T0134867

Studies in Fuzziness and Soft Computing

Volume 349

Series editor

Janusz Kacprzyk, Polish Academy of Sciences, Warsaw, Poland
e-mail: kacprzyk@ibspan.waw.pl

About this Series

The series "Studies in Fuzziness and Soft Computing" contains publications on various topics in the area of soft computing, which include fuzzy sets, rough sets, neural networks, evolutionary computation, probabilistic and evidential reasoning, multi-valued logic, and related fields. The publications within "Studies in Fuzziness and Soft Computing" are primarily monographs and edited volumes. They cover significant recent developments in the field, both of a foundational and applicable character. An important feature of the series is its short publication time and world-wide distribution. This permits a rapid and broad dissemination of research results.

More information about this series at http://www.springer.com/series/2941

Rudolf Seising · Héctor Allende-Cid
Editors

Claudio Moraga: A Passion for Multi-Valued Logic and Soft Computing

Editors
Rudolf Seising
Deutsches Museum
Munich
Germany

Héctor Allende-Cid
Pontificia Universidad Católica de Valparaíso
Valparaíso
Chile

ISSN 1434-9922 ISSN 1860-0808 (electronic)
Studies in Fuzziness and Soft Computing
ISBN 978-3-319-83912-7 ISBN 978-3-319-48317-7 (eBook)
DOI 10.1007/978-3-319-48317-7

Printed on acid-free paper

This Springer imprint is published by Springer Nature
The registered company is Springer International Publishing AG
The registered company address is: Gewerbestrasse 11, 6330 Cham, Switzerland

Foreword

This book is a tribute to Prof. Claudio Moraga on the occasion of his 80th birthday. It is a great honor for me to have the opportunity to write this foreword. I met Prof. Moraga in the late eighties when he came to the University of Chile, in Santiago, one day to visit my former supervisor, Carlos Holzmann, who had been his classmate at the Catholic University of Valparaiso. They both belong to a generation of engineers who pioneered in television, designing and building the first TV camera in Chile in 1960, as senior undergraduate students.

Professor Moraga started his academic career in Chile in the 1960s, and a few years later he became one of the founders of the Department of Computer Science (currently the Department of Informatics) at the *Technical University Federico Santa Maria* (UFSM), in Valparaiso, Chile. In 1974 he migrated to Germany, where he was Professor at the *Technical University of Dortmund* until his retirement in 2002. Soon after that he became a visiting researcher at the Technical University of Madrid. From 2006 to 2016, he became Emeritus Researcher at the *European Center of Soft Computing*, in Asturias, Spain.

Although he spent more than 40 years in Europe, he never forgot his mother country, Chile, where he had established long-lasting research collaborations with Hector Allende and Rodrigo Salas of the UFSM.

Professor Moraga is recognized worldwide as a pioneer in multiple-valued logic, a field that includes digital hardware, models, algebra, and applications. In particular, he started a research line in spectral techniques applied to multiple-valued logic, in which orthogonal transformations are applied to the analysis and synthesis of discrete functions. His seminal paper, "Complex Spectral Logic", presented at the *1978 Eighth International Symposium on Multiple Valued Logic*, opened new avenues for research that led to organizing special sessions at international conferences, workshops, and publishing two books on the topic, the latest in 2012.[1]

Although I don't know what the topic of conversation was between Prof. Moraga and my former supervisor in their meeting in the eighties, I would like to think that it was about fuzzy logic. According to Prof. Moraga's own story in *Memories of a Crisp Engineer*,[2] his work on multiple-valued switching was all about being "precise" and "crisp", because a variable may have an integral number of truth values.

The first time he heard about fuzzy logic was at a *Symposium on Multiple-Valued Logic*, where Lofti A. Zadeh was a keynote speaker in the late 1970s. Later he was systematically exposed to the world of fuzzy logic by Prof. Enric Trillas.

In fuzzy logic the truth values of variables may be any real numbers between 0 and 1. In the mid-1980, Prof. Moraga started teaching and doing research on computational intelligence: artificial neural networks, fuzzy logic, and evolutionary computation. He was immediately attracted by neuro-fuzzy systems and by the synergy between fuzzy systems and evolutionary computation. He supervised several doctoral and master theses on the theory and applications of computational intelligence. More recently he has been involved in the area of computing with words and fuzzy formal languages.

During all these prolific years, Prof. Moraga collaborated with many researchers, among them Igor Aizenberg, Jaakko Astola, Zizhong Chen, LLuis Godo, Francisco Herrera, Manuel Lozano, Rudolf Seising, Radomir Stanković, Michio Sugeno, Karl-Heinz Temme, Enric Trillas, and Gracian Triviño Barros. I would like to close this foreword by congratulating Prof. Moraga on his 80th birthday, and quoting his own words: "It has been a fuzzinating experience!".

[1]Moraga; C.: Complex Spectral Logic, in: MVL '78 Proceedings of the eighth international symposium on Multiple-valued logic, IEEE Computer Society Press Los Alamitos, CA, USA 1978, pp. 149–156.

[2]Moraga, M.: Memories of a Crisp Engineer', in: Seising, R.; Trillas, E.; Termini, S.; Moraga, C. (eds.): *On Fuzziness. A Homage to Lotfi A. Zadeh*—Vol. II, (Studies in Fuzziness and Soft Computing, Vol. 299), Berlin, Heidelberg [et al.]: Springer 2013, pp. 449–453.

Fig. 1 Prof. Dr. Pablo Estévez at July 24 opening the 2016 IEEE World Congress on Computational Intelligence in Vancouver, Canada

Pablo A. Estévez
President, IEEE Computational Intelligence
Society (2016–2017)
Professor, Department of Electrical Engineering
University of Chile

Preface

This edited book on recent advances in Multivalued Logic and Soft Computing was proposed as a Festschrift for the celebration of the 80th Birthday of Prof. Dr. Claudio Moraga, Department of Computer Science, *University of Dortmund*, Germany. The contributions of this book are written by his friends, colleagues, former students, and experts whose research interests are closely related to his work.

As the editors of this book in homage to Prof. Claudio Moraga we would like to thank all authors for their contributions, for their willingness to write their papers, to look for old or new photographs, and for their patience to us.

We are very glad that we can publish some pictures in this volume as important contemporary documents or at least nice memorabilia. We publish them with the courtesy of the owners. Other photographs we have taken from the "Fuzzy Archive" of one of the editors.

We thank Prof. Dr. Janusz Kacprzyk for accepting the book in the series *Studies in Fuzziness and Soft Computing*. We also thank Prof. Dr. Pablo Estevez, Faculty of Exact Physical and Mathematical Science, *Universidad de Chile*, for writing this foreword, and a few former students of Claudio Moraga, who have provided us with details on his academic career and other useful information. Last but not least, we thank Springer Verlag and in particular to Dr. Thomas Ditzinger, Dr. Leontina Di Cecco, and Holger Schäpe for helping this edition find its way to the publisher's list.

Fig. 2 Claudio Moraga and Héctor Allende after the latter's Thesis presentation in the *Universidad Técnica Federico Santa María* in Chile) in January 2015

Fig. 3 Claudio Moraga and Rudolf Seising during the at the "1. International Symposium Fuzziness, Philosophy and Medicine" at the *ECSC*, in Mieres, Spain, in March 2011

The invaluable help of colleagues, friends, and former students allowed us to carry out this project with ease while keeping it a secret until its publication on his birthday.

Munich, Germany
Valparaíso, Chile
September 2016

Rudolf Seising
Héctor Allende-Cid

Contents

Chapter 1
From Multi-valued Logics to Fuzzy Logic

Rudolf Seising

1.1 A Personal Introduction

When, in February 2008, I visited the *European Centre for Soft Computing* (ECSC) in Mieres Asturias, Spain, for the first time to to give a talk, I stayed for about a week. It was Enric Trillas who extended the invitation, and it was Claudio Moraga who made my first weekend in Asturias—this then unknown and foreign landscape—enjoyable. One of the first places of interest he showed me was the old church of *San Julián de los Prados*, or Santullano (built between the years 812 and 842 AD) in Oviedo's suburb Pumarí, close to the A-6 motorway (Fig. 1.1).

I remember this last point only too well. It was a Sunday morning and the square in front of this church was the meeting point; however, I missed it when I drove the rented car in this foreign city. I ended up on that highway and I had to drive almost an hour to Gijon and back. Fortunately, when I came back and reached the church, Claudio was still there. He said that he was not sure anymore whether the time of our appointment was at 11 or at 12 o'clock. On this afternoon, we drove along the coast road visiting very beautiful and interesting places, such as Colunga and Ribadesella. Regrettably, the *Jurassic Museum of Asturias* was closed in the winter season, but we had some great stops on this trip (Fig. 1.2).

Later, Claudio showed me the *Requexu Square*—the Cider Square in Mieres del Camino with the town's landmark, where we sampled a bottle of the famous *sidra* (cidre) and some tapas.

R. Seising (✉)
Geschichte der Nat, "Ernst-Haeckel-Haus",
Friedrich-Schiller-Universität Geschichte der Nat, Jena, Germany
e-mail: rudolf.seising@softcomputing.es; rudolf.markus.seising@uni-jena.de

R. Seising and H. Allende-Cid (eds.), *Claudio Moraga: A Passion for Multi-Valued Logic and Soft Computing*, Studies in Fuzziness and Soft Computing 349, DOI 10.1007/978-3-319-48317-7_1

Fig. 1.1 *San Julián de los Prados* in Oviedo; photograph by Rudolf Seising

In these days I did not yet know that I was going to have many more opportunities to enjoy Asturian cuisine, dishes and beverages in the years to come. Some months later, however, I was offered a position as a Visiting Researcher for almost a year, and, at the end of this period, I was employed as "Adjoint Researcher" in the unit "Fundamentals of Soft Computing" in the ECSC in Mieres. As I was a member of the ECSC for about 5 years, I became acquainted with many researchers, among them (in alphabetical order) Héctor Allende, Hector Allende-Cid, José María Alonso, Luis Argüelles, Christian Borgelt, Nicola Bova, Óscar Cordón, Sergio Damas, Luka Eciolaza, Itziar García Honrado, Sergio Guadarrama, Takehiko Nakama, María Navarro, Martín Pereira, David Picado, Enrique Ruspini, Adolfo Rodríguez de Soto, Daniel Sánchez, Veronica Sanz, Prakash Shelokar, Arnaud Quirin, Alejandro Sobrino, Michio Sugeno, Marco Elio Tabachhi, Settimo Termini, Gracián Triviño, Krzysztof Trawinski, Wolfgang Trutschnig, Albert van Der Heide, Juan Zamora, Fátima Zohra Hadjam, and many of us became friends. Some of these researchers contributed to this volume honouring Claudio Moraga (Fig. 1.3, left).

It was thanks to Claudio and Enric that I, as a historian and philosopher of science, could enjoy these years in this research and development centre promoted by the *Foundation for the Advancement of Soft Computing*. I was able to conduct my historical research on the history of various disciplines in Soft Computing. And, from 12 September to 14 October 2011 under the sponsorship of the Government

Fig. 1.2 Claudio Moraga at the (closed) *Jurassic Museum of Asturias* in February 2009; photograph by Rudolf Seising (see the *dropped shadow*)

of Asturias and CajAstur Savings Bank, I could organize and direct a course, with a new format for teaching in our area, on "Reflecting on fuzziness (CRF)" which had the subtitle "Philosophy, Science, Technology".

Very quickly, I became acquainted with Claudio Moraga's refinement not only in his capacity as a Computer Scientist; however, in this contribution in honour of him, I will focus on only one dimension of Claudio's cultivation, namely, his scientific knowledge. There are so many disciplines, theories and methods in science, and it seems he knows them all. He is interested in old instruments and books, in various cultures and religions, in different countries and languages. However, regarding his scientific interests, he first chose Electrical Engineering and Computer Science. He studied and taught automata theory and switching theory, and his interests lie in abstract machines. Such scientific interests have close links to various forms of logical calculus, such as propositional and predicate logic, multi-valued logic, probability logic, quantum logic and fuzzy logic. He is also a virtuoso in the concepts and calculus of artificial neural networks and evolutionary algorithms.

This paper is just a brief historical survey on these various logic concepts.

Fig. 1.3 Claudio Moraga and Rudolf Seising at the market place of Mieres in Asturias, Spain, in February 2009; photographs by Rudolf Seising

1.2 Classical Logic and Non-classical Logics

In science, the most used formal logic calculus is *classical logic* or sometimes also referred to as *standard logic*. Classical logic is as old as ancient Greek thought and thinkers of the ancient world already established some of its properties[1]:

- the **Law (or principle) of Excluded Middle (or Third)**, the *tertium non datur*: For any proposition, either that proposition is true, or its negation is true.
- the **Law (or principle) of Noncontradiction**: Contradictory statements cannot both be true in the same sense at the same time. This means that the two propositions "A is B" and "A is not B" are mutually exclusive.[2]

Many "non-classical logics" (or "alternative logics") were considered in the 20th century (see [9, 10]). "Non-classical" means that other and "special" properties

[1]There are other properties of classical logics that we don't consider here: Monotonicity and Idempotency of entailment, Commutativity of conjunction, De Morgan duality, double negative elimination, and the principle of explosion (*ex falso (sequitur) quodlibet*, "from falsehood, anything (follows)."

[2]Aristotle said in *On Interpretation* [1] that of two contradictory propositions one must be true, and the other false—(Contradictory propositions are propositions where one proposition is the negation of the other.) Russell and Whitehead stated this principle as a theorem of propositional logic in the *Principia Mathematica*.

identify these logics as extensions, variations or deviations of classical logic. Here, we consider only one of them and three of its special cases:

- **Many-valued logic**. In this logic bivalence is rejected. In many-valued logic the set of possible truth values is more than {true, false}, e.g. **three-valued logic**, **infinitely-valued logics**, and **fuzzy logic**.

Our history begins in the early years of the 20th century when the English mathematician and philosopher Bertrand Russell (1872–1970) published two important works. In 1903 *The Principles of Mathematics* appeared; in it he argued that mathematics and logic are identical:

> The fact that all Mathematics is Symbolic Logic is one of the greatest discoveries of our age; and when this fact has been established, the remainder of the principles of mathematics consists in the analysis of Symbolic Logic itself. [29, p. 5]

In the introduction to the second edition Russell wrote on the first page (Fig. 1.4):

> The fundamental thesis of the following pages, that mathematics and logic are identical, is one which I have never since seen any reason to modify. [29, p. 1]

Fig. 1.4 Bertrand Russell and the title page of the shortened version of the *Principia Mathematica*; photographs: public Domain http://russell.mcmaster.ca/~bertrand/youngbr.html and http://en.wikipedia.org/wiki/File:Pmdsgdbhxdfgb2.jpg

Russell wrote the sequel in collaboration with the English mathematician and philosopher Alfred North Whitehead (1861–1947). They originally aimed to write only a second volume to the first work; however, the subject became a very much larger than they had originally expected. The two authors therefore finally wrote a three-volume work on the foundations of mathematics, the *Principia Matematica* that was published in 1910, 1912, and 1913 [30].

Starting with usual propositional logic every proposition has one and only one of the two truth values "true" or "false"—alternatively "1" or "0". If a proposition's truth value is 'true", then we say that the proposition is true; if the proposition's truth value is "false", then we say that the proposition is false.

Because in this two-valued, or "bi-valent", logic "false" is the negation of "true" we could name these values "truth values". However, this condition does not pertain if the assignment to propositions of other values than the two values of bi-valent logic is allowed. In that case we move on to multi-valued logic where the name of the values are called "quasi truth values" or simply "values".

The Polish philosopher and logician Jan Łukasiewicz (1876–1956), who developed a three-valued system, and, the mathematician Emil Leon Post (1897–1954), (Fig. 1.5) who published his work on multi-valued logics in 1921, are known as the pioneers of many-valued logics (see [19, 28]). The latter, who was also born in Poland

Fig. 1.5 Jan Łukasiewicz and Emil Post; photographs: Public Domain https://commons.wikimedia.org/w/index.php?curid=23607576 and https://commons.wikimedia.org/w/index.php?curid=2326395

but emigrated in the USA, developed his multi-valued logic at about the same time as but independently of the former.

The mathematical methods for handling proposition logics as they were developed in the *Principia Mathematica* of Russell and Whitehead was a precondition to working out systems of multi-valued logics. This was the starting point for Post. He introduced truth matrices that allow values of the set $\{t_1, \ldots, t_m\}$ with integer $m > 2$ as values for relevant variables and expressions.

Post did not look for an interpretation of the m-valued proposition function in a multi-valued logic. This was, however, the intention of the philosopher Łukasiewicz. He questioned two-valued logic. To begin with, he replaced it by a three-valued logic using "true" (1), "false" (0) and "possible" (1/2). Later he surpassed the conception of Post when he assumed infinitely many "truth values".

Łukasiewicz was a member of the Lwów-Warsaw school, which was very active and innovative in the philosophy of science as well as in the fields of logic and semantics. The school was founded by the "father of Polish logic" Kazimierz Jerzy Skrzypna-Twardowski (1866–1938), who studied philosophy in Vienna with Franz Brentano (1838–1917) and Robert Zimmermann (1824–1898) and later lectured in Vienna (1894–1895). He became professor at the university in Lwów (Lemberg in Austrian Galicia, now Lviv in the Ukraine).

Twardowski propagated an absolute concept of truth. Every proposition is a semantic unity, in the sense that it represents an absolute truth or falsehood—independently of the point in time when the statement was made. In contrast, Kotarbinski (who later changed his mind) and Łukasiewicz resolutely denied the "pre-eternality" of propositions. It was from this perspective that Łukasiewicz studied the Aristotelian law of non-contradiction.

In search of the criteria for his indeterministic world view, he acquainted himself with the Aristotelian "Sea-battle argument". According to Łukasiewicz's interpretation, Aristotle advocated the openness and rejected the determinism of the future. Accordingly, single statements directed at the future cannot have one of the truth values "true" or "false", since the future is not predictable. Lukasiewicz wrote:

> I can assume without contradiction that my presence in Warsaw at a certain moment of next year, e.g. at noon on 21 December, neither positively, nor negatively. Hence it is possible, but not necessary, that I shall be present at Warsaw at the present time. On this assumption, the proposition 'I shall be present in Warsaw at noon on 21 December of the next year', can at the present time be neither true nor false. For if it were true now, my future presence in Warsaw would have to be necessary, which is contradictory to the assumption. If it were false now, on the other hand, my future presence in Warsaw would have to be impossible, which is also contradictory to the assumption. Therefore the proposition considered is at the moment *neither true nor false* and must possess a third value, different from "0" or falsity and "1" or truth. This value we can designate by "1/2". It represents "the possible" and joins "the true"and "the false" as a third value.
>
> The three-valued system of propositional logic owes its origin to this line of thought. [22][3]

[3]Ich kann ohne Widerspruch annehmen, dass meine Anwesenheit in Warschau in einem bestimmten Zeitmoment des nächsten Jahres, z.B. mittags den 21. Dezember, heutzutage weder im positiven noch im negativen Sinne entschieden ist. Es ist somit möglich, aber nicht notwendig, dass ich zur

Łukasiewicz assumed that the real world is an manifestation of one of all competing logical systems; in other words he thought that one of the many-valued logics is ontologically true. In the 1940s and later, he divorced himself from this assumption. At this time he declared that choosing a logic is a convention, various systems serving as the need arises. In 1951, he published *Aristotle's Syllogistic from the Standpoint of Modern Formal Logic* [21], and here he considered the 4-valued logic as perhaps being the most interesting system because in this logic we can formalize modalities like "necessary", "possible", "accidentally", "impossible". In his *Philosophical remarks on many-valued systems of propositional logic* he wrote some time earlier:

> it was clear to me from the outset that among all the many-valued systems only two can claim any philosophical significance: the three-valued one and the infinite-valued ones. For if values other than "0" and "1" are interpreted as "the possible", only two cases can reasonably be distinguished: either one assumes that there are no variations in degrees of the possible and consequently arrives at the three-valued system; or one assumes the opposite, in which case it would be most natural to suppose, as in the theory of probabilities, that there are infinitely many degrees of possibility, which leads to the infinite-valued propositional calculus. I believe that the latter system is preferable to all others. Unfortunately this system has not yet been investigated sufficiently; in particular the relations of the infinite-valued system to the calculus of probabilities awaits further inquiry. [20][4]

1.2.1 *Fuzzy Logic*

During this attempt to construe quasi truth values of multi-valued logic as probabilities in a probability logic, there arose in the mid-1960s in the USA another variant for interpreting the set of quasi truth values between "0" and "1". The name

(Footnote 3 continued)
angegebenen Zeit in Warschau anwesend sein werde. Unter diesen Voraussetzungen kann die Aussage: "Ich werde mittags den 21. Dezember nächsten Jahres in Warschau anwesend sein", heutzutage weder wahr noch falsch sein. Denn ware sie heutzutage wahr, so müsste meine zukünftige Anwesenheit in Warschau notwendig sein, was der Voraussetzung widerspricht; und wäre sie heutzutage falsch, so müsste meine zukünftige Anwesenheit in Warschau unmöglich sein, was ebenfalls der Voraussetzung widerspricht. Der betrachtete Satz ist daher heutzutage weder wahr noch falsch und muss einen dritten, von "0" oder dem Falschen und von "1" oder dem Wahren verschiedenen Wert haben. Diesen Wert können wir mit "1/2" bezeichnen; es ist eben 'das Mögliche', das als dritter Wert neben "das Falsche" und "das Wahre" an die Seite tritt. Diesem Gedankengang verdankt das dreiwertige System des Aussagenkalküls seine Entstehung. [20, p. 165].

[4]"Es war mir von vornherein klar, dass unter allen mehrwertigen Systemen nur zwei eine philosophische Bedeutung beanspruchen können: das dreiwertige und das unendlichwertige System. Denn werden die von "0" und "1" verschiedenen Werte als "das Mögliche" gedeutet, so konnen aus guten Gründen nur zwei Fälle unterschieden werden: entweder nimmt man an, dass das Mögliche keine Gradunterschiede aufweist, und dann erhält man das dreiwertige System; oder man setzt das Gegenteil voraus, und dann ist es am natürlichsten ebenso wie in der Wahrscheinlichkeitsrechnung anzunehmen, dass unendlich viele Gradunterschiede des Möglichen bestehen, was zum unendlichwertigen Aussagenkalkul führt. Ich glaube, dass gerade dieses letztere System vor allen anderen den Vorzug verdient. Leider ist dieses System noch nicht genau untersucht; insbesondere ist auch das Verhältnis des unendlichwertigen Systems zur Wahrscheinlichkeitsrechnung noch nicht geklärt." [22, p. 173].

of this attempt is "Fuzzy Logic" and the founder of the underlying mathematical theory, called "Fuzzy Set Theory", is Lotfi A. Zadeh (born in 1921), (Fig. 1.7 left) a professor of electrical engineering, who was born in 1921 in Baku, the capital of Azerbaijan, then moved to Tehran, the capital of Iran, and later settled in the USA where he continued his studies at *Massachusetts Institute of Technology* (MIT) in Cambridge, Massachusetts. In the 1950s he was a member of the Department of Electrical Engineering at *Columbia University* in New York and in 1959 he joined the Department of Electrical Engineering and Computer Science at the *University of California in Berkeley*.

The word "fuzzy" means hazy, vague, unclear, or foggy. It was the aim of Zadeh to introduce the vagueness inherent in natural languages into engineering. To this end, he established a logic of "fuzzy" entities by a generalization of the usual set's characteristic function, the so- called "membership function" μ_A of a "fuzzy set" A. The range of μ_A, i.e. all possible membership values t_i of an object x in the universe of discourse X are all numbers between 0 and 1:

$$\mu_A(x) = \begin{cases} t_1 = 1, if x \in A \\ t_i \\ t_m = 0, if x \notin A. \end{cases} \tag{1.1}$$

Zadeh founded the theory of fuzzy sets as a mathematical theory for dealing with uncertainties in the summer of 1964 [42].[5] When he analyzed the ability of conventional mathematical tools in engineering, he saw serious shortcomings: The framework was not adequate for the treatment of systems as complex as those in modern information and communication technology and even less so for those in biology and medicine.

In contrast to conventional set theory, an object is not required to be either an element of a set (membership value 1) or not an element of this set (membership value 0) but can have a membership value between 0 and 1.

Thus, he defined fuzzy sets by their membership function μ, which is allowed to assume any value in the interval [0, 1], instead of by the characteristic function of a usual set that exclusively assumes the values of 0 or 1 [42]. For fuzzy sets A and B in any universe of discourse X, Zadeh defined equality, containment, complementation, intersection, and union (for all $x \in X$), see Fig. 1.6:

- $A = B$ if and only if $\mu_A(x) = \mu_B(x)$,
- $A \subseteq B$ if and only if $\mu_A(x) \leq \mu_B(x)$,
- $\neg A$ is the complement of A if and only if $\mu_{\neg A}(x) = 1 - \mu_A(x)$,
- $A = \cup B$ if and only if $\mu_{A \cap B}(x) = max(\mu_A(x), \mu_B(x))$,
- $A = \cap B$ if and only if $\mu_{A \cap B}(x) = min(\mu_A(x), \mu_B(x))$,

The space of all fuzzy sets in X builds a distributive lattice with 0 and 1, but it is not Boolean because in this lattice there is no complementation (Fig. 1.6):

[5] For a detailed presentation of the history of Fuzzy Set Theory see [33].

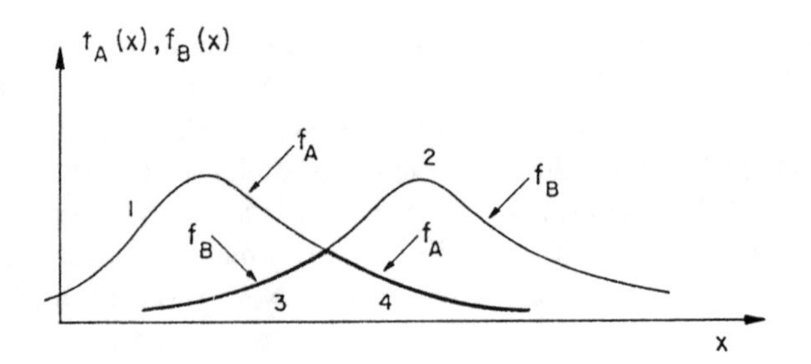

Fig. 1.6 Illustration of the union as maximum of membership functions f_A and f_B (1, 2) and the intersection as minimum of membership functions f_A and f_B (3, 4), [42]

$$\mu_A(x) \cap \mu_{\neg A}(x) \neq 0 \tag{1.2}$$

$$\mu_A(x) \cup \mu_{\neg A}(x) \neq 1 \tag{1.3}$$

A propositional logic with fuzzy concepts constitutes a "Logic of Inexact Concepts." This was demonstrated by the US-American computer scientist Joseph Amadee Goguen (1941–2006), when he was a doctoral student studying with Zadeh at Berkeley in the late 1960s [5] and in articles he published later [6, 7]. The US-American cognitive linguist George Lakoff (born 1941), a Professor at UC Berkeley, introduced the term "fuzzy logic" for this logic in 1973 [15].

Here is an interesting link concerning Lotfi A. Zadeh in the time before he founded the theory of fuzzy sets and the field of multi-valued logics in the middle of the 20th century:

In the mid-fifties Zadeh applied for a guest residency at the *Institute for Advanced Study* (IAS) in Princeton. His friend Herbert Ellis Robbins (1915–2001) was the chairman of *Columbia University*'s department of mathematical statistics at the time, and Deane Montgomery (1909–1992), another friend, was a member of the IAS. Both Robbins and Montgomery campaigned for the approval of Zadeh's IAS guest residency, even though it was rare for requests by scientists who were neither mathematicians, theoretical physicists, nor historians to receive a positive response [32]. Zadeh initially took a half-year sabbatical from *Columbia University* in 1956. During this time he wanted to learn more about logic, an interest he had cultivated since 1950, when he predicted that logic, and particularly multi-valued logic, would become increasingly more important for the problems of electrical engineering in the future [40].

In Princeton he attended lectures by Stephen Cole Kleene (1909–1994), (Fig. 1.7) who had also continued developing the multi-valued logic devised by the Polish school of logic [13, 14]. Kleene became Zadeh's mentor at Princeton: "Steven Kleene was my teacher in logic". Yes, I learned logic from Steven Kleene! [31]

Similar to the leap from two-dimensionality to n-dimensionality in mathematics, Zadeh found multi-valued logic to be a natural generalization of the conventional

Fig. 1.7 Lotfi A. Zadeh at the *World Conference on Soft Computing, San Francisco State University* in 2011 and Stephen C. Kleene; photographs: Rudolf Seising and https://en.wikipedia.org/wiki/Stephen_Cole_Kleene#/media/File:Kleene.jpg

logic of just two values into n values [32]. He was now also toying with the ideas of introducing multi-valued logic into automata theory and implementing it in electrical circuits. Once he had returned to *Columbia University* in New York he assigned two dissertations that dealt with multi-valued logic in the design of transistor circuits and with multi-valued coding:

- Werner Ulrich completed his dissertation on "Nonbinary Error Correction Codes" in 1957 [37].
- Oscar Lowenschuss completed his dissertation on "Multi-Valued Logic and Sequential Machines or Non-Binary Switching Theory" in 1958 [16–18] (Fig. 1.8).

"That's why I wanted to know about logics!" Zadeh recalled in an interview he gave for me in 2000 [31].

This history shows that Zadeh's road to fuzzy sets and fuzzy logic led across multi-valued logic but this was only half the journey. In addition and more importantly in my view, the road also led across a new mathematics of approximation that was different from probability theory and statistics; it was a road via "the mathematics of fuzzy or cloudy quantities" as he wrote already in 1962 [41, p. 856ff]!

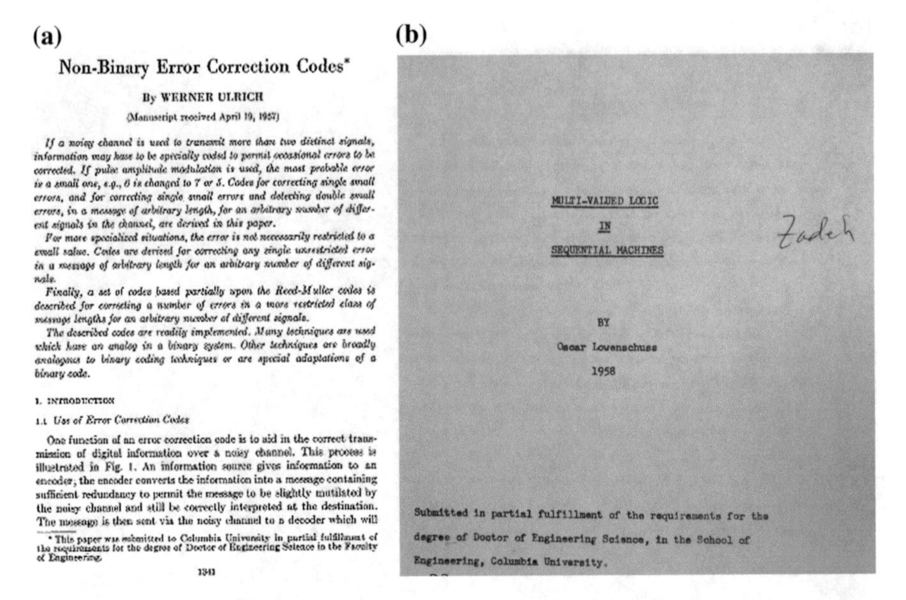

(a)

Non-Binary Error Correction Codes*

By WERNER ULRICH

(Manuscript received April 19, 1957)

If a noisy channel is used to transmit more than two distinct signals, information may have to be specially coded to permit occasional errors to be corrected. If pulse amplitude modulation is used, the most probable error is a small one, e.g., 6 is changed to 7 or 5. Codes for correcting single small errors, and for correcting single small errors and detecting double small errors, in a message of arbitrary length, for an arbitrary number of different signals in the channel, are derived in this paper.

For more specialized situations, the error is not necessarily restricted to a small value. Codes are derived for correcting any single unrestricted error in a message of arbitrary length for an arbitrary number of different signals.

Finally, a set of codes based partially upon the Reed-Muller codes is described for correcting a number of errors in a more restricted class of message lengths for an arbitrary number of different signals.

The described codes are readily implemented. Many techniques are used which have an analog in a binary system. Other techniques are broadly analogous to binary coding techniques or are special adaptations of a binary code.

I. INTRODUCTION

1.1 *Use of Error Correction Codes*

One function of an error correction code is to aid in the correct transmission of digital information over a noisy channel. This process is illustrated in Fig. 1. An information source gives information to an encoder; the encoder converts the information into a message containing sufficient redundancy to permit the message to be slightly mutilated by the noisy channel and still be correctly interpreted at the destination. The message is then sent via the noisy channel to a decoder which will

* This paper was submitted to Columbia University in partial fulfillment of the requirements for the degree of Doctor of Engineering Science in the Faculty of Engineering.

1341

(b)

MULTI-VALUED LOGIC

IN

SEQUENTIAL MACHINES

BY

Oscar Lowenschuss

1958

Submitted in partial fulfillment of the requirements for the degree of Doctor of Engineering Science, in the School of Engineering, Columbia University.

Fig. 1.8 First pages of the Ph.D. dissertation thesis by Ulrich [37] and Lowenschuss [16] in 1957 and 1958, the latter with Zadeh's signature; photographs: Rudolf seising

1.3 And Quantum Logic?

The scientific revolution brought about by the discovery of quantum mechanics in the first third of the 20th century resulted in a fundamental change in the relationship between physics and the phenomena observed in experiments. Classical experiments showed the need for a new physical theory and new physical terms to represent and predict what happens at the subatomic level. The theory of quantum mechanics was developed to fill the gap in the existing physical theories; however, quantum mechanics is completely abstract: it is a theory of mathematical state functions that have no exact counterpart in reality.

The German physicist Werner Heisenberg (1901–1976), (Fig. 1.9 left) his Danish colleague Niels Bohr (1885–1962), (Fig. 1.9 right) and others introduced new objects into the new quantum mechanics theory that differ significantly from those of classical physics. The properties of these objects are neither particles nor waves, which are the standard objects of classical physics. The properties of these objects are completely new and not comparable with those of the observables in classical theories, such as Newton's mechanics or Maxwell's electrodynamics.

However, in experimental physics classical variables like position and momentum can be observed and physicists subjected the subatomic objects of reality to experiments designed to elicit these classical variables. However, determining the state of a quantum mechanical system is much more difficult than determining the state of classical systems since sharp values for both variables cannot be measured simultaneously. This is the meaning of Heisenberg's uncertainty principle.

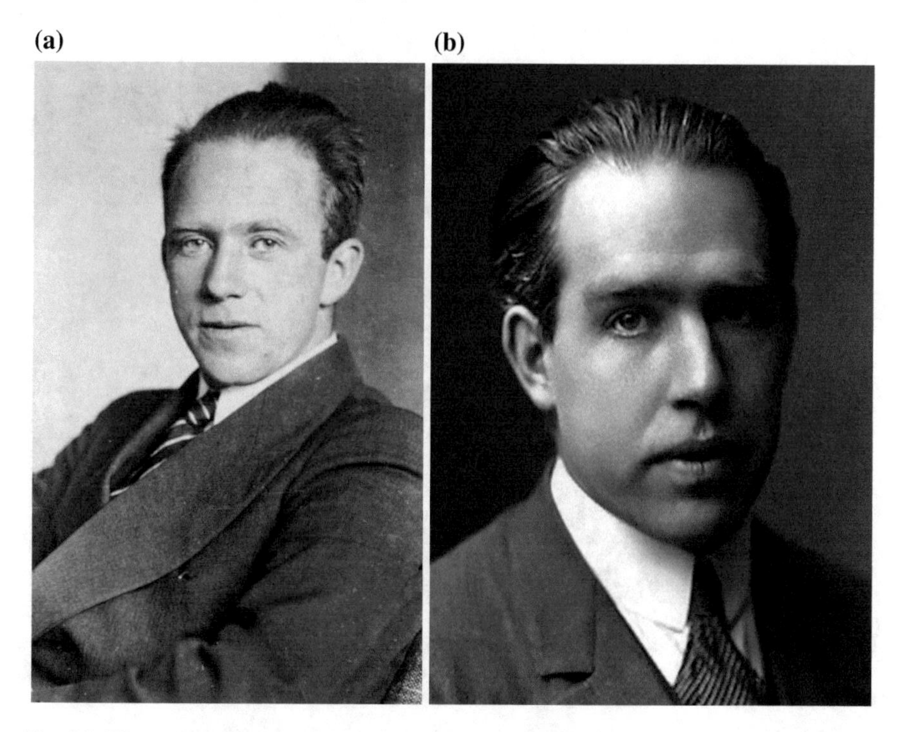

Fig. 1.9 Werner Heisenberg and Niels Bohr; photographs: Public Domain https://en.wikipedia.org/wiki/Werner_Heisenberg#/media/File:Bundesarchiv_Bild183-R57262,_Werner_Heisenberg.jpg and https://en.wikipedia.org/wiki/Niels_Bohr#/media/File:Niels_Bohr.jpg

We can perform experiments on quantum mechanical objects designed to measure a position value; we can also perform experiments with these objects in order to measure their momentum value. However, we cannot conduct both experiments simultaneously since it is impossible to measure both values for the same point in time. Nonetheless, the values at the point in time in question can be predicted on the basis of the outcomes of experiments. Since predictions are targeted at future events, we cannot give them the logical values "true" or "false."

In 1926, the German physicist Max Born (1882–1970) (Fig. 1.10 right) proposed an interpretation of the non-classical peculiarity of quantum mechanics, namely that the quantum mechanical wave function is a "probability-amplitude" [3, 4]. This is an abstract magnitude without any significance for the measurement of one of the possible observables. The absolute square of its value equals the probability of its having a certain position or a certain momentum if we measure the position or momentum respectively. The higher the probability of the position value, the lower that of the momentum value and vice versa. Thus, the absolute square of the quantum mechanical state function equals the probability density function of its having a certain position or a certain momentum in the position or momentum representation of the wave function respectively. There are varying representations of the state of a quantum mechanical object, e.g., the "position picture" and the

Fig. 1.10 George Whitelaw Mackey and Max Born; photographs: Public Domain https://commons.wikimedia.org/wiki/File:GWMackey_c1980s.jpg and https://upload.wikimedia.org/wikipedia/commons/f/f7/Max_Born.jpg

"momentum picture." These representations are complementary, which means that a subatomic object cannot be presented in both classical pictures at the same time.

However, from the mathematical point of view it has to be stated that no joint probability for the event in which both variables have a certain value exists since there is no classical probability space (no Boolean lattice) that constitutes such events. Therefore we need a radically different kind of uncertainty theory that is not describable in terms of classical probability distributions.

The mathematician John von Neumann (1903–1957), who was originally born in Hungary, then adopted German citizenship and later settled in the USA, had been collaborating with Max Born on the mathematical foundation of quantum mechanics since 1927 [24]. In 1932 he published his famous book *Mathematical Foundations of Quantum Mechanics* [25], in which he defined the quantum mechanical wave function as a one-dimensional subspace of an abstract Hilbert space, which is defined as the state function of a quantum mechanical system.

In 1936, (Fig. 1.11 left) not long after founding the mathematical theory of quantum mechanics John von Neumann together with the US-American mathematician Garett Birkhoff (1911–1996) (Fig. 1.11 right) proposed the introduction of a "quantum logic". The set of quantum mechanical predictions constitutes a non-distributive lattice; this lattice is, therefore, non-Boolean [2]. The distributive law, says that, if x, y and z are propositions we have

$$x \wedge (y \vee z) \neq (x \wedge y) \vee (x \wedge z) \tag{1.4}$$

Fig. 1.11 Garett Birkhoff and John von Neumann; photographs: Public Domain https://commons.wikimedia.org/wiki/File:Garrett_Birkhoff.jpegbyKonradJacobs,Erlangen and https://en.wikipedia.org/wiki/John_von_Neumann#/media/File:JohnvonNeumann-LosAlamos.gif

Birkhoff and Neumann remarked that "the study of mechanics points to the distributive identities as the weakest link in the algebra of logic" [2, p. 837]. It is possible to find an even weaker law that is satisfied in the case of a finite-dimensional Hilbert space. In that case, the lattice of quantum mechanical prediction has the property of "modularity"; the modular law is formulated as follows:

$$x \leq y \Rightarrow x \vee (y \wedge z) = y \wedge (x \vee z) \tag{1.5}$$

Later, in 1963, the US-American mathematician George Whitelaw Mackey (1916–2006) (Fig. 1.10 left) attempted to provide a set of axioms for the propositional system of predictions of experimental outcomes. He was able to show that this system is an orthocomplemented partially ordered set [23]. In this logico-algebraic approach, the set of predictions of the properties of a quantum mechanical object is not additive and not distributive.

In the 1960 s the US-American philosopher Patrick Colonel Suppes (1922–2014) discussed the "probabilistic argument for a non-classical logic of quantum mechanics" [35, 36]. He introduced the concept of a "quantum mechanical σ-field" as an "orthomodular partial ordered set" covering the classical σ-fields as substructures.

Eventually, in the 1980s, a "quantum probability theory" was proposed and developed by authors such as the US-American mathematician Stanley P. Gudder and the Israeli historian and philosopher of science Itamar Pitowski (1950–2010) [8, 27]. These developments regarding a theory of probabilistic structures of quantum mechanics became very complex. The quantum mechanical lattice of predictions is Suppes' "quantum mechanical σ-field," which can be restricted to Boolean lattices, each corresponding to a given observable and those observables compatible with it. In every case of one of these restrictions, the quantum probabilities became classical probabilities again that only apply to predictions of compatible observables.

1.4 Outlook

Soon after "Fuzzy Sets" appeared in the journal *Information and Control* the Japanese physicist Satosi Watanabe (1910–1993) wrote an article for the same journal entitled "Modified Concepts of Logic, Probability, and Information Based on Generalized Continuous Characteristic Function", which was published in 1968 [39]. Watanabe had studied at *Tokyo Imperial University*, and the university had sent him to Europe, where he studied under the French physicist Luis de Broglie (1892–1987). In 1937, he moved to Leipzig to study nuclear theory under Heisenberg. He left Germany because of World War II, but he stayed in Copenhagen for some time with Niels Bohr before returning to Japan.

In the short abstract to this paper he stated: "All the basic laws of the traditional logic can be derived from the characteristic function $f(A/a)$ which is 1 or 0 according as object a satisfies predicate A or not. There is good reason to believe that it is worthwhile to extend this formalism to the case where $f\ (A/a)$ can take any value in the continuous domain [0, 1]." [39, p. 1].

In his introduction, Watanabe stated that the basic postulate—he called it "the postulate of fixed truth set"—saying "that a and A determine the value of $f\ (A/a)$ is not always satisfied" [39, p. 1]. Watanabe emphasized that this postulate is valid if the characteristic function is two-valued, and in that case "everything with regard to predicates can be reduced to the set theory of objects, and it is easy to derive directly therefore logical operations such as conjunction, disjunction, negation etc." However, his intension was to "reconstruct logic, probability and information theory on the basis of irreducibly continuous characteristic functions" in this article; in that case the characteristic function is multi-valued and the postulate is not valid; therefore "we have to derive these logical operations of predicates independently of the set-theoretical operations applied to objects." [39, p. 2].

Watanabe had read "Fuzzy Sets", and he distanced his approach from Zadeh's theory "which uses from the beginning the notions of set, conjunction, disjunction, etc. as if they were already known to us. His determination of the values of the membership function for conjunction and disjunction is arbitrary." [39, fn. 2].

Interestingly, when Watanabe described the background of the emergence of his "stream of thought" in the last section of his article, he highlights the fact that this was not Zadeh's theory of fuzzy sets or fuzzy logic but quantum logic. He referred to the "theoretical structure of quantum mechanics" "on what Birkhoff and Neumann called quantum logic" [39, p. 19].

Watanabe pointed to the deep mathematical formulation of quantum logic in terms of the infinitely-dimensional Hilbert Space, on the one hand and the Japanese physicist Kodi Husimi (1909–2008), on the other, who tried already in 1937, the year after Birkhoff and von Neumann's publication [11], "to found the non-distributive modular logic directly on the experimental basis of atomic phenomena without relying on the Hilbert space" [39, p. 19].

Husimi demonstrated that a weaker law than the modular law, the orthomodular law, is satisfied in the lattice of quantum mechanics[6]:

$$x \leq y \Rightarrow x \vee \left(x^{\perp} \wedge y\right) = y \tag{1.6}$$

As the Israeli physicist, philosopher and historian of science Max Jammer (1915–2010) wrote in *The Philosophy of Quantum Mechanics*, Husimi presented this attempt at a meeting of the Physico-Mathematical Society in Japan in 1937:

> Defining the implication by the statement that the numerical probability of the antecedent is smaller or equal to that of the consequent Husimi showed that the existence of the meet requires the sum of two quantities whose existence he derived from the correspondence principle according to which 'every linear relation between the mean values in the classical theory is conserved in the process of quantization'. [12, p. 354]

At the end of his article Watanabe stated his belief "that the modular logic would have a wide application outside atomic physics". Already in 1959 he taught on non-Boolean logic in a lecture series at *Yale Graduate School of Physics* where he introduced the characteristic function $f(A/a)$.[7]

His goal was to derive from this characteristic function $f(A/a)$ the logical relation of implication. Here, he followed the dictum of the American philosopher, logician and mathematician Charles Sanders Peirce (1839–1914), who wrote in his essay "The New Elements of Mathematics" [26] on the Boolean mathematics and on a sign that he called "illation". Watanabe referred to Peirce as follows: "there is one primary and fundamental logical relation, that is illation [implication]." [39, p. 19].

Watanabe argued that we have to move on from the sharp dichotomy between "true" and "false" to a fluid decision from "Yes" to "No". One of his reasons for this "blasphemous sentence" was related to logics and probabilities: "it seems more natural and truthful to our human experience to derive logic form a probability-like function such as μ_A. Probability precedes logic"!

---

[6]Here we need the definition of the "orthocomplement" $x^{\perp}$ of the element x: This element satisfies (1) the *complement law* $x^{\perp} \vee x = 1$ and $x^{\perp} \wedge x = 0$; (2) the involution law $x^{\perp\perp} = x$; (3) the *order reserving law* if $x \leq y$ then $y^{\perp} \leq x$.

[7]The lecture notes of this course *Physical Information Theory* (1959–60) were published under the title "Algebra of Observations" but not until 1966 unfortunately [38].

Fig. 1.12 Claudio Moraga and Rudolf Seising had an hour's rest during Saturday's Scientific Conversations "Thinking and Fuzzy Logic" in Palermo (Sicily, Italy), 14th May, 2011, photograph by Rudolf Seising

The "Peirce-principle" signifies that the basic operation of human deductive reasoning is implication. However, the appropriate implication is an implication based on our everyday-usage of vague or inexact reasoning and decisions.

Acknowledgments I would like to thank Mark Winstanley for proofreading and suggestions for improvement.

References

1. Aristotle: On Interpretation, Translated by E. M. Edghill: https://ebooks.adelaide.edu.au/a/aristotle/interpretation/.
2. Birkhoff, Garett, von Neumann, John: The Logic of Quantum Mechanics, *Annals of Mathematics*, Series 2, vol. 37, 1936, p. 823 ff.
3. Born, Max: Zur Quantenmechanik der Stoßvorgänge, *Zeitschrift für Physik*, vol. 37, 1926, pp. 863–867.
4. Born, Max: Das Adiabatenprinzip in der Quantenmechanik, *Zeitschrift für Physik*, vol. 40, 1926, pp. 167–191.
5. Goguen, Joseph A.: L-Fuzzy Sets, *Journal of Mathematical Analysis and Applications*, vol. 18, 1967, pp. 145–174.

6. Goguen, Joseph A.: Categories of Fuzzy Sets: Applications of a Non-Cantorian Set Theory, Ph.D. Dissertation, University of California at Berkeley, June 1968.
7. Goguen, Joseph A.: The Logic of Inexact Concepts, *Synthese*, vol. 19, 1969, pp. 325–373.
8. Gudder, Stanley P.: *Quantum Probability*, Academic Press: San Diego, 1988.
9. Haack, Susan: *Deviant Logic*, Chicago, Ill.: Cambridge University Press, 1974.
10. Haack, Susan: *Deviant Logic, Fuzzy Logic: Beyond The Formalism*, Chicago, Ill.: University of Chicago Press, 1996.
11. Husimi, Kodi: Studies in the Foundations of Quantum Mechanics, *Proceedings of the Physico-Mathematical Society of Japan*, vol. 19, 1937, pp. 766–789.
12. Jammer, Max: *The Philosophy of Quantum Mechanics. The Interpretations of QM in Historical Perspective*, John Wileys and Sons, 1974.
13. Kleene, Stephen C.: On a notation for ordinal numbers, *The Journal of Symbolic Logic*, vol. 3, 1938, pp. 150–155.
14. Kleene, Stephen C.: *Introduction to Metamathematics*, Amsterdam: North-Holland, 1952.
15. Lakoff, George: Hedges: A Study in Meaning Criteria and the Logic of Fuzzy Concepts, *Journal of Philosophical Logic*, vol. 2, 1973, pp. 458–508.
16. Lowenschuss, Oscar: Multi-Valued Logic in Sequential Machines. Ph.D. Thesis. School of Engineering, Columbia University, New York, 1958.
17. Lowenschuss, Oscar: A Comment on Pattern Redundancy, *IRE Transactions on Information Theory*, December 1958, p. 127.
18. Lowenschuss, Oscar: Restoring Organs in Redundant Automata, *Information and Control*, Vol. 2, 1959, pp. 113–136.
19. Łukasiewicz, Jan: O logice trójwartościowej (in Polish), *Ruch filozoficzny*, vol. 5, 1920, pp. 170–171. English translation: On three-valued logic, in: Borkowski, L. (ed.), *Selected works by Jan Łukasiewicz*, Amsterdam: North–Holland, 1970, pp. 87–88.
20. Łukasiewicz, Jan: Philosophische Bemerkungen zu mehrwertigen Systemen des Aussagenkalküls, *Comptes rendus de séances de la Société de Sciences et des Lettres de Varsovie*. Cl. iii, vol. 23, 1930, pp. 51–77.
21. Łukasiewicz, Jan: *Aristotle's Syllogistic from the Standpoint of Modern Formal Logic*. Oxford: Clarendon Press, 1951. 2nd, enlarged ed., 1957.
22. Łukasiewicz, Jan: *Selected Works*, ed. L. Borkowski. Amsterdam: North-Holland, 1970.
23. Mackey, George W. *Mathematical Foundations of Quantum Mechanics*, New York: W. A. Benjamin, 1963.
24. von Neumann, John: Mathematische Begründung der Quantenmechanik, Güttingen: Nachr. Ges. Wiss., Math.-phys. Klasse, 1927, pp. 1–57.
25. von Neumann, John: *Mathematical Foundations of Quantum Mechanics*, Princeton Univ. Press, Princeton, New Jersey, 1955.
26. Peirce, Charles S.: *The New Elements of Mathematics*, Eisele, Carolyn (ed.): Volume IV – Mathematical Philosophy, Chapter VI – The Logical Algebra of Boole, The Hague: Mouton Publishers, Paris: Humanities Press, N. J.: Atlantic Highlands, 1976, pp. 106–115.
27. Pitowski, Itamar: *Quantum Probability, Quantum Logic*, Lecture Notes in Physics, Berlin: 1989.
28. Post, Emil L.: Introduction to a General Theory of Elementary Propositions, *American Journal of Mathematics*, vol. 43, 1921, pp. 163–185.
29. Russell, Bertrand: *Principles of Mathematics*, 1903; 2nd edition: W. W. Norton & Company 1938.
30. Whitehead, Alfred N.; Russell, Bertrand: *Principia mathematica* 1 (1 ed.) 1910; 2 (1 ed.), 1912; 3 (1 ed.), 1913, Cambridge: Cambridge University Press.
31. Seising, Rudolf: Interview with L. A. Zadeh on July, 26, 2000, University of California, Berkeley, Soda Hall.
32. Seising, Rudolf: Interview with L. A. Zadeh on June 16, 2001, University of California, Berkeley, Soda Hall.
33. Seising, Rudolf: *The Fuzzification of Systems. The Genesis of Fuzzy Set Theory and Its Initial Applications – Developments up to the 1970s*, Berlin, New York, [et al.]: Springer 2007.

34. Seising, Rudolf: General Systems, classical systems, quantum systems, and fuzzy systems: an introductory survey, *International Journal of General Systems* (Special Issue: Fuzzy and Quantum Systems, Rudolf Seising (guest editor), vol. 40 (1), 2011), pp. 1–9.
35. Suppes, Patrick: Probability Concepts in Quantum Mechanics, *Philosophy of Science*, vol. 28, 1961, pp. 378–389.
36. Suppes, Patrick: The Probabilistic Argument for a Non-Classical Logic of Quantum Mechanics, *Philosophy of Science*, vol. 33, 1966, pp. 14–21.
37. Ulrich, Werner: Nonbinary Error Correction Codes, *Bell Systems Technical Journal*, November 1957, pp. 1341–1142.
38. Watanabe, Satosi: Algebra of Observation, *Progress of Theoretical Physics*, Suppl. Nos. 37-38, 1966, pp. 350–367.
39. Watanabe, Satosi: Modified Concepts of Logic, Probability, and Information Based on Generalized Continuous Characteristic Function, *Information and Control*, vol. 15, 1069, pp. 1–21.
40. Zadeh, Lotfi A.: Thinking Machines. A New Field in Electrical Engineering, *Columbia Engineering Quarterly*, January 1950, pp. 12–30.
41. Zadeh, Lotfi A.: From Circuit Theory to System Theory, Proceedings of the IRE, vol. 50, 1962, pp. 856–865.
42. Zadeh, Lotfi A.: Fuzzy Sets, *Information and Control*, vol. 8, 1965, pp. 338–353.
43. Zadeh, Lotfi A.: Fuzzy Sets and Systems. In: J. Fox (ed.): *System Theory*, Microwave Res. Inst. Symp. Series XV, Brooklyn, New York: Polytechnic Press, 1965, pp. 29–37.

Chapter 2
A Dialogue Concerning Contradiction and Reasoning

Enric Trillas

To Claudio Moraga, who, joyously for me, is more than a colleague, is a loved friend.

Carla. The understanding of an idea, once it is linguistically expressed, requires moving back up to reduce it to something that is currently considered clear enough. For instance, and concerning precise predicates, the Aristotle's principle of contradiction expressed by 'nothing can and cannot be simultaneously', is often understood by mathematicians [1] passing through the following steps:

1. Question: Can it be that x is P and is not P, for some elements x into consideration?
2. Suppose that X is the set containing all those elements x.
3. Let it be $\mathbf{P}$ the subset of X specified by the use of P.
4. The subset $\mathbf{P}^c$ of X, the complement of $\mathbf{P}$, just contains those elements in X to which P is not applicable.
5. The principle of contradiction is just represented by the set's law $\mathbf{P} \cap \mathbf{P}^c = \emptyset$;
6. This law is 'identified' with the contradiction principle.
   But this is only acceptable for predicates specifying a set, and fuzzy logic deals with those that can't specify a set where the principle, if identified in an analogous form, fails.

Karl. Well, it depends on the forms of expressing both the 'and', and the 'not'. For instance, if the 'and' is represented by a continuous t-norm T, and the 'not' by a strong negation function N, the corresponding functional equation $T(a, N(a)) = 0$, with a in [0, 1], has the solutions $\mathrm{T} = \mathrm{W}_f$, $\mathrm{N} \leq \mathrm{N}_f$, with f any order-automorphism of the unit interval. Hence, it is not true that the principle always fails, but it is true

E. Trillas (✉)
European Centre for Soft Computing, Mieres (Asturias), Spain
e-mail: etrillasetrillas@gmail.com

R. Seising and H. Allende-Cid (eds.), *Claudio Moraga: A Passion for Multi-Valued Logic and Soft Computing*, Studies in Fuzziness and Soft Computing 349, DOI 10.1007/978-3-319-48317-7_2

Fig. 2.1 Claudio Moraga, Enric Trillas, and Fatima Zohra Hadjam at the "1. International Symposium Fuzziness, Philosophy and Medicine" at the *European Centre for Soft Computing*, Mieres (Asturias), Spain, March 23–24, 2011; in the background, Mila Kwiatkowska; photograph by Rudolf Seising

that it fails in many cases; it holds with $f = id$, $T = W$ and $N = 1 - id$, and fails with $f = id$, $T = prod$, $T = min$, and the same negation $1 - id$ [2].

Carla. Yes, and without considering strong negations there are also positive and negative cases. For instance, with the order-automorphism $f(x) = x^2$, and the non-strong but continuous negation $N(x) = 1 - x^2$, it is $W_f(a, 1 - a^2) = [max(0, a^4 - a^2)]^{1/2} = 0$, but with $f = id$, it is $W(a, 1 - a^2) = max(0, a - a^2) = 0 \Longleftrightarrow a \leq a^2 \Longleftrightarrow a = 0$, or $a = 1$. In the first case the principle holds, but it does not hold in the second.

Karl. Yes, by keeping a continuous t-norm for representing the 'and', and independently of the strong character of the negation, the solutions of the functional equation $T(a, N(a)) = 0$, are $T = W_f$ and $N \leq N_f$, with independence on the properties of N. Obviously, with $f(x) = x^2$, it is $N(x) = 1 - x^2 \leq N_f(x) = [1 - x^2]^{1/2}$. Actually, even without strong negation, there are cases in which the principle fails, and cases in which it holds.

Carla. Even if it is true that the preservation of the Aristotle's principle is bounded, in the standard algebras of fuzzy sets, to expressing the conjunction by a 'f-modification' of the Łukasiewicz t-norm $W(a, b) = max(0, a + b - 1)$, that is, to continuous t-norms in the 'family' of Łukasiewicz, it is also true that the principle not always fails. Hence, it cannot be properly said that fuzzy sets violate the principle [3].

Karl. Let me add that all that comes from understanding the principle *more clasico*, that is, in the form $\mu \cdot \mu' = \mu_0$ that, when $\mu \in \{0, 1\}^X$, reduces to the classical, or crisp, 'empty intersection'. It is a view directly coming from the theory of crisp sets. Are there other views?

Carla. Yes, of course. For instance, by understanding "it is $\mu \cdot \mu'$ impossible" as "$\mu \cdot \mu'$ is not deducible from μ" and the relation of deduction as the pointwise ordering between fuzzy sets, if it were $\mu \cdot \mu$' deducible from μ, or $\mu \leq \mu \cdot \mu'$, since it is clear that the principle holds. But notice that this validity depends on a particular form of interpreting deduction, and only for those μ such that its intersection with μ' is not coincidental with μ; certainly, for a lot of fuzzy sets [4].

Karl. Ok, but I doubt that Aristotle could think on 'impossible' as a synonym of 'deductible'. Even if, perhaps, he could thought on the absurdity of a syllogism with premises p, and p & p'; that any conclusion of such syllogism seems to be contradictory.

Carla. Yes, this is the right concept, self-contradictory! The greatest 'sin' that can be considered. Something that nobody will accept.

Karl. Although Aristotle did state that the principle cannot be submitted to proof, you should have just proven in which cases it holds in the standard algebras of fuzzy sets, and also its validity under the equivalence of impossible and non-deductible. Is it possible to prove it under the interpretation of impossible as self-contradictory?

Carla. Well, it is indeed possible by taking into account some definitions and by supposing some properties of what is defined. What is interesting to note is that the proof can be achieved under a simple symbolism verifying very few and soft suppositions.

Karl. It seems to mean that the principle holds in very general settings. Go ahead, please!

Carla. Suppose it exists a 'natural' relationship of inference that, even produced in the brain as a part of the natural phenomenon called *thinking*, can be represented by the symbol $\leq$, with which $p \leq q$ represents that the statements p and q are linked in the form 'q is inferred from p'. Two statements p and q are called inferentially equivalent whenever $p \leq q$ and $q \leq p$, and can be written $p = q$. Of this symbol we need to just supposing it is, in itself, transitive, that is, verifies: $\mathrm{p} \leq \mathrm{q}, \mathrm{q} \leq \mathrm{r} \Longrightarrow \mathrm{p} \leq \mathrm{r}$. In addition, it is required to count with the symbols (&) for 'and', and (′) for 'not', of which it will be supposed they verify the simple laws $p \,\&\, q \leq p$, $p \,\&\, q \leq q$, and $p \leq (p')'$. Finally, a definition: p is contradictory with q if and only if $p \leq q'$; hence, p is self-contradictory if and only if $p \leq p'$.

Karl. Wow, I see the proof! It is very simple [5]:

$$p \,\&\, p' \leq p,$$
$$p \,\&\, p' \leq p' \Longrightarrow (p')' \leq (p \,\&\, p')',$$
$$p \leq (p')' \Longrightarrow p \leq (p \,\&\, p')':$$
$$p \leq p' \leq (p \,\&\, p')'.$$

That is, p & p' is self-contradictory, or impossible!

Carla. Yes, that's all. Under very few hypotheses on the behavior of the conjunction, and the negation, a symbolic proof of the principle follows. And such proof holds when the symbols & and $'$ are interpreted in different forms; for instance, when taking the relation $\leq$ as the natural ordering of the lattice, as the connectives in an Ortholattice, or a De Morgan algebra, and of course, in a Orthomodular lattice, a Boolean algebra, and, of course, as those in a Basic algebra of fuzzy sets and, in particular, in a standard algebra of them with, always, the inference relation identified with the pointwise ordering of fuzzy sets. Actually, the proof covers a big number of frameworks where statements are typically represented, and, in particular, all cases where the negation is strong, that is, verifies $p \leq (p')'$ and $(p')' \leq p$, or p and $(p')'$ are inferentially equivalent.

Karl. Really nice! All that means that in classical, quantum, intuitionistic, and fuzzy logics, the contradiction principle actually holds, and thinkers can be relaxed for what concerns the solid character of their grounds. The fact that the principle can be proven is, in my view, something new.

Carla. Notice that neither commutative nor associative laws for &, no duality law for both & and $'$, no functional expressibility of the connectives in the fuzzy case, etc., are supposed. The theorem is free from most of the constraints that are typically considered. Anyway, the theorem can fail in those cases in which it is not $p \leq (p')'$, that is, the double negation of a single statement p cannot be inferentially obtained from it; nevertheless, in any case, the principle will fail for this statement p, or for all those statements whose double negation will be not reachable from them.

Karl. Turning back to the concept of contradiction, it should be noticed that in the setting of Ortholattices it is $p \leq q' \Longrightarrow p \cdot q = 0$, since the inequality implies $p \cdot q \leq q' \cdot q = 0$. In the case of Orthomodular lattices, and provided p and q commute, that is, $p = p \cdot q + p \cdot q'$, the reciprocal holds since $p \cdot q = 0$ implies $p = p \cdot q'$, or $p \leq q'$. In particular, since in Boolean algebras all pair of elements commute, it is $p \leq q' \Longleftrightarrow p \cdot q = 0$. Thus, in Boolean algebras, it is $p \leq p' \Longleftrightarrow p = 0$: the only self-contradictory element is 0; this is perhaps a reason for which it is difficult, at least for children, to accept the empty set, $\emptyset$, as a set.

Carla. But this is not the case in De Morgan algebras, since, for instance, in the case of the De Morgan algebra in the unit interval $[0, 1]$ with $\cdot = min$, $+ = max$, and $' = 1 - id$, it is $0.3' = 1 - 0.3 = 0, 7$, but $0.3 \cdot 0.7 = 0.3 \neq 0$, even if $0.3 < 0.7 = 0.3'$. In this example, there are a lot of self-contradictory elements since $p \leq p' = 1 - p \Longleftrightarrow p < 0.5$, that is, all numbers in the sub-interval $[0, 0.5]$, and only they, are self-contradictory. Anyway, if $p \cdot q = min(p, q) = 0$, since it implies $p = 0$, or $q = 0$, it follows $0 = p \leq q'$, or $0 = q \leq p'$.

Karl. Contradiction is a symmetrical property, $p \leq q' \Leftrightarrow q \leq p'$, provided the negation verifies $q \leq (q')'$, since $p \leq q' \Rightarrow (q')' \leq p' \Rightarrow q \leq p'$.

Carla. Boolean algebras are endowed with so a lot of laws that many theorems can be proven in them. Boolean algebras cannot allow, for instance, to distinguish

Fig. 2.2 Claudio Moraga and Ramon López de Mántaras during the First "Alfredo Dean" Seminar on Ordinary Reasoning, to be held at the Cultural Center "Muralla Romana" of Gijon (Asturias), Spain, April 29, 2011.; photograph by Rudolf Seising

between incompatibility and contradiction. In the classical setting contradiction and incompatibility are indistinguishable concepts, but not in the quantum, nor in the fuzzy case, and less again in language. That is, Boolean algebras are not well suited for representing commonsense reasoning; they have too much constrictions, and it analogously happens with Ortholattices even if the particular case of Orthomodular lattices are sometimes taken to represent and to study some aspects of quantum reasoning. But in commonsense reasoning there are laws common to both Ortholattices and De Morgan algebras, that are not always valid like it appears, for instance, when studying the validity of the laws $(\mu \cdot \sigma)' = \sigma + (\mu' \cdot \sigma')$, and $\mu = \mu \cdot \sigma + \mu \cdot \sigma'$, with fuzzy sets [6]. Let me say that what cannot be supposed for the second law when dealing with imprecise predicates is duality!

Karl. And what on the principle of excluded middle? Since it does not hold in intuitionistic logic, I suppose no straightforward proof is available.

Carla. Well, excluded middle, or *tertium non datur*, was not so clearly explicitly posed by Aristotle as it was contradiction, but, and from very early in the Middle

Age, it is understood by 'either the affirmative, or the negative, always hold', that in set theory is represented by $\mathbf{P} \cup \mathbf{P}^c = X$, if the statements are on a universe of discourse X, and are specified by its subsets. In this case, it straightforwardly follows from contradiction because of duality: $\mathbf{P} \cup \mathbf{P}^c = (\mathbf{P} \cap \mathbf{P}^c)^c = \emptyset^c = X$, and reciprocally. In classical logic both principles are equivalent. They correspond to stating "'x is P' or 'x is not P'" for all x in X.

Karl. In the case of lattices with a negation it corresponds to $p + p' = 1$, with the disjunction $+$ and the maximum, 1, of the natural order, and that, in Ortholattices is a structural law, but not in De Morgan algebras. Of course, since in these lattices the negation is strong, the disjunction is commutative, and the duality law $p' + q' = (p \cdot q)'$ holds, it is clear that, like with sets, it is $p + p' = p'' + p' = (p' \cdot p)' = (p \cdot p')' = 0' = 1$. The two laws of contradiction and excluded middle are equivalent, and both do hold.

Carla. With fuzzy sets in a Basic algebra, the principle can be posed in the same form $\mu + \mu' = \mu_1$, and in the very particular case of a standard algebra its validity comes from solving the functional equation $S(a, N(a)) = 1$, for all a in [0, 1], with S a continuous t-conorm, N a negation either strict, or strong, and whose solution is $S = W*_f$, and $N_f \leq N$, for all order-automorphisms f of the unit interval. Hence, there are cases in which the principle holds and others that not as, for instance and respectively, with $S = W*$, or with $S = max$, and in both cases with $N = 1 - id$[2].

Karl. Thus, the two principles jointly hold in the standard algebras of fuzzy sets with $T = W_f$, $S = W*_g$, and $N_g \leq N \leq N_f$, where f and g are order-automorphisms of the unit interval [2]. Of course, if S is the N-dual of T, $S = N \circ T \circ (N \times N)$, the two principles are equivalent like it happens in lattices.

Carla. Analogously to the first principle, the second can be analogously posed by asking if "p or p'" is deductible from the single premise p, and each time the disjunction $+$ verifies $p \leq p + q$, it is $p \leq p + p'$. That is, $p + q$ is actually deductible from p.

Karl. By interpreting $\leq$ as the 'natural' relation of inference, and accepting it verifies $p \leq p + q$, what can be simply said is that $p + q$ is inferable from p. Is not it?

Carla. Yes, it is, and 'inferable from p' either can mean that $p + q$ can be conjectured from p, or that it can be deduced from p depending from which properties can have the relation $\leq$.

Karl. And, concerning the beforehand interpretation of impossible by self-contradictory, what can be said?

Carla. The problem is not identical to that of contradiction, since what should be here represented is 'always holds'. It is an affirmative argument for which it seems that a constructive argument could be in order. Trying to prove it through a reduction to absurd reasoning is another possibility that, notwithstanding, has the trouble of representing the 'always'.

In the case of the unit interval before mentioned, notice that the numbers that are not self-contradictory are those in (0.5, 1), coming from the inequality $p' \leq p \Leftrightarrow$

$1 - p \le p \Leftrightarrow 0.5 \le p$, equivalent to $p' \le (p')'$. This last expression shows that the numbers p not self-contradictory are those whose negation is self-contradictory; these numbers, being not self-contradictory, and after being understood those in $[0, 0.5]$ as ‘impossible’, can be seen as ‘possible’. Thus, and since in presence of duality it is $p + p' = (p \cdot p')'$, it appears the possibility of identifying the excluded middle with the inequality $(p + p')' \le ((p + p')')'$. Can it be proven?

Karl. I see. Under duality and double negation, last inequality reduces to $p' \cdot p \le p + p'$, whose validity is obvious. I suppose, by analogy with the proof of contradiction, that it should come from accepting the inferences $p \le p + q$, and $q \le p + q$. Is not it?

Carla. Yes, it seems so. They imply $(p + q)' \le p'$, $(p + q)' \le q'$, and then the question is how to continue for arriving at a proof.

Karl. Very easy once accepted that the negation verifies $p \le (p')'$. The proof is:

$$p \le p + p' \Rightarrow (p + p')' \le p' \le p + p' \le ((p + p')')'.$$

Carla. Yes, but there is a problem in it. The step $p' \le p + p'$, is allowed?

Karl. You are right ... we did not yet accept the commutative property of the operations. It is $p' \le p' + p$.

Carla. There is the exit’s way of supposing not only $p \le p + q$, but also $p \le q + p$, that does not seem anything bizarre. What do you think?

Karl. Certainly it is not a bizarre supposition that can be also accepted for conjunction: $p \cdot q \le p$, and $q \cdot p \le p$. Hence, with it the proof is ok.

Carla. In the case of the unit interval, the inequalities translating the two principles should be verified by all the numbers between 0 and 1. Since there is duality and the negation is strong it suffices to see it for just one of them, for instance, $p \cdot p' \le (p \cdot p')' \Leftrightarrow min(p, 1 - p) \le 1 - min(p, 1 - p) = max(1 - p, p)$, obviously true.

Karl. What turns in my mind is that intuitionistic logic not accepts excluded middle, but that in the form we established the principle, it should hold. There seems to be something rare.

Carla. I don’t think so since no constructive procedure asserts here that $p + p'$ is true. We just prove the principle as a theorem by saying that “not (p or not q)” is self-contradictory that, in some way, means that “p or not p” is not self-contradictory. Passing from this to truth, it lacks a form of asserting what is true, and we don’t entered in such a question that, I think, is in the center of intuitionism. We left constructive truth aside. What we add is just that in all those cases in which the inference relation, the negation, the conjunction, and the disjunction, satisfy a few and very soft properties, the principles hold as theorems on the basis of considering self-contradiction as a death ‘sin’. We escaped from truth and the ways for asserting it. But nothing else!

Karl. May be ours is merely an idealistic way of posing the principles ...

Carla. Why? I am not sure. In the classical case $p \cdot p' \leq ((p \cdot p')'$ just means $p \cdot p' = 0$, and $(p + p')' \leq ((p + p')')'$ means $p + p' = 1$. That is, our way of posing the principles reduces to the common way in which they are classically posed, with the difference that there the principles are axioms and here theorems. I see the classical axiomatic way more idealistic than is ours. We find minimal conditions supporting the principles, and these conditions seem to be of a so great generality that allow the validity of both principles in many settings.

Karl. The Aristotle's idea that the principle of contradiction cannot be submitted to proof seems to be challenged by an interpretation of the word impossible he did use, as the concept self-contradictory, and in a framework of great amplitude, where the two principles of contradiction and excluded middle, appear as theorems. These theorems reduce, in the classical case, to the set interpretation of the principles, and shows that both the algebras of fuzzy sets and also De Morgan algebras satisfy these principles: It cannot be properly said that De Morgan algebras and fuzzy sets violate the two principles.

Carla. There are indeed some multiple-valued logics also verifying them, although some others did not because of the properties their negations 'enjoy'. Nevertheless, what still is not fully clear is which Is the relationship between what has been presented with both the concept of truth, and, specially, with the constructivism for asserting truth and that intuitionists consider essential for assuming mathematical results. In fact, from what we escape is from constructivism.

Karl. This, jointly with the analysis of the two principles for consequences operators whatsoever, that is, for any kind of deductive inference, remains open to scrutiny.

Carla. Anyway, all that on which we jointly debated here cannot be seen as something belonging to just the province of logic, but to the larger domains of natural language and commonsense reasoning. And, at this respect the words of Russell and Whitehead, that mathematics and logic are merely concerned with 'the correct use of a certain small number of words', should be specifically taken into account. Common reasoning concerns all words, and in all contexts.

Karl. All right. May be, and provided our discussion could have any value at all, it could lie in a change of perspective for the study of the reasoning's phenomena by placing at its center the ideas of a natural linking between statements, and contradiction, instead of truth and the representation of conditional, If/Then, statements by any sort of material conditional.

Carla. Also, and from my point of view, to abandon the expression of the conditional 'If p, then q' coming from, for instance and in Boolean algebras, the logical equivalence between $p' + q = 1$ and $p \leq q$, by substituting the partial order by a just transitive 'natural' relation that links statements, could arrive to be fruitful. Who knows!

Fig. 2.3 Claudio Moraga, at the "1. International Symposium Fuzziness, Philosophy and Medicine" at the *European Centre for Soft Computing*, Mieres (Asturias), Spain, March 23–24, 2011, in the background, Clara Barroso; photograph by Rudolf Seising

References

1. P. R. Halmos: *Naïve Set Theory*, Van Nostrand, 1961.
2. E. Trillas, C. Alsina, J. Jacas: *On contradiction in fuzzy logic*, *Soft Computing*, Vol. 3/4, 1999, pp. 197–199.
3. E. Trillas, C. Alsina, A. Pradera: *Searching for the roots of non-contradiction and excluded-middle*, *International Journal of General Systems*, Vol. 31/5, 2002,pp. 499–513.
4. E. Trillas, I. García-Honrado, S. Termini: *Some algebraic clues towards a syntactic view on the principles of Non-Contradiction and Excluded-Middle*, *International Journal of General Systems*, Vol. 43/2, 2014, pp. 162–171.
5. E. Trillas: *Non-contradiction, excluded-middle, and fuzzy sets*, in: Vito Di Gesù, Sankar K. Pal, Alfredo Petrosino: Fuzzy Logic and Applications, 8th International Workshop, WILF 2009, Palermo, Italy, June 9–12, 2009, Proceedings. Lecture Notes in Computer Science 5571, Springer, pp. 1–11.
6. E. Trillas, C. Alsina: *Standard theories of fuzzy sets with the law* $(\mu \cdot \sigma)' = \sigma + (\mu' \cdot \sigma')$, *International Journal of Approximate Reasoning*, Vol. 37/2, 2004, pp. 87–92.
7. A. N. Whitehead, B. Russell, *Principia mathematica*, Cambridge University Press, 1910.

Chapter 3
Some Entertainments Dealing with Three Valued Logic

Itziar García-Honrado

3.1 Introduction

I met Professor Claudio Moraga when I was carrying out my Ph.D. in the *European Centre for Soft Computing* (ECSC) in Mieres (Spain), I was a member of the Unit of Fundamentals of Soft Computing and he was working there as an Emeritus Researcher.

During that time he was a great supporter, he helped me in my researches correcting some papers and commenting me some ideas to enlarge them. I am grateful for his work and his kindness.

In 2011, we attended together to the *IEEE International Symposium on Multiple-Valued Logic* in Tuusula, Finland. I presented there a study elaborated with Professor Enric Trillas [1] about symmetric difference. This work studied the behavior of symmetric difference found out in which cases disjunctive syllogism for a exclusive *or* is verified. The study was done in three valued logic, and in Standard Algebras of fuzzy sets through functional equations. So, in that paper an idea which I want to show in this writing is collected, it is that in the path to generalize some classical rules to fuzzy logic, an intermediate steep could be considered: to analyze the behavior in multiple-valued logic, for instance in three valued logic.

This is the reason to propose along this chapter some "entertainments" with some well known three valued logic: Łukasiewicz, Gödel, Kleene, Bochvar and Post. Through the truth tables of basic operation $(+, \cdot, ')$ in all these logics. I propose to study the behavior of Sheffer stroke and its classical properties (following the methodology done for symmetric difference) and from new conception of the principle of no contradiction to study of different set of conjectures [2]. Therefore, some examples of how Conjectures models works in these three valued logic will be shown.

I. García-Honrado (✉)
Facultad de Ciencias, Departamento de Estadística, Investigación Operativa y Didáctica de las Matemáticas, Universidad de Oviedo, Oviedo, Spain
e-mail: garciaitziar@uniovi.es

R. Seising and H. Allende-Cid (eds.), *Claudio Moraga: A Passion for Multi-Valued Logic and Soft Computing*, Studies in Fuzziness and Soft Computing 349, DOI 10.1007/978-3-319-48317-7_3

3.2 Three-Valued Logic

Classical logic allows elements to totally verify or not a proposition, deals with values 0 and 1. In the case of three-valued logic a new value is introduce, $\frac{1}{2}$, this is the simplest way to break the duality of classical logic and what makes difficult translating directly our knowledge of classical logic.

Along this chapter operations in three-valued logic are considered. All these operations are based in the basic operation: negation ($'$), intersection or product ($\cdot$), and union or addition ($+$). Behind, tables [3] defining these basic operations are collected for the logic of: Łukasiewicz, Gödel, Kleene, Bochvar and Post.

- Łukasiewicz

	$'$
1	0
$\frac{1}{2}$	$\frac{1}{2}$
0	1

$\cdot$	1	$\frac{1}{2}$	0
1	1	$\frac{1}{2}$	0
$\frac{1}{2}$	$\frac{1}{2}$	$\frac{1}{2}$	0
0	0	0	0

$+$	1	$\frac{1}{2}$	0
1	1	1	1
$\frac{1}{2}$	1	$\frac{1}{2}$	$\frac{1}{2}$
0	1	$\frac{1}{2}$	0

- Gödel

	$'$
1	0
$\frac{1}{2}$	0
0	1

$\cdot$	1	$\frac{1}{2}$	0
1	1	$\frac{1}{2}$	0
$\frac{1}{2}$	$\frac{1}{2}$	$\frac{1}{2}$	0
0	0	0	0

$+$	1	$\frac{1}{2}$	0
1	1	1	1
$\frac{1}{2}$	1	$\frac{1}{2}$	$\frac{1}{2}$
0	1	$\frac{1}{2}$	0

- Kleene

	$'$
1	0
$\frac{1}{2}$	$\frac{1}{2}$
0	1

$\cdot$	1	$\frac{1}{2}$	0
1	1	$\frac{1}{2}$	0
$\frac{1}{2}$	$\frac{1}{2}$	$\frac{1}{2}$	0
0	0	0	0

$+$	1	$\frac{1}{2}$	0
1	1	1	1
$\frac{1}{2}$	1	$\frac{1}{2}$	$\frac{1}{2}$
0	1	$\frac{1}{2}$	0

- Bochvar

	$'$
1	0
$\frac{1}{2}$	$\frac{1}{2}$
0	1

$\cdot$	1	$\frac{1}{2}$	0
1	1	$\frac{1}{2}$	0
$\frac{1}{2}$	$\frac{1}{2}$	$\frac{1}{2}$	$\frac{1}{2}$
0	0	$\frac{1}{2}$	0

$+$	1	$\frac{1}{2}$	0
1	1	$\frac{1}{2}$	1
$\frac{1}{2}$	$\frac{1}{2}$	$\frac{1}{2}$	$\frac{1}{2}$
0	1	$\frac{1}{2}$	0

- Post

	$'$
1	$\frac{1}{2}$
$\frac{1}{2}$	0
0	1

$\cdot$	1	$\frac{1}{2}$	0
1	0	0	$\frac{1}{2}$
$\frac{1}{2}$	0	1	$\frac{1}{2}$
0	$\frac{1}{2}$	$\frac{1}{2}$	$\frac{1}{2}$

$+$	1	$\frac{1}{2}$	0
1	1	1	1
$\frac{1}{2}$	1	$\frac{1}{2}$	$\frac{1}{2}$
0	1	$\frac{1}{2}$	0

3.3 Translating Terms and Operations from Classical Logic

The presence of a third element, the undecided one, make difficult translating all principles or laws that are verified in classical logic. The conflict starts with principles that are considered the basis of the logic, as it is the case of the Aristotelian Principles of Non-Contradiction and Excluded-Middle [4]. The Principle of Non-Contradiction formulated by 'It is impossible for anything at the same time to be and not to be' it is usually translated as $a \cdot a' = 0$. It is clear that, for instance, in the three-valued logic of Łukasiewicz, taking $a = \frac{1}{2}$, it is obtained that $a \cdot a' = \frac{1}{2} \neq 0$. Then, dealing with multi-valued logic some conflicts happened: our knowledge of classical logic translated to multiple-valued logic fails (or mostly, fails) or we have the challenge of rethink how the solid bases of classical logic could work.

Therefore, in this chapter to proceed with three valued logic, to different aspects are considered:

- To analyze the behavior of some classical laws directly translated into the operations $(', \cdot, +)$ as it is done in classical logic.
- To find some translations that covered the case of classical logic and englobe the verification of some three-valued logics.

Regarding these second vision, the Aristotelian principle of Non- Contradiction could be translated as 'a and not a' is 'self-contradictory', that is equivalent to "If 'a and not a', then not (a and not a)" [5, 6]. With this translation three-valued logic of Łukasiewicz verifies that principle, but it is not the case of the three-valued logic of Post, in which taking $a = \frac{1}{2}$, it is obtained that $a \cdot a' = \frac{1}{2} \cdot 0 = \frac{1}{2} \not\leq 0 = (\frac{1}{2})' = (a \cdot a')'$.

Along this writing, with respect to the first point, two classical laws will be named:

- Symmetric difference, $a \cdot b = (a + b) \cdot (a \cdot b)'$, and the exclusive syllogism disjunctive $(a \cdot b) \cdot a' \leq b$ and $(a \cdot b) \cdot a \leq b'$.
- Sheffer stroke, $a|b = (a \cdot b)'$, and the relation of sheffer stroke and its classical equivalences $a' = a|a$, $a \cdot b = (a|b)|(b|a)$ and $a + b = (a|a)|(b|b)$

The representation of both of them with Venn diagrams is shown in Fig. 3.1. Both of them get rid of the intersection. To translate both operations in language; symmetric difference could be understood as either a or b, while Sheffer stroke could be understood as neither a nor b.

On other hand, considering different interpretation of the term *not inconsistent*: not incompatible, not self-contradictory or not contradictory, an analysis of sets of conjectures depending on these interpretations in all the previous three-valued logic will be done.

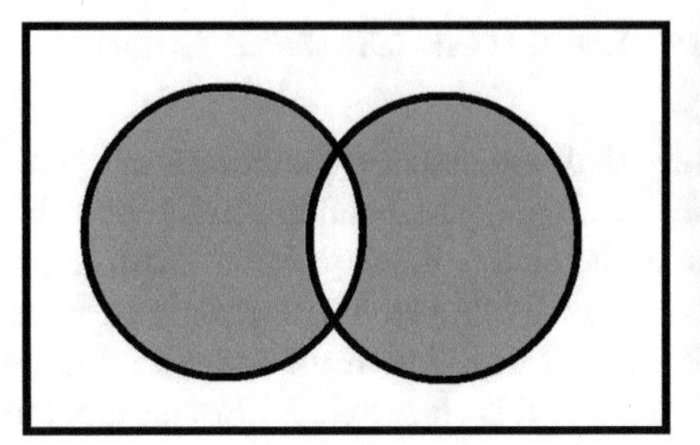
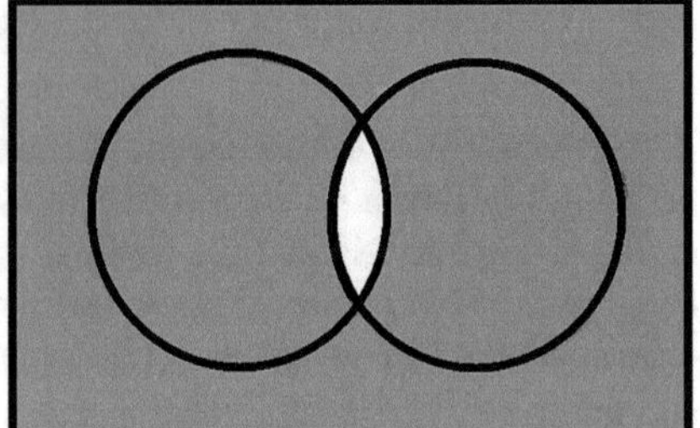

Fig. 3.1 Venn diagrams of symmetric difference and Sheffer stroke

In classical logic, the term not incompatible $a \cdot b \neq 0$ is equivalent to not self-contradictory $a \cdot b \not\leq (a \cdot b)'$ of to not contradictory $a \not\leq b'$. If $a \cdot b = 0$ and $0 \leq 1$, it is $a \cdot b \leq (a \cdot b)'$. In the same way from $a \cdot b \leq (a \cdot b)'$, it is obtained $a = a \cdot b \cdot b' \leq (a \cdot b)' \cdot b'$. Finally, if $a \leq b'$, it is $a \cdot b' \leq b \cdot b' = 0$.

These equivalences are not verified in three-valued logic. For instance, in three valued logic of Post, for $1 \cdot \frac{1}{2} = 0$ but $1 \not\leq \frac{1}{2}' = 0$. Therefore, the variety of these logics allow us to study the behavior of those interpretations of *not incompatible* through the set of conjectures.

Conjectures are those elements that are *not incompatible* with the information available that is, with the information contents in the set of premises. That is the relation, which shows the reason to obtain three different set of conjectures depending of the previous translations of the term, not incompatible.

Let me deals with a resumé of Premises $r(P)$. Framing set of conjectures in three valued logic, the resumé will be considered an element in $L = \{0, \frac{1}{2}, 1\}$.

Then, these three set of conjectures will be considered:

- $Conj_1(P) = \{q \in L; r(P) \cdot q \neq 0\}$
- $Conj_2(P) = \{q \in L; r(P) \cdot q \not\leq (r(P) \cdot q)'\}$
- $Conj_3(P) = \{q \in L; r(P) \not\leq q'\}$

3.4 Two Operations: Symmetric Difference and Sheffer Stroke

In a previous publication [7], it is analyzed the behavior of symmetric difference in those three valued logics. Firstly, tables of the operation $a \Delta b = (a + b) \cdot (a \cdot b)'$ are obtained one for Łukasiewicz, Kleene and Bochvar logics, other for Gödel's, and other for Post's.

- Łukasiewicz, Kleene and Bochvar

Δ	1	$\frac{1}{2}$	0
1	0	$\frac{1}{2}$	1
$\frac{1}{2}$	$\frac{1}{2}$	$\frac{1}{2}$	$\frac{1}{2}$
0	1	$\frac{1}{2}$	0

- Gödel

Δ	1	$\frac{1}{2}$	0
1	0	0	1
$\frac{1}{2}$	0	0	$\frac{1}{2}$
0	1	$\frac{1}{2}$	0

- Post

Δ	1	$\frac{1}{2}$	0
1	0	0	1
$\frac{1}{2}$	0	1	$\frac{1}{2}$
0	$\frac{1}{2}$	$\frac{1}{2}$	$\frac{1}{2}$

Then, the schemes of inference '$(a \cdot b) \cdot a' \leq b$ and $(a \cdot b) \cdot a \leq b'$' are studied, getting as results that only in the three valued logic of Gödel those schemes are verified.

To proceed with Sheffer stoke the same steps will be follows.

Four different tables of the operation $a|b = (a \cdot b)'$ are obtained, one for Łukasiewicz and Kleene, and the other three for: Gödel, Bochvar, and Post.

- Łukasiewicz and Kleene

$\mid$	1	$\frac{1}{2}$	0
1	0	$\frac{1}{2}$	1
$\frac{1}{2}$	$\frac{1}{2}$	$\frac{1}{2}$	1
0	1	1	1

- Gödel

$\mid$	1	$\frac{1}{2}$	0
1	0	0	1
$\frac{1}{2}$	0	0	1
0	1	1	1

- Bochvar

$\mid$	1	$\frac{1}{2}$	0
1	0	$\frac{1}{2}$	1
$\frac{1}{2}$	$\frac{1}{2}$	$\frac{1}{2}$	$\frac{1}{2}$
0	1	$\frac{1}{2}$	1

- Post

$\mid$	1	$\frac{1}{2}$	0
1	1	1	0
$\frac{1}{2}$	1	$\frac{1}{2}$	0
0	0	0	0

Note that the values hold in classical logic when dealing with elements $\{0, 1\}$ are preserved in Łukasiewicz, Kleene, Gödel and Bochvar but not in Post. And in the case of three valued logic of Gödel Sheffer stroke does not allow undecided elements as result.

With the tables of operations it is easy to analize what happens with the classical equivalences of Sheffer stroke: $a' = a|a$, $a \cdot b = (a|b)|(b|a)$ and $a + b = (a|a)|(b|b)$. In the case of Łukasiewicz, Kleene, and Bochvar those equivalences are hold, but it is not the case of Gödel where only the equivalence related to the negation is hold, the others do not hold since through Sheffer stroke all undecidable elements become 0 or 1.

- Gödel

$(a\mid b)\mid(b\mid a)$	1	$\frac{1}{2}$	0
1	1	1	1
$\frac{1}{2}$	1	0	1
0	1	1	0

$(a\mid a)\mid(b\mid b)$	1	$\frac{1}{2}$	0
1	1	1	0
$\frac{1}{2}$	1	1	0
0	0	0	0

Regarding to Post, any equivalence is hold, in fact the table of Sheffer stroke is coincidental with the one of $(a|b)|(b|a)$, and with the table of the negation of the operation $a \cdot b$.

- Post

$(a\|b)\|(b\|a)$	1	$\frac{1}{2}$	0
1	1	1	0
$\frac{1}{2}$	1	$\frac{1}{2}$	0
0	0	0	0

$(a\|a)\|(b\|b)$	1	$\frac{1}{2}$	0
1	$\frac{1}{2}$	0	1
$\frac{1}{2}$	0	0	0
0	1	0	1

3.5 Conjectures

As it is indicated before, this section is devoted to study how the sets of conjectures previously defined behave in the collected three valued logic.

First of all, to obtain conjectures we need an algebraic structure. In [2] it appears the definition of Basic Flexible Algebra (BFA), this structure collects the minimum characteristics an algebraic structure should verify in order to build a Conjectures models.

Definition 3.1 A Basic Flexible Algebra (BFA) is a seven-tuple $\mathscr{L} = (L, \leq, 0, 1; \cdot, +, ')$, where L is a non-empty set, and

1. $(L, \leq)$ is a poset with minimum 0, and maximum 1.
2. The binary operations $\cdot$ and $+$ are mappings $L \times L \to L$, such that:
   (a) $a \cdot 1 = 1 \cdot a = a$, $a \cdot 0 = 0 \cdot a = 0$, for all $a \in L$
   (b) $a + 1 = 1 + a = 1$, $a + 0 = 0 + a = a$, for all $a \in L$
   (c) If $a \leq b$, then $a \cdot c \leq b \cdot c$, $c \cdot a \leq c \cdot b$, for all $a, b, c \in L$
   (d) If $a \leq b$, then $a + c \leq b + c$, $c + a \leq c + b$, for all $a, b, c \in L$
3. The operation $' : L \to L$ verifies
   (a) $0' = 1$, $1' = 0$
   (b) If $a \leq b$, then $b' \leq a'$
4. It exists L_0, $\{0, 1\} \subset L_0 \subsetneq L$, such that with the restriction of the order and the three operations $\cdot$, $+$, and $'$ of $\mathscr{L}$, $\mathscr{L}_0 = (L_0, \leq, 0, 1; \cdot, +, ')$ is a Boolean algebra

Many well known algebraic structures [8] are BFA: Boolean algebras, ortholattices, De Morgan algebras,...

In this section, set of conjectures are built from a BFA using the operations defined in three-valued logic. The poset is $L = \{0, \frac{1}{2}, 1\}$ with the linear order $0 \leq \frac{1}{2} \leq 1$. Then, the operations $(+, \cdot, ')$ defined in Łukasiewicz, Kleene and Gödel three valued logic verify items 2 and 3 in the definition. It is not the case of the operations defined in Bochvar or Post three valued logic, since it is not verified the monotonicity of $\cdot$ or $+$. In the case of Bochvar it is enough to take $0 \leq \frac{1}{2}$, and then $1 = 0 + 1 \leq \frac{1}{2} + 1 = \frac{1}{2}$. Analogously, in the case of Post, it is enough to take $0 \leq \frac{1}{2}$ and then $\frac{1}{2} = 0 \cdot 1 \not\leq \frac{1}{2} \cdot 1 = 0$.

Finally, regarding item 4 of the definition, the operations defined in Łukasiewicz, Kleene and Gödel verify the classical operations in $\{0, 1\}$. Therefore, since operations in Łukasiewicz, Kleene are coincidental, two different BFA can be considered, let me referee to each one as Łukasiewicz and Gödel BFA.

To obtain a conjecture, it is necessary to have a body of information, that is called set of premises, P. In this section, it is dealt with elemental sets of premises, those that have only one element. If it were a set of two or more premises, these premises would be summarized in the résumé, that could be the intersection of all of them.

This premise is selected from the set $L = \{0, \frac{1}{2}, 1\}$, the element 0 is not considered as a premise since 0 do not give any information.

So, a table of conjectures from the element of total knowledge (1), and another table for conjectures from the undecidable element ($\frac{1}{2}$) are obtained for each different sets of Conjectures depending of the translation of *not inconsistent*:

$P = \{1\}$	$Conj_1(P)$	$Conj_2(P)$	$Conj_3(P)$
Łukasiewicz and Kleene	$\{1, \frac{1}{2}\}$	$\{1\}$	$\{1, \frac{1}{2}\}$
Gödel	$\{1, \frac{1}{2}\}$	$\{1, \frac{1}{2}\}$	$\{1, \frac{1}{2}\}$
$P = \{\frac{1}{2}\}$	$Conj_1(P)$	$Conj_2(P)$	$Conj_3(P)$
Łukasiewicz and Kleene	$\{1, \frac{1}{2}\}$	$\emptyset$	$\{1\}$
Gödel	$\{1, \frac{1}{2}\}$	$\{1, \frac{1}{2}\}$	$\{1, \frac{1}{2}\}$

In the case of Gödel BFA, all sets of conjectures are equal, all elements different to 0 are conjectures. In Łukasiewicz BFA different sets of conjectures depending the on the interpretation of *not inconsistent* and the premise used are obtained.

Accordingly, let us considered some properties that conjectures verify in orthomodular lattices [2]:

1. $Conj_i(p) \neq \emptyset$, for $p \in L$ and $i = 1, 2, 3$.
2. $Conj$ is expansive: $P \subset Conj(P)$
3. It exists an operator C_i such that $Conj_i(P) = \{q \in L; q' \notin C_i(P)\}$, for $i = 1, 2, 3$.

Let us check if the sets $C_i(P)$ (for $P = \{1\}, \{\frac{1}{2}\}$ and $i = 1, 2, 3$) obtained above verify that properties:

1. In the case of Łukasiewicz BFA, it is $Conj_2(\frac{1}{2}) = \emptyset$.
2. In the case of Łukasiewicz BFA, it is $\{\frac{1}{2}\} \notin Conj_i(\frac{1}{2})$ for $i = 2, 3$.
3. In orthomodular lattices the operators C are operators of consequences in the sense of Tarski: extensive ($P \subset C(P)$), monotonic (If $P \subset Q$, then $C(P) \subset C(Q)$) and a clausure ($C(C(P)) = C(P)$). For each set of conjectures they are defined as:

- $C_1(p) = \{q \in L;\ p \cdot q' = 0\}$
- $C_2(p) = \{q \in L;\ p \cdot q' \leq (p \cdot q')'\}$
- $C_3(p) = \{q \in L;\ p \leq q\}$.

To define sets of conjectures as $Conj_i(P) = \{q \in L; q' \notin C_i(P)\}$ (for $i = 1, 2, 3$), it is necessary that the negation is involutive, it is $(a')' = a$. But this property is not hold on Gödel BFA.

Therefore, let's calculate sets of $C_i(\{1\})$, and $C_i(\{\frac{1}{2}\})$, for $i = 1, 2, 3$ in Łukasiewicz BFA.

Łukasiewicz	$C_1(P)$	$C_2(P)$	$C_3(P)$
$P = \{1\}$	$\{1\}$	$\{\frac{1}{2}, 1\}$	$\{1\}$
$P = \{\frac{1}{2}\}$	$\{1\}$	$\{0, \frac{1}{2}\}$	$\{1, \frac{1}{2}\}$

In that case, it is not possible that C_i is an operator of consequences since it is not extensive, since for $i = 1, 2, 3$. For instance, taking $P = \{\frac{1}{2}\}$, it is $\{\frac{1}{2}\} \not\subseteq C_1(P) = \{1\}$.

Moreover, in orthomodular lattices $C_i(P) \subset Conj_i(P)$, but in Łukasiewicz BFA is not hold since taking $P = \{\frac{1}{2}\}$, $C_2(P)\{\frac{1}{2}\} = \{0, \frac{1}{2}\} \not\subseteq \emptyset = Conj_2(\{\frac{1}{2}\})$.

These problems are solved in [9] by avoid set of premises, or elements in premises p that are self contradictory in the sense that $p \leq p'$. In that way, the premise $P = \{\frac{1}{2}\}$ could not be considered. What is more, in [9] operators of consequences considered are consistent operator of consequences, so it is impossible the existence of an element and its negation in the same set of conjectures, therefore self contradictory elements could not be considered consequences. It is easy to cheek that in Łukasiewicz BFA the operator C_2 is not consistent since $\frac{1}{2} = (\frac{1}{2})' \in C_2(\{1\})$.

As conclusion, all this operations done with three valued logic under the definition of BFA, allows us to corroborate that the problems that appears when trying to enlarge Conjectures models into Standard Algebras of Fuzzy Sets, that is the set of fuzzy set defined in a Universe of discourse X, and endowed with a t-conorm, a t-norm and a negation $([0, 1]^X, T, S, N)$, actually until now only some results are obtained for $([0, 1]^X, T, S, N)$, which is a De Morgan algebra. Notwithstanding, some definitions that look for avoid self contradictory elements in those models solve many problems, but are there other ways to solve them or at least control these problems? I would like this study could highlight these conflicts and maybe inspire future studies.

3.6 To End Up

This is not a research work, it is nothing else than an entertainment to work with three valued logic. I do not work in the field of multi-valued logic, nor in the field Professor Claudio Moraga works. But since Professor Claudio Moraga is strongly related with multi-valued logic, and some times I look for some strange behavior of classical world in three valued logic, I consider that it is interesting to devote this chapter to look over what happens in three valued logic. Therefore, in this writing I do some games with three valued logic. I considered two types of games: The first one is relating to translate a operation, the Sheffer stroke, and trying to cheek whether classical properties are hold. And the second one, through the new interpretations of *not inconsistent* some curiosities are shown related to models of Conjectures using the operations defined in some three-valued logic.

Fig. 3.2 Professors Rudolf Seising, Claudio Moraga and Enric Trillas during the second edition of Saturday's Scientific Conversations in Parlermo, 2011

Finally, I want to remark that the years spent working in the Unit of Fundamentals of Soft Computing in ECSC, was a very nice period to approach to Research, and I should thanks all my mates in that unit the great atmosphere of research generated. In the photogram 3.2 some of them appears.

References

1. I. García-Honrado and E. Trillas. (2011). Notes on the exclusive disjunction. In *ISMVL'11*, pages 73–77.
2. E. Trillas and I. García-Honrado. *On an Attempt to Formalize Guessing*, chapter 12, pages 237–255. Studies in Fuzziness and Soft Computing. Springer-Verlang, 2011.
3. L. Bolc and P. Borowik. (1992). *Many-valued Logics*. Springer-Verlag, Berlin.
4. Aristotle. *Metaphysics*. Book IV, Translated by W. D. Ross, eBooks@Adelaide, 2007.
5. I. García-Honrado and E. Trillas. Characterizing the principles of non contradiction and excluded middle in [0,1]. *Internat. J. Uncertainty Fuzz. Knowledge-Based Syst.*, 2:113–122, 2010.
6. E. Trillas, I. García-Honrado, and S. Termini. Some algebraic clues towards a syntactic view on the Principles of Non-Contradiction and Excluded-Middle. *Int. J. General Systems*, 43(2): 162–171, 2014.

7. I. García-Honrado and E. Trillas. (2015). Remarks on Symmetric Difference from an inferential Point of View. In *Journal of Multiple-Valued Logic and Soft Computing*, 24(1-4):35–51.
8. G. Birkhoff, *Lattice Theory*, Amer. Math. Society, Colloq. Publs, 3th. edition, (7th. print.), 1993.
9. E. Trillas, I. García-Honrado, and A. Pradera. Consequences and conjectures in preordered sets. *Information Sciences*, 180(19):3573–3588, 2010.

Chapter 4
Fuzziness as an Experimental Science: An Homage to Claudio Moraga

Marco Elio Tabacchi and Settimo Termini

4.1 Introduction

To choose a topic to homage Claudio is at the same time very easy and very difficult. Easy as, due to his widespread scientific interests, we will be spoiled for choice of a topic. Difficult since every specific choice would impinge on a very subtle and tiny aspects of his omni-comprehensive interests, overlooking many others as prominent as the chosen one.

Our first reaction, was then to pick up a very specific topic: the idea that Church-Turing thesis can serve as a meeting point to establish a dialogue of ideas between a few considerations briefly touched upon in our previous work [1–7] and Claudio's considerations in his fuzzy computation paper [8], in the hope that this "dialogue at a distance" could subsequently produce a real, verbal dialogue, possibly as a basis for approaching in a new way what is now the twenty years old notion of "Computing With Words"; a very challenging idea but still immature with regards to its theoretical and conceptual aspects.

However, after ruminating a little on this project, we concluded that this was not the right way to homage Claudio. And the reason was the following: it is possible to begin a true dialogue either in the case in which one is affording a very specific topic with definite tools or in the case in which one has in mind a few general questions but it is clear the context in which the discussion can be developed (in other words, the conceptual space in which we want to explore the ideas is clearly defined). The reflection on possible connections between Church-Turing Thesis and CWW does not belong to either of the two previously mentioned cases. Much work is still

M.E. Tabacchi (✉) · S. Termini
Universitá degli Studi di Palermo, Palermo, Italy
e-mail: metabacchi@gmail.com

S. Termini
e-mail: settimo.termini@gmail.com

R. Seising and H. Allende-Cid (eds.), *Claudio Moraga: A Passion for Multi-Valued Logic and Soft Computing*, Studies in Fuzziness and Soft Computing 349, DOI 10.1007/978-3-319-48317-7_4

Fig. 4.1 Enric Trillas, Settimo Termini and Claudio Moraga attending the Saturday's Scientific Conversations "Thinking and Fuzzy Logic" in Palermo (Sicily, Italy), 14th of May, 2011; photograph by Rudolf Seising

necessary to understand what is the "conceptual space" in which these connections can be seen and studied. So we decided that this could be the occasion for beginning giving a personal small contribution to the clarification of which is the nature of this conceptual space.

More or less we shall, then, move around such general questions as: "Are there connections between the research program of Cybernetics and the birth of Fuzzy Sets theory?", "Why the new large field of challenging investigations moving around the notion of information has not yet found a stable name?", "The successions of names Cybernetics, General Systems, Cognitive Sciences indicates only changes of names or reflect a deep change of the focus?", "In which sense approximation and uncertainty are essential parts of this field?" and we fear that we should ask again—50 years after Zadeh's seminal paper: "What is fuzziness?"

Very big questions, indeed. We are listing them not hoping to provide answers but just to stress the horizon in which one should move when facing fundamental (and foundational) problems regarding Fuzzy Sets. In the following pages we shall limit to ask which can be the best way of formally studying fuzziness expressing some doubts about the fruitfulness of taking as a unique possible paradigm classical mathematical logic. In this direction (and influenced by cybernetic suggestions), we shall comment on an old paper by Norbert Wiener on the nature of Mathematics (Sect. 4.2). We shall then ask whether it is not the case of trying to understand fuzziness going back to a sort of experimental study as advocated by Trillas (Sect. 4.3). Subsequently, in light of the conceptual connections existing between Cybernetics and the birth of FST, we shall ask which kind of constraints Cognition (the present avatar—or reincarnation?—of Cybernetics) poses on the previously discussed questions (Sect. 4.4).

4.2 Mathematizing Fuzziness

Some authors [9–12] have outlined the analogy and similarity of development of Fuzzy logic and Cybernetics (innovative initial ideas, crucial new questions posed and asked but not completely answered, difficulty of going on by developing the original program along the same conceptual lines presented at the beginning). In our view these analogies are rooted in some of the central questions of the new informal field that begin to grow in the mid of last Century and which, in our view, has not yet reached its point of equilibrium. This is also the reason why we realized that an old paper by Norbert Wiener can be useful for pinpointing some questions.

4.2.1 The Resistible Legacy of Mathematical Logic in the Mathematization of Uncertainty

The authors of a relatively recent paper [13], mathematical logicians by trade, write that: "our results appear to document the fact that fuzzy logic, taken seriously, is not just applied logic but may well be considered a branch of philosophical logic (since it offers a rich formal model of consequence under vagueness) as well as of mathematical logic (since it brings problems demanding non-trivial mathematical solutions)".

Let us remember that "fuzzy logic, taken seriously", is what is usually called "fuzzy mathematical logic". So what the authors affirm is that if we want to "take seriously" fuzzy logic we cannot but take as reference and model the problems and questions of classical mathematical logic. We are not questioning here the importance of the results of "fuzzy logic, taken seriously" in itself. What we want to ask is whether such a theory can be considered an applied logic and, more importantly, a general model of reasoning. We think that—independently from the value of the obtained results—it cannot provide any true help for applications in situations where uncertainty and fuzziness play a crucial role (see [14]. And this for very simple and fundamental reasons; namely the central notions of mathematical logic, coherence and completeness, in the first place, lose their crucial role when we are concerned from the start with uncertainty and imprecision. And its complex (and wonderful) machinery (if taken seriously) complicates any approach both to solving concrete problems and understanding "reasoning" as it is intended and used in natural language. We know that we are not saying something new, mathematical logicians have always been perfectly aware of the points remembered above. And the same holds also for the work done in fuzzy, and in general, in non standard logics.

However, since these points, although known, are rarely discussed, we would call, with a particular urgency, the attention to them, asking to draw the necessary conclusions. From a practical, or could we say, procedural point of view the urgency about the centrality of these reflections can only stress the necessity of paying a different attention to what is involved. In the long run this may perhaps lead to a paradigm shift, or, on a lesser scale, to a change of perspective. Let us clarify our point. We are not affirming that we must not afford the problem of modelling reasoning with rigor and with sophisticated mathematical tools; what we are saying is that we have to look for this modelling forgetting the burdensome legacy of the (wonderful) achievements of mathematical logic of last Century. If we look, in fact, at the true questions staying behind the informal notion of reasoning (permeated, imbedded, as it is, with approximations, uncertainty and ambiguity and—notwithstanding all these things—miraculously working) we realize that we are moving in a universe completely different from the one inhabited by Hilbert and Gödel.

4.2.2 From a Norbert Wiener's Early Idea

We chose, then, to introduce the argument of looking at and seeing fuzziness as an experimental science, by reading again a quite dated paper, written before the great results in logic of the Thirties of last Century [15]. In our search, in fact, to go back to more informal uses of the term reasoning we found interesting a few simple considerations done by such a mathematician as Norbert Wiener in 1923; before than the complete outline (as well as the decline and fall) of Hilbert program (and the brilliant gems produced in the course of this gigantic struggle). What Wiener does is to describe in a very plain style what—according to him—is the Nature of Mathematics. We want, then, to recall these remarks by Wiener on the empirical nature of math which are very consonant and tuned with the ones heralded by Enric Trillas at the end of his long scientific engagement in doing research in FST (a position we completely and wholeheartedly share). But let us now try to report, briefly, some of the content of Wiener's remarks. After observing that the way in which something new is obtained ("order of being") is very different from the way in which the new findings are presented ("order of thinking") he continue by affirming: "Now there is perhaps, no place where the order of being and the order of thinking need to be differentiated with such care as in mathematics. It needs but little reflection to see that any account of mathematics which makes logic not only the norm of the validity of its processes but also its chief heuristic tool is absurd on the face of it", and subsequently: "Logic will never answer a question for you until you have put it a definite question ... logic is a critic, not a creator, even as regards its own laws of criticism". (ibid., p. 268) Later, after observing that "Mathematics is every bit as much an imaginative art as a logical science and the art of mathematics is the art of asking the right questions", he asks the question of recognizing the interesting results among all that can be mechanically derived from the premises and answer by affirming that "This charge is entirely beyond the jurisdiction of logic, but the ability to discriminate between such trivial theorems and the really vital conclusions of a mathematical science is precisely that quality which the competent mathematician has and the incompetent mathematician lacks". The emphasis of the style can be, perhaps, related to the appearance of *Principia mathematica* from 1910 to a decade before.

In fact, "In order to do good mathematical work, then, and in fact to do any mathematical work, it is not enough to grind out mechanically the conclusions to be derived from a given set of axioms, as by some super-Babbage computing machine. We must select". (ibid., p. 269) The reference to Babbage seems particularly meaningful, since we must remember that his paper appeared in 1923, thirteen years before the appearance of Turing's (and Church's) paper. This shows, in our view, the fact that—notwithstanding the fact of writing more than a decade before the birth of the Theory of Computation, Wiener had clear in mind what should be avoided by simplistic mechanizations. Which does not, of course, imply that we cannot look for sophisticated mechanical (algorithmic) models of human activities. In fact, he writes: "The imagination is the mainspring of mathematical work, while logic is its balance-wheel". Parallel to the imagination an important role is played by the "habits of thought" which represent a new acquisition of the human mind: "Habits of thought—it is these rather than the sensory and imaginational content of the mind which constitute what is vital in mathematical imagining. Inasmuch as the mathematical imagination must sooner or later submit to the criticism of logic, it is essential that these habits should accord with logic." (ibid., p. 270) All this machinery will help along the mathematical investigation guiding to find the right level at which it is better to move: "In every branch of mathematics there is one plane of generality on which the theorems are easiest to prove and needless complication arises as quickly by falling short of this as by exceeding it". And the same habits of thought allow the great mathematician to take "a number of separate theories, fragmentary, intricate and tortuous, and by a profound perception of the true bearing and weight of their methods to have welded them into a single whole, clear, luminous, and simple." And he continues by affirming that: "Mathematics is an experimental science. The formulation and testing of hypotheses play in mathematics a part not other than in chemistry, physics, astronomy, or botany ... An experiment is the confronting of preconceived notions with hard facts, and the notions of the scientist are just as much the result of preconceptions, the facts just as hard, in mathematics as anywhere else" (ibid., p. 271).

To sum up Wiener's ideas as presented in this essay, we can say that mathematics is seen as an experimental science (a bold statement explicitly done by him), the use of logic in math is crucial but it does not play a role at the beginning of a process which can lead to a new theorem or a new theory. However, at the beginning, there is not a free imagination alone seen as a sort of dream; there is imagination regimented by the habit of knowing how a certain field is or should be structured, of understanding the connections among the new ideas and suggestions: a very complex conception, indeed. In this perspective he can conclude that "it is just these waifs of notions that may furnish the new point of view which will found a new discipline or reanimate an old. He who lets his sense of the mathematically decorous inhibit the free flow of his imagination cuts off his own right hand" (ibid., p. 272).

4.3 Fuzziness as an Experimental Science

Now we shall try to present—albeit in a very synthetic way—Enric Trillas' arguments, championed at least since 2006 [16, 17] and, recently, in two complementary books, respectively, in Catalan [18] and Spanish [19] on the importance of considering Fuzziness as an experimental science.

Fuzziness is usually seen as just an extension of two values (true/false) logic aimed at a better treatment of missing information, imprecision and uncertainty, but in any respect an extension of traditional mathematical means, built on axioms and rules. In Trillas' opinion, which we make our own, Fuzziness is also something more: a paradigm shift that highlights new ways to implement the treatment of imprecision, which is worthwhile to pursue in a separate but intertwined path as the rest of "traditional" Fuzzy Logic. In his 2006 paper [16] Trillas traced a possible pathway: "[R]ethinking fuzzy logic [will lead to] a ramification of current fuzzy logic in three branches: An experimental science of fuzziness, mainly dealing with imprecision in natural and specific languages; theoretical fuzzy logic, dealing with mathematical models and their linguistic counterparts as well as the necessary computing tools for their computer implementation, and a broad field of new practical applications to a multiplicity of domains, like internet, robotics, management, economy, linguistics, medicine, education, etc." The first two versions of fuzziness, the formal/mathematical approach and the experimental one, seem—at a first reading to have a strong analogue in what Lotfi Zadeh termed respectively Narrow and Wide Fuzzy Logic [20]. One side a toolbox for reasoning about uncertainty and vague predicates using a formal logical system, the other a complex "system of reasoning and computation" made of many components, that should allow to create a description of the real world that enclose its inherent imprecision. From one perspective a toolset living in the formal realm of logic whose predicates need a constant translation to and from the real world, in a constant switching from precisiations to vague descriptions and back; from the other an all-encompassing system, self-sufficient in its abilities of reality description and control, where fuzzy concepts are the only needed citizens. All this justified by the fact that "Despite its name, fuzzy logic is, in the first place, used for representing some reasonings involved with imprecision and uncertainty [and as such] closer to an experimental science than to a formal one" [16].

We have discussed further consequences of this idea in [4, 7], noting that due to the normal mechanisms of science advancement, development of Fuzziness has been rich in dichotomies and strong contraposition, mainly due to the fact that the intrinsic innovativeness of the basic ideas rendered a normal process of assimilation impossible. Trillas managed to take away a portion of the development of FST usually devoted to the paradigm of mathematical logic, and to aim at a general strategy centered instead on the nuances of the concepts related to topics such as language and cognition; the experimental model seems more at ease with problems of such nature, more at least than the strict realm of mathematical logic. But let us to say more. Perhaps, we could also speak of a pernicious role of mathematical logic in the assessment of the mathematical foundations of fuzzy set theory, in the sense that the paradigm of mathematical logic was seen as the unique, the only possible to which refer for providing formally acceptable and respectable mathematical developments. As we have briefly indicated at the beginning of this Section and argued, in more detail, in [14], in order to obtain good mathematical modellings of phenomena connected to an unavoidable presence of fuzziness, the attitude to follow is to go back to the attitude of mathematicians before the storm of the Thirties of last Century and the discovery of the jewels of XX Century mathematical logic. They are incomparable jewels, but to acknowledge this does not imply that every attempt at model reasoning must take it as the unique possible paradigm. The correct rules of reasoning in everyday language full of precious slippery notions (ambiguity, fuzziness, imprecision, locality, context dependence, irony etc.) cannot coincide with the ones needed to construct what we could call Gödel-Hilbert Cathedral.

In this perspective, we can read Trillas' threefold division in a different sense. "Theoretical fuzzy logic", by using Trillas words does not coincide with "Fuzzy logic in the narrow sense", in Zadeh's words, if this "narrow sense" is interpreted, as it is usually done, as Fuzzy mathematical logic. This last one is only a tiny fragment of what is and will be "Theoretical fuzzy logic" which will try to model in a mathematical reasonable way aspects of human reasoning for which such notions as soundness and coherence have no role to play. Results, we maintain, will show that considering Fuzziness in a very general sense as an experimental science and looking for an "old style" mathematization of relevant classes of phenomena not only has a specific place in the history of the discipline, but reinforces the idea that the new developments in the field, if supported by the experimental checking, will preserve all the conceptual innovativeness of the original idea. The idea of Fuzziness as an experimental science offer a solution to the problem of further development of fuzziness; by following the experimental approach, the existence of a "coherent something" which has many distinguishing and specific features which prevent its total collapse into other, less appropriate and already established paradigms could be usefully employed. We think that this idea is in synch with the discipline founders', and strongly reinforces the possibility that the new developments will merge in a coherent framework all the conceptual originality of the primitive vision.

4.4 Cognitive Aspects and the Constraints Induced on "Reasoning"

In this last Section we shall consider and review some cognitive aspects which reinforce the previous analysis. It would be interesting to investigate whether there are fundamental reasons explaining the conditions under which the notion of rational behavior can (and cannot) be captured by a (finite) set of mechanical rules.

As a good starting point for such analysis we could consider a recent paper on the subject by Pere Julià [21]. Julià starts from the idea that "Experience is fuzzy", and as such Fuzzy Logic and even more Soft Computing are disciplines whose contribution to cognition should be valued more, and said contributions should also be the catalyst for a continuous interchange that could foster both the development of cognitive sciences and computational intelligence. He then proceeds to lament a predominance of the engineering approach to fuzziness, which brings with it a primacy of the machine view of cognition; in this context cognitive research in fuzziness seems to be driven by the quality of achievement in the AI domain over the understanding of human performances, a direction which in any sensible cognitive research program should be reversed, and akin to championing the unicity of machines over the specificities of human reasoning when delimiting the area of action of cognitive fuzziness—especially when language is concerned. From here Julià moves to pinpoint the forced identification of reasoning with written language as the main cause for the failure of properly achieving understanding of cognition, and while voluntarily glossing over the conative aspects of language, a debate that would have brought to the table long debated and unresolved questions about state of minds, qualia and the likes, he embarks on a detailed discussion of how "confusing constructs and events" brings hard sciences, although at different degrees, to the risk of valuing more accuracy than the integration of empirical data when natural language is involved. Conclusions show an apparent and sincere optimism for the future of cognitive research in fuzziness, well represented by a citation of Klir and Yuan [22] on the importance of computational intelligence in cognition (the 'psychology' and 'FST' labels sounds like an anachronism due to the age of the cited paper) and on the bidirectional relationship they should enjoy: "Psychology is not only a field in which it is reasonable to anticipate profound applications of fuzzy set theory, but also one that is very important to the development of fuzzy set theory itself."; this is reinforced by Smithsons' observation that [applications of FST to humanities] are creatively stimulating for both researchers and theorists" [23].

But in fairness Julià's reflection seems to end on some sour notes: the centrality of natural language in fuzziness and Soft Computing is strongly dependent on the use of Language Variables and Rules, which, according to the author, are semantic insufficient, relying on a symmetry between the speaker and the listener that exists only in computer models and theories; knowledge cannot be represented by rules alone, and this is pointed as a recurring error in computational intelligence, already beleaguering cybernetics and various system theories; the actual process of "smartification" of machines seems to be leaving behind the behaviour of humans themselves, and with it the ability to learn how to comprehend and interact; even worse, the lack of a theory of mind approach based on emotions and intentions highlights the presumed impossibility of a specific machine effort (as intended today). In Julià's words, "[t]he fundamental fact remains: machines just do not have the right nerves going to the right places. In their case, we do not have to worry about all-too-human factors like frustration, fatigue, self-doubt, cognitive dissonance, absent stimulation, delayed action in the sense described, the perceptual/verbal/motor triad, not to speak of the automatic/non-automatic dichotomy that proves pivotal in human self-knowledge and self-regulation." [21] Some of the argumentation previously discussed is, in our opinion, less cogent and central to the debate today as it was in the past. A certain vein of biologicism seemingly in accordance with the original Searlian ideas of single realisation has been one of the strong contenders in the AI debate of the eighties, but it is a much less tenable position today. There is a large agreement nowadays that multiple realisation of intelligence is possible, and as such the community at large has accepted the idea that even the simulation of human behaviour should not necessarily depend by physical structures that are isomorphic to the ones human beings have been blessed with. Research has recently reached a number of milestones in human performance replication [24], and the fact that such results have been obtained employing means that are certainly different both in development and design from the ones at disposal in humans is a clear sign of how the intelligence target can be aimed at from different perspectives and achieved in different guises; reasoning in particular is concerned here, as it would be hardly opinable that it is not an important, if not altogether essential, part of intelligent behaviour. It is also significative that such results are not simply, to use Julià's wording, "spectacular achievements in engineering", devoid of any context and place in human experience—as indeed many similar feats were in the eighties and nineties. When they are of the competitive type (such as IBM Watson applied to Jeopardy! or chess, or Google Deep Mind) their reasoning power is surely based on a vast amount of data, but at the same time the choice to operate using spoken language is indication of a stern intention to confront the subject on her own terms (perhaps an instance of the kind of rigor we mentioned in the introduction. A further proof of this is the recent inclusion in Watson's speech recognition algorithms of a module that includes emotion detection, or with its technical name a "tone analyzer". This is a clear first attempt at bringing into the picture the conative factors, without doubt in order to learn from humans and one day include as well in a production paradigm the emotional layer. When of the collaborative type (of which autonomous vehicles are probably the most advanced example) reasoning must necessarily involve the subjects, and a continuous interchange of signals and information could not avoid all the dimensions of communication—something stressed out by the non verbal nature of the task itself. In time any autonomous agent, in order to be really useful and integrated, must assume at least partially the role of Subject, using its knowledge as Expert to gain an acceptable level of social behaviour. This is in no way an attempt to emulate Mamdani's "human controllers", but a specific attempt at integrating human and artificial reasoning in search of solutions for common problems.

But when Julià goes right at the core of the reasoning question, his argumentation really hits a target. We have already noted [3, 25] that there really is an unfortunate evidence of the fact that due to its roots in engineering [10] soft computing community tends to give special evidence to result obtained in control theory, reinforcing the idea that fuzziness is mainly a formal science, akin to mathematics or propositional logic (which, in turn, explain the particular affection for the use of the original term FST, instead of the more current Soft Computing, Fuzziness or Computational Intelligence). As already discussed in Sect. 4.2, and again following Enric Trillas [11, 26] we have championed the necessary evolution of Fuzziness toward an experimental science [4]. This approach, especially when applied to methodologies that are already experimental in essence, such as Computing with Words and Perception (CWP), can bring to the already rich plate of Fuzziness the aspect of human experience that is so sorely missed. In an experimental approach to CWP, Rules are no more "orders/commands to be executed if other well-specified variables and constraints concur" [21], but they evolve naturally from the observation of the surrounding environment, including the contribution of human actors, and the behaviour inspired by them is constantly changing and adapting, its only guide and inspiration being the evaluation of the agent's performance and the integration with what is acting around and above it. Experience is fuzzy indeed, and can be obtained by learning—we do this every day, constantly, with a certain nonchalant indifference, something which is deeply ingrained in human nature. By favouring an experimental approach to Fuzziness the same mechanism can be implemented in machines, not necessarily in the same physical structure, through evolution. We do not know beforehand if there are and what are the intrinsic limits to what is obtainable by an experimental approach that employs a CWP paradigm in terms of reasoning, but are optimistic for the future, as this approach would see the conjunction of an approach that keeps first and foremost the very nature of human reasoning, and means that intrinsically value the imperfection and uncertainty of the real word.

4.5 Final Remarks

We must confess that the fact that the situation appears to be so intricate is unpleasant. However epistemological analyses have the vantage point of focusing critical aspects. Now it seems to us that substantial parts of the problems appear to emerge from the (uncritical) use of technical tools and of previous results which looked as the right instruments for solving the open questions.

We must be brave enough to recognize that some of the questions asked in the midst of last Century are both more innovative and slippery than one thought. Think about the notion of intelligence as presented—smartly—by Turing in order to have a tractable notion. Think at the way in which the problem of mechanical translation was presented and afforded in the fifties of last Century (without doing any successful step forward) and the brutal way in which it is afforded today (with moderately broad results). Think to the recent success of obtaining unforeseeable and unpre-

Fig. 4.2 Enric Trillas, Settimo Termini and Claudio Moraga and students Valerio Perticone, Sergio Perticone and Fabio D'Asaro attending the Saturday's Scientific Conversations "Thinking and Fuzzy Logic" in Palermo (Sicily, Italy), 14th of May, 2011; photograph by Rudolf Seising

dictable results in a "mechanical" playing of GO. Everything should be reviewed and reconsidered. We have already a context that has been successfully laid down decades ago (and by such giants as Norbert Wiener). However, this context must be refined and adapted according to the indications grown by the work done in the last decades. The trick, if we may use this word, is to use new and traditional methodologies, trying to control what happens in these new fields to be explored, without any dogmatism, ready to adapt and change when it seems desirable to do so. And, maybe, also the problem of reflecting on possible (unusual) connections between Church-Turing Thesis and Computing with Words, can be taken into account.

References

1. D'Asaro, F.; Perticone, V.; Tabacchi, M. E.; Termini, S.: Reflections on technology and human sciences: rediscovering a common thread through the analysis of a few epistemological features of fuzziness, *Archives for the Philosophy and History of Soft Computing*, vol. 1, 2013.
2. Dasaro, F.A.; Perticone, V.; Tabacchi, M. E.; Termini, S.: Technology and human sciences: a dialogue to be constructed or a common tread to be rediscovered? In: Pedrycz, W.; Reformat, M. (eds.): Joint IFSA World Congress NAFIPS Annual Meeting, IEEE SMC, 2013.

3. Tabacchi, M.; Termini, S.: Experimental modeling for a natural landing of fuzzy sets in new domains. In: Magdalena, L.; Verdegay, J. L.; Esteva, F. (eds): *Enric Trillas: A Passion for Fuzzy Sets*, Springer International Publishing (Studies in Fuzziness and Soft Computing, vol. 322), 2015, pp. 179–188
4. Tabacchi ME, Termini S (2014) Some reflections on fuzzy set theory as an experimental science. In: Laurent A, Strauss O, Bouchon-Meunier B, Yager RR (eds) Information Processing and Management of Uncertainty in Knowledge-Based Systems, Springer International Publishing, Communications in Computer and Information Science, vol. 442, pp. 546–555
5. Tabacchi M. E.; Termini, S.: Fifty fuzzily gone, many more to go. An appreciation of fuzziness' present and an outlook to what may come. *Informatik Spektrum* vol. 38(6), 2015, pp. 484–489.
6. Tabacchi, M. E.; Termini, S.: Future is where concepts, theories and applications meet (also in fuzzy logic). In: Kacprycz, J.; Trillas, E.; Seising, R. (eds): *Towards the Future of Fuzzy Logic*, Springer (Studies in Fuzziness and Soft Computing, vol. 325), 2015, pp. 323–339.
7. Termini, S.; Tabacchi, M. E.: Fuzzy set theory as a methodological bridge between hard science and humanities, *International Journal of Intelligent Systems* vol. 29(1), 2014, pp. 104–117.
8. Moraga, C.: Towards a fuzzy computability? Mathware & Soft computing, vol. 6 (2–3), 1999.
9. Montagnini, L:; Tabacchi, M. E.; Termini, S.: Out of a creative jumble of ideas in the middle of last century: Wiener, interdisciplinarity, and all that. Biophysical Chemistry, 2015.
10. Seising, R.; Tabacchi, M. E.: The webbed emergence of fuzzy sets and computer science education from electrical engineering. In: Ciucci, D.; Montero, J.; Pasi, G. (eds): Proceedings of the 8th conference of the European Society for Fuzzy Logic and Technology, European Society for Fuzzy Logic and Technology, Atlantis Press, Advances in Intelligent Systems Research, 2013.
11. Seising R, Tabacchi ME, Termini S, Trillas E (2015) Fuzziness, cognition and cybernetics: a historical perspective. In: Alonso, J; Bustince, H.; Reformat, M. (eds.): Proceedings of the 2015 Conference of the International Fuzzy Systems Association and the European Society for Fuzzy Logic and Technology, Atlantis Press, Advances in intelligent System Research, vol. 89, 2015, pp. 1407–1412
12. Tamburrini, G.; Termini, S.: Do cybernetics, system science and fuzzy sets share some epistemological problems? i. an analysis of cybernetics. In: Proc. of the 26th Annual Meeting Society for General Systems Research, Washington, D.C., 1982, pp. 460–464.
13. Hajek, P.; Paris, J.; Shepherdson, J.: The liar paradox and fuzzy logic, *The Journal of Symbolic Logic* vol. 65 (01), 2000, pp. 339–346.
14. Tabacchi, M. E.; Termini, S.: Back to "reasoning". In: Proceedings of SMPS 2016, in publishing
15. Wiener, N.: On the nature of mathematical thinking, *The Australasian Journal of Psychology and Philosophy* vol. 1 (4), 1923, pp. 268–272.
16. Trillas, E.: On the use of words and fuzzy sets, *Information Sciences* vol. 176 (11), 2006, pp. 1463–1487.
17. Trillas, E.; Moraga, C.: Reasons for a careful design of fuzzy sets. In: 8th conference of the European Society for Fuzzy Logic and Technology (EUSFLAT-13), Atlantis Press, 2013.
18. Trillas, E.: *En defensa del raonament*, Universitat de València, 2015.
19. Trillas, E.: *Razonamiento, significado, incertidumbre y borrosidad*, Universidad Pública de Navarra, 2015.
20. Zadeh, L. A.: Foreword. In: Trillas, E.; Bonissone, P. P.; Magdalena, L., Kacprzyk, J. (eds): *Combining Experimentation and Theory: A Hommage to Abe Mamdani*, Springer (Studies in Fuzziness and Soft Computing, vol. 271), 2012.
21. Julià, P.: On reasoning with words and perceptions. In: Kacprycz, J.; Trillas, E.; Seising, R. (eds): *Towards the Future of Fuzzy Logic*, Springer (Studies in Fuzziness and Soft Computing, vol. 325), 2015, pp. 1–20.
22. Klir, G.; Yuan, B.: Fuzzy sets and fuzzy logic, vol. 4, 1995, New Jersey: Prentice Hall, 1995.
23. Smithson, M.: *Fuzzy set analysis for behavioral and social sciences*, Springer Science & Business Media, 2012.

24. Dasaro, F. A.; Perticone, V.; Tabacchi, M. E.: L'obiezione di una lady ed il computer che vince ai telequiz. come la flessibilità ha consentito all'intelligenza artificiale di superare un limite immaginario. In: Giunta, I. (ed.): *FlessibilMENTE - Un modello sistemico di approccio al tema della flessibilità*, La Società Formativa, Pensa Multimedia, pp. 379–398
25. Tabacchi ME (2015) In the future everyone will be a fuzzy set. In: Argüelles Méndez, L.; Seising, R. (eds.): *Accuracy and Fuzziness – A Life in Science and Politic*, Springer (2015), pp. 149–158
26. Trillas, E.; Termini, S.; Tabacchi, M: E:; Seising, R.: Fuzziness, cognition and cybernetics: an outlook on future. In: Alonso, J.; Bustince, H.; Reformat, M. (eds): Proceedings of the 2015 Conference of the International Fuzzy Systems Association and the European Society for Fuzzy Logic and Technology, Atlantis Press, Advances in intelligent System Research, vol. 89, 2015, pp. 1413–1418.

Chapter 5
Some Reflections on the Use of Interval Fuzzy Sets for Dealing with Fuzzy Deformable Prototypes

José A. Olivas

5.1 Introduction

I had the pleasure of meeting Prof. Moraga in a Workshop at Santiago de Compostela, organized by Prof. Alejandro Sobrino in mid-nineties. He gave very valuable advice of my doctoral work [1] (advised by Prof. Trillas), that I finished in the year 2000 where I introduced the Fuzzy Deformable Prototypes (from now on FDPs), which can provide a formal framework for working with prediction systems and, in general, representing and dealing with prototypical knowledge in every application.

FDPs come from the confluence of two interesting approaches to the concept of prototype: Bremermann's "deformable prototypes" [2], introduced in the late seventies from the field of pattern recognition, and Zadeh's fuzzy prototypes [3], as a result of his controversy with cognitive psychologists [4]. This proposal firstly summarize some concepts initially presented in author's doctoral work and the application of such ideas to several real fields and problems such as Forest Fires Prediction and Prevention, Software Engineering, Medicine, etc.

In early seventies, Zadeh [5] said that the construction of a fuzzy set, with the membership degrees of each element to the fuzzy set is the biggest problem for using fuzzy sets in real applications. So he introduced the concept of a type-2 fuzzy set: A type-2 fuzzy set is a L-fuzzy set over a referential set X for which the membership degrees of the elements are given by fuzzy sets defined over the referential set $[0, 1]$, which is a lattice with respect to Zadeh's union and intersection operators. Starting with this idea many works and concepts had been developed, which will be later on described.

J.A. Olivas (✉)
Department of Information Technologies and Systems, SMILe Research Group (Soft Management of Internet and Learning), Escuela Superior de Informática, University of Castilla-La Mancha, Paseo de la Universidad 4, 13071 Ciudad Real, Spain
e-mail: JoseAngel.Olivas@uclm.es

R. Seising and H. Allende-Cid (eds.), *Claudio Moraga: A Passion for Multi-Valued Logic and Soft Computing*, Studies in Fuzziness and Soft Computing 349, DOI 10.1007/978-3-319-48317-7_5

Fig. 5.1 Prof. Sobrino and Prof. Moraga, University of Santiago de Compostela, June 2016

The aim of this paper is to show some thoughts on how the application of the proposals based on Zadeh's initial idea, such as interval fuzzy sets, to Fuzzy Deformable Prototypes management could improve the performance of real applications. The work will be organized as follows: A short introduction, definition and applications of Fuzzy Deformable Prototypes will be presented in Sect. 5.2. Section 5.3 briefly introduces interval fuzzy sets and in Sect. 5.4 some reflections on the use of interval fuzzy sets for dealing with Fuzzy Deformable Prototypes are presented at the end of the chapter.

5.2 Fuzzy Deformable Prototypes

Fuzzy Deformable Prototypes model tries to be a closer representation than the standard AI ones on how humans take decisions. Many AI tools use patterns, but most of them adopt the behavior of the most similar pattern with a real situation. But each real scenario is different to any other one (mainly when dealing with natural or human phenomena, it is not possible that a past situation occurs again). Then, it seems adequate trying to describe any new situation deforming the patters abstracted of known occurred ones (Data+knowledge), basing this deformation on the similarity of the real situation with the old data patterns.

In the framework of Bremermann's 'deformable prototypes' a real element is classified according to the minimum energies required for physically deforming the closest pattern. Zadeh criticized the classical prototype theories from the point of view of psychology, due to the fact that these theories do not fit the function that a prototype should have: A fuzzy prototype is not an element—usually the best representative of a set or class for the classical prototype theories—, but fuzzy schemas of good, bad and borderline elements of a category. Zadeh's idea suggests a concept that encompasses a set of patterns, which represent the high, medium, or low compatibility of the instances with the concept. So FDPs can be defined as a linear combination of Fuzzy Prototypical Categories (described as tables of attributes) able to be adapted to any real situation, where the coefficients are the degrees of membership to each of these Fuzzy Prototypical Categories, represented by standard fuzzy sets.

For the combination to the case of affinity with more than one Fuzzy Prototypical Category, the definition of a real situation would be:

$$C_{real}\left(w_1,\ldots,w_n\right)=\left|\sum \mu p_i\left(v_1,\ldots,v_n\right)\right|. \quad (5.1)$$

where:

C_{real} Real case.
$(w_1, \ldots, w_n)$ Parameters describing the real case.
μp_i Degrees of compatibility with Fuzzy Prototypical Categories different from 0.
$(v_1, \ldots, v_n)$ Parameters of these Fuzzy Prototypical Categories.

Usually, these Fuzzy Prototypical Categories are defined using a Data Mining/KDD inspired process (which we called FPKD: Fuzzy Prototypical Knowledge Discovery, Fig. 5.2).

The prototypes are represented as fuzzy numbers with the aim of obtaining the degrees of affinity of a real situation with the prototypes. Taking into account these degrees of affinity with several prototypes, the prototypes with degree of affinity

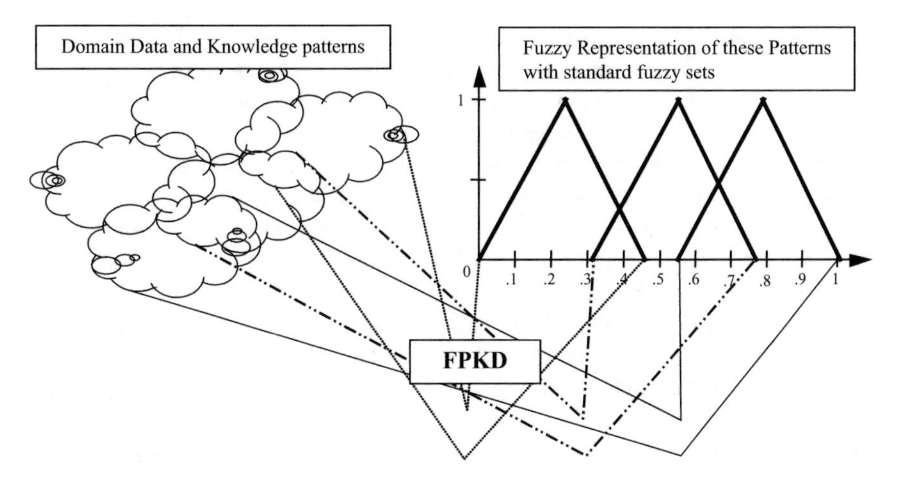

Fig. 5.2 Data/knowledge, patterns and fuzzy representation

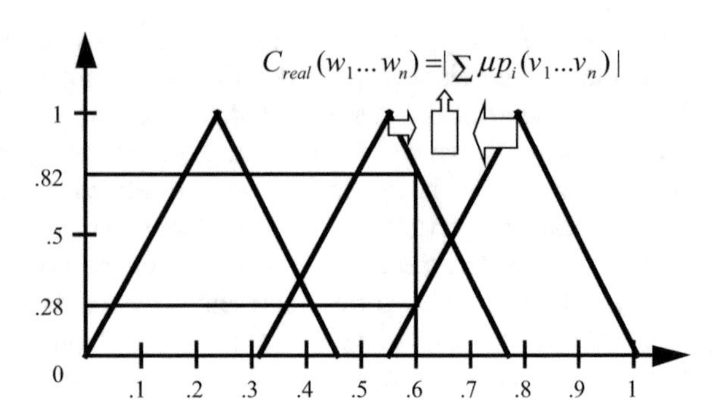

Fig. 5.3 Deformation of the closest fuzzy deformable prototypes for describing uniquely the new real situation

different from zero are deformed for describing the real situation, as it is shown in Fig. 5.3 with an arbitrary example.

Applying these ideas to the problem of assigning and optimizing resources in the daily fight against forest wild fires (usually due to the frequency and simultaneity of the fires together with the limited resources available), the INCEND-IA [6, 7] system was developed for predicting the evolution of the forest fire occurrence-danger rate for a given area in the short and medium term. Many other applications of this approach had been further developed, such as the ones related to Information Systems and Software Engineering [8, 9], traffic control [10], health records and documents management [11], social sciences [12] or Information Retrieval and Web Search [13].

5.3 Interval Fuzzy Sets

Karnik and Mendel [14] provided this definition of a type-2 fuzzy set: "A type-2 fuzzy set is characterized by a fuzzy membership function, i.e. the membership value (or membership grade) for each element of this set is a fuzzy set in [0, 1], unlike a type-1 fuzzy set where the membership grade is a crisp number in [0, 1]". Membership functions of type-2 fuzzy sets are three dimensional. It is the third dimension of type-2 fuzzy sets that provides additional degrees of freedom that make it possible to model uncertainties in complex systems. There are several options like interval and general Type-2 fuzzy sets including geometric Type-2 fuzzy sets and non-stationary fuzzy sets [15–17]. The concept of an interval-valued fuzzy set was presented in 1975 by Sambuc [18] and Jahn [19] in mid-seventies. At the end of this decade the definition of an interval-valued membership function (Grattan-Guinness [20]) was published, but it was in the next decade, starting with some works of Gorzalczany [21], when these interval-valued fuzzy sets took importance in the scientific community. Later, there are many relevant works, mainly the Karnik et al. [14], Liang and Mendel [22], Mendel et al. [23] ones.

Finally, it is important to underline the works of Bustince et al. [24] where all these concepts are widely described and the Multi-valued Logic Descriptive Power Hierarchy is shown, where: Type-1 fuzzy sets $\rightarrow$ interval-valued fuzzy sets $\rightarrow$ interval $T2$ fuzzy sets $\rightarrow$ $T2$ fuzzy sets

5.4 Some Reflections on the Use of Interval Fuzzy Sets for Dealing with Fuzzy Deformable Prototypes

A short introduction, definition and applications of Fuzzy Deformable Prototypes were presented and also a brief description of interval fuzzy sets. The main shortcoming to solve for the using of the type-2 and interval fuzzy systems models is the computational complexity of the inference mechanisms. The use of interval-valued type-2 fuzzy sets (IVT2FS) can solve this problem because this representation is a simplified version of full type-2 fuzzy sets. Concepts like embedded type-2 fuzzy sets [22], and the Footprint of Uncertainty [23] could allow us to manage the different points of view of the patterns/prototypes.

But perhaps the most important that it can bring the use of interval-valued fuzzy sets for dealing with Fuzzy Deformable Prototypes is the ability to represent more fully the uncertainty and vagueness (noise) inherent to large databases of that are currently often to extract patterns and all the problems associated with Big Data. We have applied some of these ideas to work with big databases of the field of Intelligent Tutoring Systems [24] with very promising results.

Fig. 5.4 Prof. Trillas and Prof. Moraga after Trillas' honorary doctorate ceremony, University of Santiago de Compostela, June 2016

Finally, I fondly acknowledge Prof. Moraga his support and high quality (from all points of view) advice and ideas in all my research fields. But the most important is our friendship over the years.

References

1. Olivas, J. A.: A contribution to the experimental study of prediction based on Fuzzy Deformable Categories. PhD Thesis, University of Castilla-La Mancha, 2000.
2. Bremermann, H.: Pattern Recognition. In: Bossel, H.: *Systems Theory in the Social Sciences*, Birkhäuser Verlag, 1976, pp. 116–159.
3. Zadeh, L. A.: A note on prototype set theory and fuzzy sets, *Cognition* vol. 12, 1982. pp. 291–297.
4. Osheron D. N., Smith, E. E.: On the adequacy of Prototype Theory of Concepts, *Cognition* vol. 9, 1981, pp. 35–58.
5. Zadeh, L. A.: Quantitative fuzzy semantics, *Information Sciences* vol. 3, 1971, pp. 159–176.
6. Olivas, J. A., Sobrino, A.: An Application of Zadeh's Prototype Theory to the Prediction of Forest Fire in a Knowledge-based System. Proc. of the 5th. International IPMU, vol. II, 1994, pp. 747–752.
7. Olivas, J. A.: Forest Fire Prediction and Management using Soft Computing. Proc. of the 1st IEEE-INDIN'03, IEEE International Conference on Industrial Informatics, 2003, pp. 338–344.
8. Genero, M., Olivas, J. A., Piattini, M., Romero, F. P.: Using metrics to predict OO information systems maintainability. In Dittrich, Geppert and Norrie (eds): Advanced Information Systems Engineering CAISE 2001, Springer, LNCS 2068, 2001, pp. 388–401.
9. Peralta, A., Romero, F. P., Olivas, J. A., Polo, M.:. Knowledge extraction of the behaviour of software developers by the analysis of time recording logs. Proc. of the 2010 IEEE International Conference on Fuzzy Systems (FUZZ-IEEE'10), 2010.
10. Angulo, E.; Romero, F. P.; García, R.; Serrano-Guerrero, J.; Olivas, J. A.: An adaptive approach to enhanced traffic signal optimization by using soft-computing techniques, *Expert Systems with Applications* 38 (3), 2011, pp. 2235–2247.
11. Sobrino, A., Puente, C., Olivas, J. A.: Extracting Answers from causal mechanisms in a medical document, *Neurocomputing* vol. 135, 2014, pp. 53–60.
12. Romero F. P.; Caballero, I.; Serrano-Guerrero, J.; Olivas, J. A. (2012). An approach to web-based Personal Health Records filtering using fuzzy prototypes and data quality criteria, *Information Processing & Management* vol. 48 (3): 451–466.
13. Garcés, P., Olivas, J. A., Romero, F. P.: Concept-matching IR systems versus Word-matching IR systems: Considering fuzzy interrelations for indexing web pages, *Journal of the American Society for Information Science and Technology* JASIST vol. 57 (4), 2006, pp. 564–576.
14. Karnik, N.N., Mendel, J.M., Liang, Q.: Type-2 fuzzy logic systems, *IEEE Transactions on Fuzzy Systems*, 7(6), 1999, pp. 643–658.
15. John, R. I.: Embedded interval valued type-2 fuzzy sets. In: Proc. of the 2002 IEEE International Conference on Fuzzy Systems, Honolulu, HI, USA, 2002, pp. 1316–1320.
16. Coupland, S., John, R. I.: Geometric type-1 and type-2 fuzzy logic systems, *IEEE Transactions on Fuzzy Systems*, 2007, pp. 3–15.
17. Garibaldi, J. M., Musikasuwan, S., Ozen, T.: The association between non-stationary and interval type-2 fuzzy sets: A case study. In Proc. of the 2005 IEEE Int. Conf. on Fuzzy Systems, 2005, pp. 224–229.
18. Sambuc, R.: Function ϕ-Flous, Application a l'aide au Diagnostic en Pathologie Thyroidienne, These de Doctorat en Medicine, University of Marseille, 1975.
19. Jahn, K. U.: Intervall-wertige Mengen, *Mathematische Nachrichten* vol. 68, 1975, pp. 115–132.

20. Grattan-Guinness, I.: Fuzzy membership mapped onto interval and many-valued quantities, *Zeitschrift für mathematische Logik und Grundladen der Mathematik* vol. 22, 1976, pp. 149–160.
21. Gorzalczany, M. B.: A method of inference in approximate reasoning based on interval-valued fuzzy sets, *Fuzzy Sets and Systems* vol. 21, 1987, pp. 1–17.
22. Liang, Q., Mendel, J. M.: Interval type-2 fuzzy logic systems: Theory and design, *IEEE Transactions on Fuzzy Systems* vol. 8, 2000, pp. 535–550.
23. Mendel, J. M., John, R. I., Liu, F.: Interval type-2 fuzzy logic systems made simple, *IEEE Transactions on Fuzzy Systems*, vol. 14(6), 2006, pp. 808–821.
24. Bustince, H., Fernandez, J., Hagras, H., Herrera, F., Pagola, M., Barrenechea, E.: Interval Type-2 Fuzzy Sets are Generalization of Interval-Valued Fuzzy Sets: Toward a Wider View on Their Relationship, *IEEE Transactions on Fuzzy Systems* vol. 23(5), 2014, pp. 1876–1882.

Chapter 6
The Way to the BliZ

Erdmuthe Meyer zu Bexten

6.1 Introduction

6.1.1 Our First Meeting

Dear Claudio,

We came to know each other in the winter term 1986/87, so this year marks our thirtieth anniversary! As I remember, you and Christian Krieb held the project group 108 (PG 108) together at the University of Dortmund (todays TH Dortmund) (Fig. 6.1). Here is a reminder of that time:

If my memory serves me correctly, at that time you just arrived from the University of Bremen back to Dortmund. Our work on the projects was a very fulfilling experience for all of us; not only from a professional point of view but on a social level as well.

6.1.2 Transition from Studying to Teaching

Later, after my studies were finished we led both project groups together: The PG „Entwicklung eines symbolischen Simulators für Signalverarbeitungssysteme" in the summer term 1991 and winter term 1991/92 (Fig. 6.2).

On the following winter term 92/93 and summer term 93 we led the PG 212 titled „CASSY 2: Weiterentwicklung eines Simulationssystems".

E.M. zu Bexten (✉)
Zentrum für blinde und sehbehinderte Studierende (BliZ),
Technische Hochschule Mittelhessen (THM), Wiesenstrasse 14, 35390 Giessen, Germany
e-mail: emzb@bliz.thm.de

R. Seising and H. Allende-Cid (eds.), *Claudio Moraga: A Passion for Multi-Valued Logic and Soft Computing*, Studies in Fuzziness and Soft Computing 349, DOI 10.1007/978-3-319-48317-7_6

Fig. 6.1 PG 108 group shot

You took also took care as my superviso over my diploma thesis that I wrote at Siemens, Munich and together with my future husband Dr. Volker Meyer zu Bexten, in 1988, titled: „Testbarkeitskriterien und Testmustererstellung für iterative logische und systolische Felder" (Fig. 6.3).

I held my very first class during the summer term in 1992 (seen on the picture) (Fig. 6.4).

To this time you introduced me to the topic of handicapped accessible user interfaces („Behindertengerechte Benutzeroberflächen"). This topic is today known as barrier freedom („Barrierefreiheit") and is still my field of research which I convey to my students in the classes called „Software-Ergonomie".

6.1.3 Graduation

From this field I drew the topic for my doctor thesis „Eine Simulationsumgebung für Signalverarbeitende Systeme" which I worked on at the Frauhofer Institute for Microelectronic Circuits and Systems in Duisburg, Germany and the examination was on the 20th August in 1992. You and Prof. Dr. Franz Pichler from the Johannes Kepler University at Linz, Austria were my doctoral thesis supervisors (Fig. 6.5).

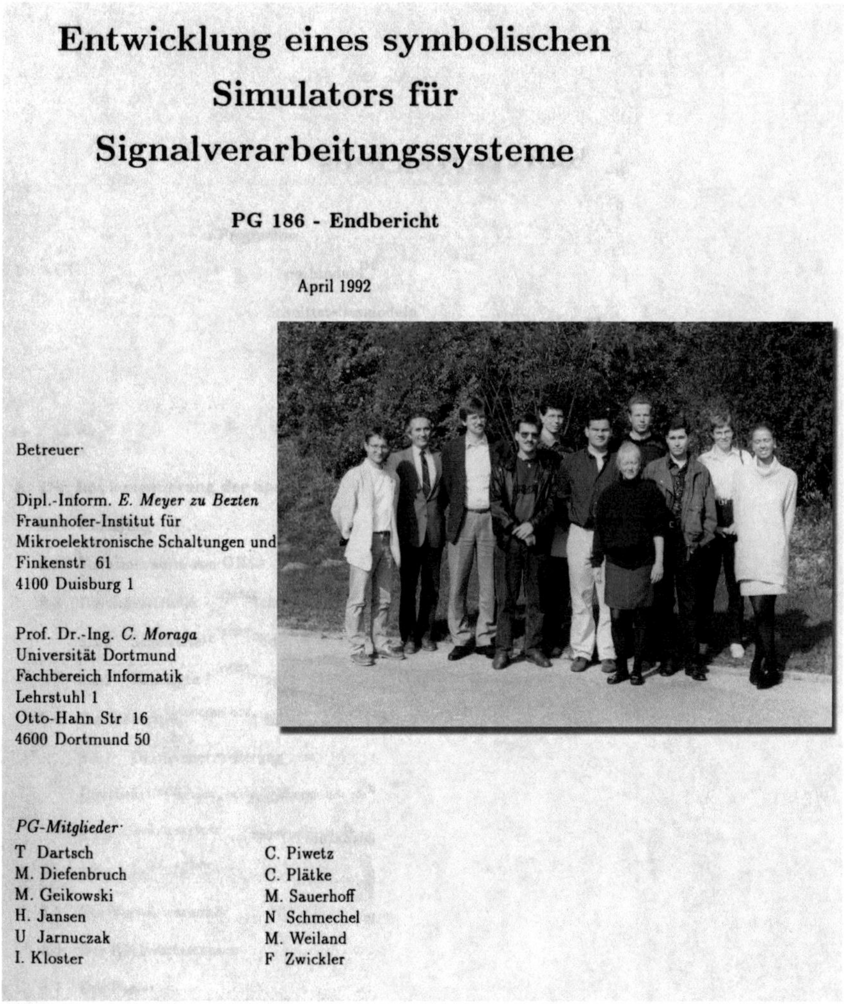

Entwicklung eines symbolischen Simulators für Signalverarbeitungssysteme

PG 186 - Endbericht

April 1992

Betreuer·

Dipl.-Inform. *E. Meyer zu Bexten*
Fraunhofer-Institut für
Mikroelektronische Schaltungen und
Finkenstr 61
4100 Duisburg 1

Prof. Dr.-Ing. *C. Moraga*
Universität Dortmund
Fachbereich Informatik
Lehrstuhl 1
Otto-Hahn Str 16
4600 Dortmund 50

PG-Mitglieder·

T Dartsch
M. Diefenbruch
M. Geikowski
H. Jansen
U Jarnuczak
I. Kloster
C. Piwetz
C. Plätke
M. Sauerhoff
N Schmechel
M. Weiland
F Zwickler

Fig. 6.2 Final report PG 186—1991/1992

6.1.4 The Years After Graduation

After my graduation we worked together on many occasions. I remember two instances very well: the conference in Iizuka in 1994 and the Workshop in Mangalia 1996, shortly after this I got appointed as professor to Gießen. In Iizuka, Japan we gave a presentation on Knowledge-based Genetic Algorithms with Fuzzy Fitness. Later, in 1996 we went together with the professorship on a workshop for some days, which I remember fondly, as the following pictures show (Figs. 6.6, 6.7, 6.8 and 6.9):

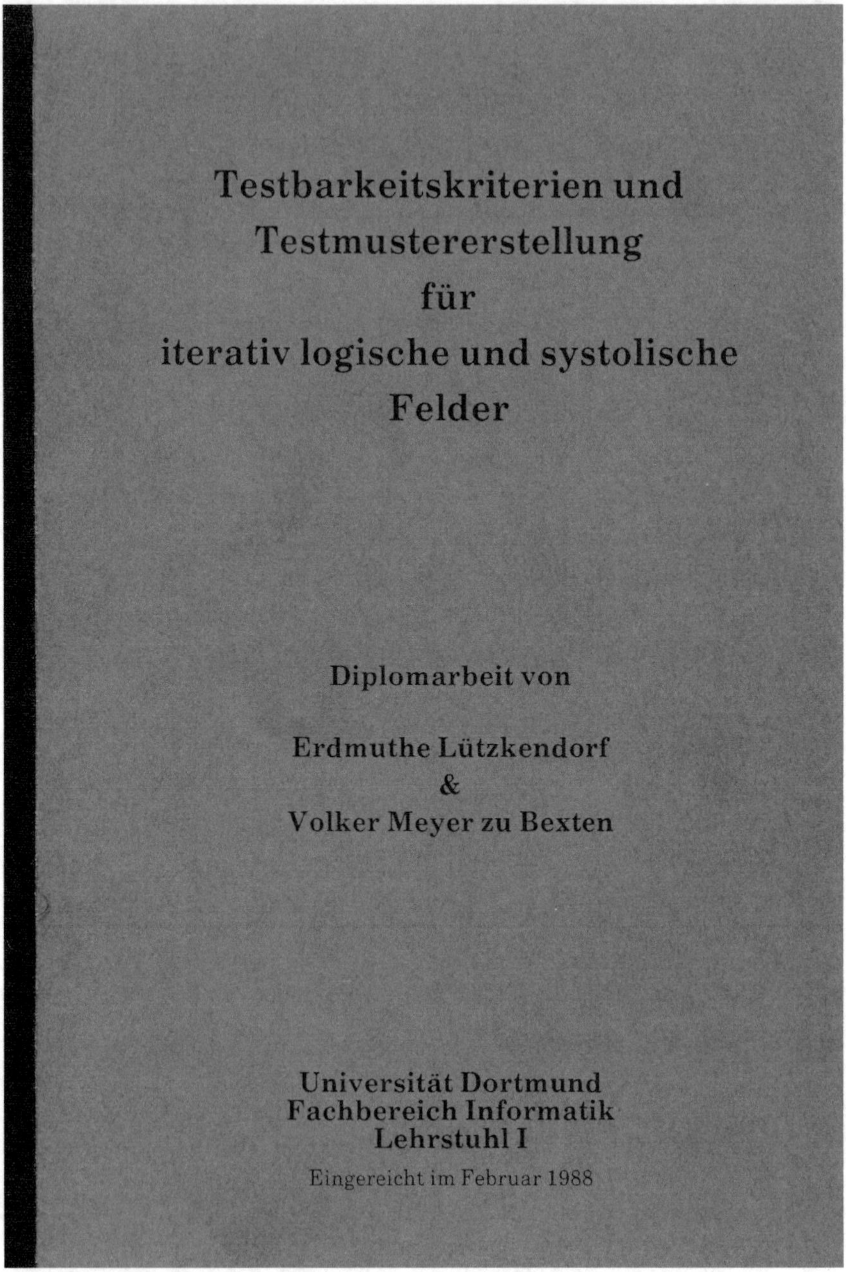

Testbarkeitskriterien und
Testmustererstellung
für
iterativ logische und systolische
Felder

Diplomarbeit von

Erdmuthe Lützkendorf
&
Volker Meyer zu Bexten

Universität Dortmund
Fachbereich Informatik
Lehrstuhl I

Eingereicht im Februar 1988

Fig. 6.3 Diploma, 1988

Vorlesungskommentar

des Fachbereichs
Informatik

der
Universität Dortmund

für das
Sommersemester 1992

Brailletastaturen

Zusätzlich zu normalen Tastaturen können Blinde auch Dateneingaben über spezielle Brailletasttaturen vornehmen. Mittels weniger Tasten werden die Braillezeichen zusammengesetzt und an den PC geschickt, der normale ASCII-Zeichen empfängt.

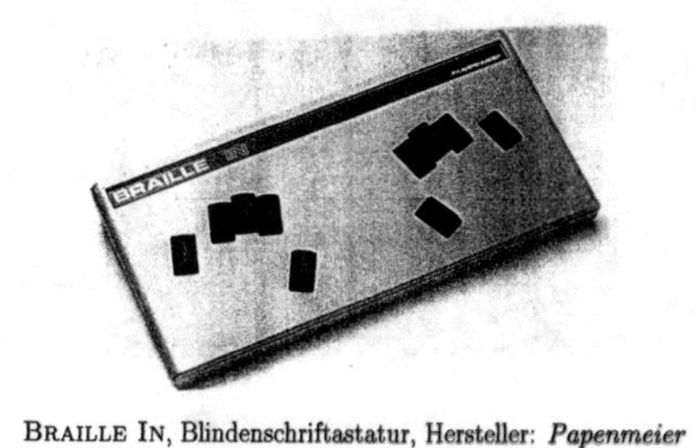

BRAILLE IN, Blindenschriftastatur, Hersteller: *Papenmeier*

FhG IMS/UniDo

Spezialvorlesungen mit Übungen

Grundlagen zur Entwicklung graphischer Benutzeroberflächen (4 V, 2 Ü)

042339	4 V	Di	8.15 - 10.00 GB V/HS 113	Moraga/
		Do	8.15 - 10.00 HG I/HS 2	Meyer zu Bexten
042340	2 Ü	n.V.		Meyer zu Bexten

Im Rahmen dieser Vorlesung wird zunächst die Entwicklungsgeschichte von Benutzeroberflächen (BO) erläutert und Beispiele von BO für Klein- und Großrechner präsentiert. Anschließend werden Richtlinien zur Entwicklung und Gestaltung von BO vorgestellt und diskutiert sowie Werkzeuge vorgestellt, die den Programmierer bei der Entwicklung von BO unterstützen (sogenannte UIM-Systeme). Das bedeutet, es wird das X-Window-System, OSF/Motif, Openlook, InterViews, Motifation, Serpent, TeleUse usw. vorgestellt. Danach werden Ein- und Ausgabegeräte präsentiert und anhand von Ergonomierichtlinien untersucht und bewertet. Weitere Themen sind Graphik und Text-Editoren, Adaptives Lernen sowie Behindertengerechte BO. Im Rahmen der Übungen sollen verschiedene Hilfssysteme entwickelt werden. Ein Prototyp davon wird dann in der Vorlesung vorgestellt und diskutiert.

Fig. 6.4 First class in the summer term 1992

Fig. 6.5 Prof. Dr. Claudio Moraga (2nd from the *left*) and Prof. Dr. Franz Pichler (on the *right*)

6.1.5 Appointment to Gießen and the Development of the BliZ

It was thanks to your supervision and guidance that I learned not only the practice of good science. You also taught me how to give good lectures. On a more important note you prepared me for my job as a professor and did spark my interest for the field of study I am engaging. It was you who introduced me to the topics of medicinal engineering, which I still adhere today. In the end it was you who led me on the path I followed and on its end I developed the BliZ.

„*You see more blind persons walking alone, not because there are more of us these days but because we have learned to make our own way*" to quote Malinao [5]. I stumbled upon this quote and found it to be quite inspiring, as it sums up in easy words what the original idea of the BliZ was. To help blind and visual impaired people in their education and empower them to lead their life as normal as possible.

We in our society tend to envision disabled people as helpless persons, victim to their inborn or acquired conditions. This thinking has long been the dominant mindset and is inherent to the so called medical model, one of the two dominant models to define disability in the academic discourse. The medical model views disability as a problem that is belonging to the individual and of no concern to any other person, hence the way of thinking that blind people have to overcome their two major'problems' on their own:

First, they must learn the skills and techniques which will enable them to carry on as a normal, productive citizen in their community; and second, they must become aware of and learn to cope with the public attitude and misconceptions about

Proceedings of the 3rd International Conference on Fuzzy Logic, Neural Nets and Soft Computing (Iizuka, Japan, August 1–7, 1994) p.p.473–474

C. Moraga, et al.

Knowledge-based Genetic Algorithms with Fuzzy Fitness

Claudio Moraga, Erdmuthe Meyer zu Bexten

University of Dortmund, Department of Computer Science
44221 Dortmund, Germany
{moraga, erdmuthe}@LS1.informatik.uni-dortmund.de

Abstract

In this paper three models are introduced to work with genetic algorithms when no fitness function is available, but the performance of the system to be optimized can be measured with variable confidence in a significant part of the problem space. Two of the models simulate the experience of experts by means of fuzzy linguistic variables whereas the third model suggests an interpolation method to cover measurement gaps. Simulation results are positive.

1. INTRODUCTION

Genetic Algorithms [Hol71, Hol75, Hol92] ("GA") are computational models of nat use operations known in natural optimizing process in a comple work with sets (*Populations*) o properly coded represent poten problem. Populations are pr following the cycle *selection, m mutation*. The selection is based the individuals of a given pop solutions) by means of a *Fitn* processing cycle will be repeate condition (e.g. remaining erro computing time exhausted) is r Theorem [Hol75] guarantees, th optimal solution will be reached problems. GAs do not pose spec problem to be solved (e.g. contin other than the existence of an function to evaluate the suitability be a solution of the problem. It b the fitness drives the selection.

In this paper two methods [Zah65] will be disclosed, whicl the case where no fitness f measurements of the performance carried out. These measureme homogeneously reliable, but experienced operators. Moreover there might be some areas of the search space where no measurements are possible. A typical example of such a situation is the problem of optimal tuning of a complex (possibly old) industrial plant. A mathematical model does not exist or, if it exists, it does not hold any longer due to changes introduced by aging of subsystems. The only remaining alternative is to measure the performance of different configurations -(tuning states)- of the system. Due to the above mentioned reasons it is possible that the measurements are not equally reliable for different configurations of the system. It is assumed that some experienced operators of the system are knowledgeable enough to judge the measurements. It may also be the case that some configurations which are theoretically possible, lead to severe overload of the system and cannot be activated. In such a case, no measurement of performance is available.

2. A CONSERVATIVE APPROACH

Assume that an industrial plant has the performance

judgement. The system does not converge to the configuration with the maximum measured value of performance, since this was judged to be only *acceptable*. The GA does however not find the theoretically best configuration, since the corresponding measurement even though *acceptable with some confidence*, was judged to have a deviation and accordingly was penalized leading to a lower fitness value.

Fig. 6.6 *Center* The 3rd international conference on fuzzy logic, neural nets and soft computing. Iizuka, August 1–7, 1994. *Right* Prof. Dr. Erdmuthe Meyer zu Bexten, Dr. Volker Meyer zu Bexten, Prof. Dr. Claudio Moraga

Fig. 6.7 FU Dortmund, FB Informatik, Lehrstuhl I, Mangalia, Romania at the Black Sea, September 27–29, 1996

Fig. 6.8 Prof. Dr. Claudio Moraga giving a presentation during the workshop in Mangalia

Fig. 6.9 Mangalia: Dinner with the group, Dr. Prof. Claudio Moraga, Prof. Dr. Erdmuthe Meyer zu Bexten, Dr. Jens Hiltner

blindness which go deep into the very roots of our culture and permeate every aspect of social behavior and thinking [3].

Our culture is heavily build around the ability to see and so blind people are confronted with a number of visual challenges everyday—be it reading the label of canned food or figuring out if they're at the right bus stop.

The medical approach would suggest that the problem exists because of their blindness, rather than normal eyesight being the only possible way to decipher what is included in the frozen dinner. Nowadays the favors shift more towards the second big defining model: The social model. This model draws on the idea that it is actually the society that disables people, through designing everything to meet the needs of the majority of people who are not disabled. There is a recognition within the social model that there is a great deal that society can do to reduce, and ultimately remove, some of these disabling barriers, and that this task is the responsibility of society as a whole, rather than the disabled person alone. An important principle of the social model is that the individual is the expert on their requirements in a particular situation, and that this should be respected, regardless of whether the disability is obvious or not [4].

The advancement of the social model is associated with the rapid development of the Information and Communication Technology (ICT). As mentioned, our culture is heavily build around the ability to see. When the internet got more and more developed it slowly penetrated every aspect in our modern day lifes; everything is managed online relaying more on text and video than ever before, so people with a visual impairment may find different online environments can be significantly disabling. This, together with a lecture I attended a long time ago, led me to the thought: „Does not anybody feel disabled some-times in his or her life? How easy it is to lose eyesight or suffer from any other disability. Is it therefore not in all our interests to help those, who have difficulties?"

As I was—and am—a professional in the medical information sciences I began to work on handicapped accessible workplaces. Since the late 80s I worked in cooperation with medical scientists and patients on handicapped accessible user interfaces. As time went by and I got to work as a professor at the THM Gießen I realized firsthand the one thing that Dobransky and Hargittai [1] warn for.

„The increasing spread of the Internet holds much potential for enhancing opportunities for people with disabilities. However, scarce evidence exists to suggest that people with disabilities are, in fact, participating in these new developments. Will the spread of information technologies (IT) increase equality by offering opportunities for people with disabilities? Or will a growing reliance on IT lead to more inequality by leaving behind certain portions of the population including people with disabilities?" [1].

To include these portions of the populace it is important to enable them to the right education. As a result more young people aim for higher education to participate in this new environment, but at the same time they are faced with new unique barriers and challenges at the university.

They might face difficulty in keeping up with the curriculum due to the visual nature of the classes, e.g. power point presentations, the requirement of reading specific literature which might not be available in embossed printing and so on. Much of the progress shown by blind pupils in the field of higher education stems from their interaction with the environment. Therefore, it is extremely important for the student to have a good learning environment with the necessary amount of support.

The responsibility of the learning and teaching support for blind students rests in the hands of the institution staff, the student body, the academic staff (that is the teaching staff, programme managers and module leaders) and the support staff. They must provide the optimum environment setting required by visually disabled students for education by working together in the best way possible and provide the technology required for learning. So many different institutions that need to work together can only result in chaos. My position as professor in medical computer science allowed me to change that. As such my first step was to establish a study course for blind students. Once this course was running, it was only a small step to the next big project: An institution that combines all the different services needed to enable—at first—blind and visually-impaired people normal academically studies.

All this specifications are met by the THM Gießen in one facility: with the BliZ, headed by me. In my responsible for practical computer science and medical computer science I came up with the idea to design a facility with the sole purpose to aid disabled persons with their academic carrier. This text is about this facility and the great opportunities it offers to those who need it. Let me introduce you to the …

The BliZ that is short for „Zentrum für Blinde und Sehbehinderte Studierende" was brought to life in 1998 at the THM Gießen (formerly known as Fachhochschule Gießen-Friedberg) under my supervision. It is in Germany the first and up to this date the only institution.

The next few pages are about this BliZ, a project I am very proud of. Then you will read the following passage always keeping in mind that this was only possible through all the teaching and guiding you offered to me.

6.2 The BliZ

The first year started with five students, three of them studied computer sciences while the other two were enrolled in economics. Today in the year 2015/16 there are approximately 40 students and a staff of 32 people (Fig. 6.10).

Fig. 6.10 The BliZ today—picture from the official homepage

Besides supporting the handicapped students, the main objectives are R&D of software and hardware for blind and visually impaired people. The BliZ is also involved in many projects to improve the situation of persons with low vision or without any residual vision in the professional context. Nowadays most of our visually impaired professionals have to use computers in their daily work. They have to deal with document management, video conferencing, unified messaging, application sharing or even teleworking. Teleworking in particular offers an interesting future perspective for people with disabilities. It is a good starting point for equal opportunities in professional work. In an environment of up to date information technology combined with the respective adapted access, it is possible to work effectively and with good efficiency.

But, alas, the available solutions for adapted access are far from being perfect. As of this, the BliZ is in close contact with the manufacturers of these devices to give advice and assistance for testing and improving ergonomics. If a problem is experienced with the adapted access technology, then new alternative working strategies will be developed, evaluated and finally implemented.

Nonetheless, the main objective lies in supporting the students and offer them the best study environment possible.

6.2.1 Services for Students

Deciding to undertake academic studies and visit a college for at least three years is one of the most important decisions a person has to make in his or her lifetime. It is only natural that a number of questions arise before, during and after college. The people working at the BliZ are well aware of this and therefore developed a wide program to assist young people during their academic life and beyond.

Starting life at the university can be a daunting experience, especially if one has a special condition to consider. Therefore, it is one of the most important tasks to make this transition as easy and successful as possibly. The most crucial task is to find the field of study that pleases oneself the most. Therefor the BliZ provides educational and career guidance for young adults who are interested in taking up their studies at the THM or just want information on their general options. The Staff of the BliZ will assist students with their orientation around the campus and their way to the different facilities and identifying external agencies.

The workroom (as depicted in the following photographs) is fitted with six different computers, each with a different monitor of varying size. Additionally available are two notebooks for presentations out of the THM. To be able to work the room also holds three Braille-output devices and diverse screen readers. Together with a number of other assistive technologies the BliZ offers a wide range of possibilities for the students to work on their curricula (Figs. 6.11 and 6.12).

To give a better understanding here you can find a short description of all the devices that can be used by the students:

Fig. 6.11 Workstation with Braillekeyboard and -display

Fig. 6.12 The workroom

Fig. 6.13 Braille display (from our partner Papenmeier)

Braille Displays

This device displays information on a computer screen by raising and lowering combinations of braille pins. It typically sits underneath the user's key-board and refreshes in real time as the user moves the cursor on the screen (Fig. 6.13).

Braille Printers

Also known as „braille embossers", these printers translate print text in Braille format onto specialized paper.

Screen Readers

Screen Readers are software programs that interface with the computer's operating system and provides the user full control over reading and interacting with their computer (e.g. using a navigation menu, highlighting text, using a spell checker, etc.). Jobs Access with Speech (JAWS)[1] is arguably the most widely used screen-reading program. It is a huge help for students in using any software which is displayed on the screen.

Screen Magnification

Screen Magnification Systems allow the user to enlarge the graphic, media, and text on a computer screen. Similar to a magnifying glass, the user can control what gets

[1] http://www.freedomscientific.com/Products/Blindness/JAWS.

Fig. 6.14 Screen magnification as used in our workrooms

magnified (e.g. text cursor, mouse pointer, icons, title bars, etc.). ZoomText[2] is an example of a screen-reading and magnification program that provides students with access to visual and auditory translation for what's appearing on their computer screen.

Video Magnifiers/CCTV

Also known as a closed-circuit television system, video magnifiers use stand-alone cameras to project magnified images onto a television screen, computer monitor or video monitor (Fig. 6.14).

Additional to these assistive technologies the BliZ offers a wide range of different services:

Individual Coaching

The staff assists the students based on their individual needs. They will find a tutor to teach the student the subjects he might fall behind or edit the script used in lectures, so that it can be read by the student.

Text in Alternative Formats

Most visually impaired students rely on texts in an alternate format like large printing or in Braille. Only a small number of books used in university are available

[2]http://www.zoomtext.de/.

in electronic formats and the presentations are oftentimes not co-rectly read by a screenreader. The BliZ assists students by reformatting these texts as needed.

Test Format

Tests can be administered to students with visual impairment in a number of ways. In consultation with the lecturer students can take their tests fitted to their special needs. Tests may be converted to Braille or texts in alternate for-mat or read by a computer with voice output (see JAWS further up). Faculty members also allow extra time for exams. Tests are taken in the workroom of the BliZ as it helds the assistive devices needed for the students. They are supervised by members of the staff.

Punkt Bilder

To enable blind and visual impaired students to write their tests the BliZ developed its own unique solution. The project Punkt Bilder (which roughly translates to „point picture") was programmed to make tests „readable" by transferring them into Braillegraphics which can directly viewed on the screen or can be printed in Braille.

E-Learning/HeLB

The HeLB Project (short for Hessisches elektronisches Lernportal für chronisch Kranke und Behinderte) is the heart of the BliZ. It is its central selfdeveloped e-learning platform. All the aforementioned activities are made possible through HeLB. The reduced access to information technology experienced disabilities (see [1]) creates an initial barrier to this type of learning. An elearning platform therefore has to address both the technical issues and pedagogical aspects of accessibility and inclusion.

For the development of HeLB the BliZ adheres to the seven basic rules, the so called „seven pillars of barrier-freedom", formulated by Hellbusch [2]

1. Text orientation
2. Contrast and colors
3. Scalability
4. Linearizability
5. Device-independence and dynamic
6. Comprehensibleness, navigation and orientation
7. Structured contents

Following these basic guidelines enables the BliZ to deliver software which is tailored to the specific needs of each individual student or professional and allows for intuitive use during the course of their academic carrier or later in their field of work as a professional. The portal is specially tailored to meet the specific needs of the individual student/user by adjusting the graphical interface, diagrams, video- audio- and live streams to the specific form of disability. Additionally, HeLB serves as an online-learning platform by offering complete e-lectures and online exams. With its services HeLB contributes to the principle of equal opportunities by enabling people to get an academic degree and start with the best chances in their life after graduation (Fig. 6.15).

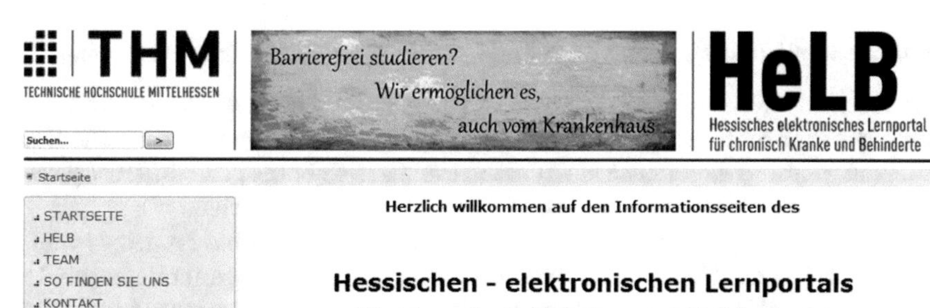

Fig. 6.15 The Information page of HeLB, the online Portal, which connects student to the university and the outside world

Being able to actually utilize their newly learned skills and techniques and be a productive part of the community is the ultimate goal of every student. Even after their studies we remain in contact with our former students. The curriculum of the BliZ therefore does not end with the final exam. The transition program focuses on life after being a student at the THM and is help-ing each graduate to make his or her own personal transition into a professional carrier. To realize its agenda the BliZ is working together with different partners in industry, other colleges and teaching hospitals in Research and Development on different projects to further the development in assistive technologies.

6.2.2 Projects

While primary focusing on the students, the BliZ is not only working in the academically field. Its expertise proofs valuable for a wide range of professional fields. This paragraph introduces some of the bigger projects. They encompass a wide area of uses and range from assistance in healthcare to different online services but in the end they all serve the bigger goal to include visual impaired people into the society.

Barrier Free Information

In cooperation with TransMIT GmbH, which is one of the biggest german corporations dealing in transfer of technology, the BliZ developed the TransMIT Project sector for barrier free information and rehabilitation technology. The projects intentions are to provide customers with an array of solutions to barrier free information access.

Appraisal of Webpages, Software and Creation of Barrier Free Documents

Far too many Webpages and are hard to navigate for someone who is blind. The same goes for Software which is needlessly complicated in its structure. The BliZ offers its yearlong experience and if asked evaluates webpages and different pieces of software, giving suggestions how to make them more accessible.

Likewise, many important Documents are not easily accessible for people with a visual impairment. This is especially frustrating if they are losing information vital to their life. The BliZ therefore is editing this documents in a way, that it can be accessed by anyone and in different formats like HTML, PDF and the DAISY-Standard.

Counseling and Workshops

Besides evaluating the existing products, the BliZ offers counseling on topics like barrier free documents, barrier free web design and inclusion into education. Especially educational institutions profit from this offer, as they can build on the long years of experiences gathered in the BliZ.

Diverse Literatures

The BliZ also aims to provide its patrons with literature of many diverse topics. To make this possible, several contracts with different publishers were made. If asked for a certain book, the people at the BliZ will rework it in a way, that it can be accessed by blind people in a comfortable way.

BAUM

BAUM is a abbreviation for **B**arrierefreies **A**ufbereiten/**U**msetzen von **M**aterialien. Many barrier free documents come with the problem that way have no unifying structure like for example no alternative description for pictures which can be read by screen reading software. These documents need to be reworked to be more accessible. BAUM acts as a guideline for the BliZ staff how to work with different formats like HTML, Word or PDF and make them accessible for everyone (Fig 6.16).

Healthcare

Vital areas, like hospitals, rely more and more on new technology to simplify the process of determining the cause of the patient. Mostly the solicited in-put is done via touchscreen, which is a real hindrance for blind and visually impaired people. The BliZ, together with its partners, is working on possible solutions to this problem, like equipping the terminals with assistance soft-ware. The center also offers trainings for emergency aid personal to increase their knowledge in critical situations. For better results, the ambulance assignments will be monitored with interactive mobile software and through a special developed controlling system.

Linux for Blind and Visual Impaired People

Another project specifically aims to make the operation System Linux usable for blind people. Linux is divided into a text oriented and a graphical user interface. Solutions for blind people to work with the text based interface exist since 1996, the

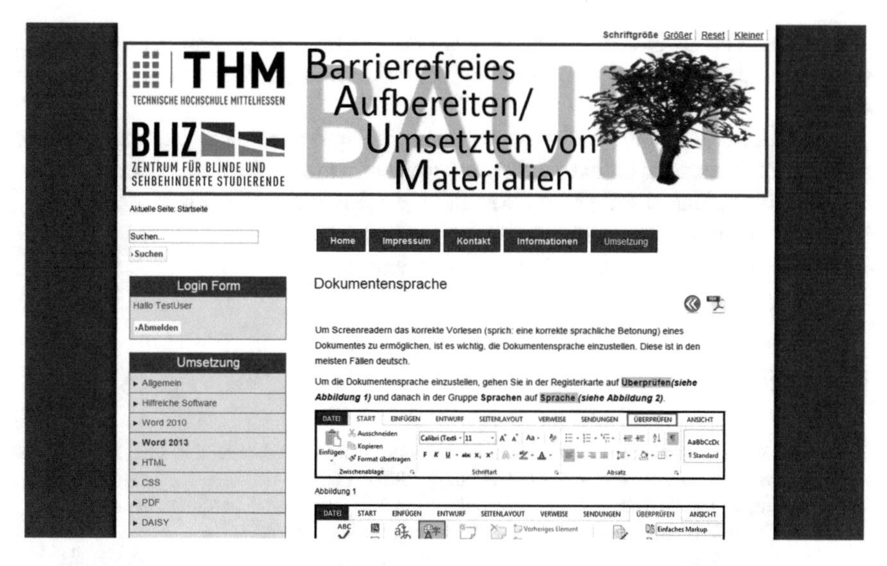

Fig. 6.16 The Interface for BAUM. By choosing a format on the right the user will get a description on how to work with it

most known screenreader is Suse-Blinux. It does recognize Braille-displays and is as such easily accessible by blind people. Over the course of the last years graphical interfaces, like the ones used by Windows, become more and more important. The members of the BliZ are working on projects which will make this graphical interface usable to the blind and visually impaired people.

Arbe-IT

Arbe-IT is a project developed especially to make the entrance into the working live as easy as possible. The goal is to link the BliZ, the THM and the job market. It is a web-portal that provides barrier free information and working offers. It works both ways, meaning that students can look for work and employers, which are interested in employing disabled persons, can look up information on the students too. The project is a co-venture with the Agentur für Arbeit Gießen (Jobcenter Gießen) and the Landeswohlfahrtsverband Hessen.

6.3 Closing Words

As you can see the BliZ is a smashing success story. Its services are well received among the students, the facilities and the job market. The combination of practical computer science with the individual, personal care from the staff helps to keep barriers for blind people to a minimum. Over the last few years we managed to work together with contributors from many areas, from publishers to healthcare and IT companies. Thanks to our commitment we opened up a whole new area for people who suffer from an otherwise disabling condition like blindness in the higher job market.

Dear Claudio, I thank you again for your great and exemplary patronage during and after my academic studies, during my time with the doctoral thesis and none the less for your guidance and preparation into my academic carrier. I wish you the very best for your future but healthiness above everything else.

Sincerely yours, Erdmuthe Meyer zu Bexten.

References

1. Dobransky, K.; Hargittai, E.: The Disability Divide in Internet Access and Use, *Information Communication and Society* vol. 9 (3), 2006, pp. 309–311.
2. Hellbusch, Jan Eric: Barrierefreies Design – eine Aufgabe für Webarchitekten, http://www.barrierefreieswebdesign.de/knowhow/architekten/gestaltung.html, 2001.
3. Jernigen, Kenneth: http://www.blind.net/a-philosophy-of-blindness/individual-articles/blindness-concepts-and-misconceptions.html, 1969.
4. Kent, Mike: Disability and eLearning: Opportunities and Barriers, *Disability Studies Quarterly*; vol. 35 (1)
5. Malinao, Reynaldo: http://www.benilde.edu.ph/article.aspx?nid=114.
6. Meyer zu Bexten, Erdmuthe: Autismus, *Deutsches Polizeiblatt*, vol. 6, 2014, pp. 6–9.
7. Meyer zu Bexten, Erdmuthe: Umgang mit beeinträchtigten Menschen im Studium, www.unidue.de/imperia/md/content/diversity/web_handicap_261115.pdf, 2015.
8. Papenmeier: Rehatechnik, http://www.papenmeier.de/rehatechnik-203.html

Chapter 7
Milestones of Information Technology—A Survey

Franz Pichler

7.1 Electrical Transmission of Written Documents: Telegraphy

7.1.1 Needle Telegraphy

After the discovering of the fact that an electrical current causes a movement of a magnetic needle (compass) by Oersted (1820) and the development of the "multiplier" by Schweigger which consists simply by a winding of wires (solenoid) it was straightforward to use this for the invention of a device for the electrical transmission of information. Gauß and Weber, both professors at the University of Göttingen, Germany are considered of being the first which designed a system for the transmission of numerical data between the astronomical Observatory and the University. In England Cooke and Wheatstone developed on the same basis their system of "needle telegraphy" with great success (Fig. 7.1). English railways used needle telegraphs until the midst of the 20th century.

7.1.2 ABC-Telegraphy

After the invention of the electromagnet by Henry and the use of it to construct clockworks to move a pointer it was possible to realize telegraphs which made directly use of the normal alphabet. Werner von Siemens, the famous German inventor and founder of the Siemens Company was one of the first to construct such an "ABC-

F. Pichler (✉)
Johannes Kepler University Linz, Linz, Austria
e-mail: telegraph.pichler@aon.at

R. Seising and H. Allende-Cid (eds.), *Claudio Moraga: A Passion for Multi-Valued Logic and Soft Computing*, Studies in Fuzziness and Soft Computing 349, DOI 10.1007/978-3-319-48317-7_7

telegraph" which found application in the German railway system (Fig. 7.2a). Besides of Germany and England ABC-telegraphs were widely used in France replacing the optical telegraph of Chappe. A major company in France which manufactured such telegraphs was Breguet of Paris (Fig. 7.2b).

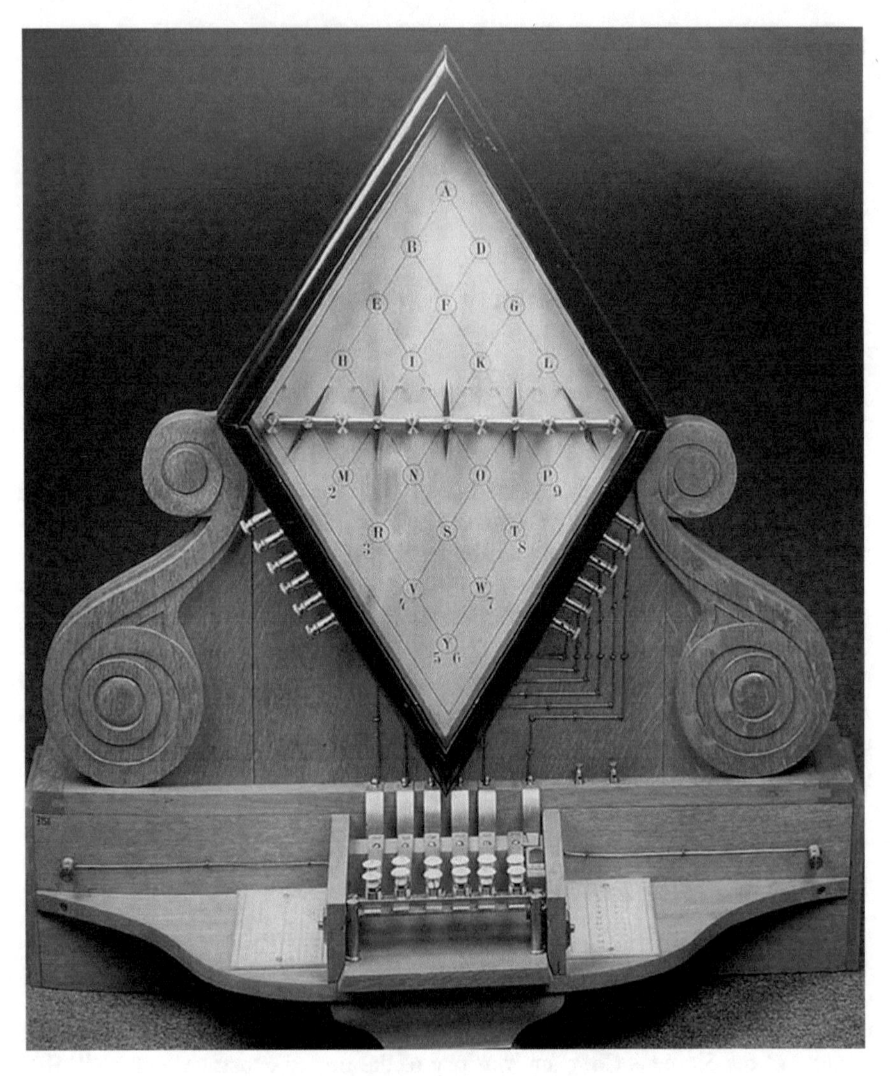

Fig. 7.1 5-needle telegraph by Cooke and Wheatstone (1837)

7.1.3 Morse-Telegraphy

The American inventor Samuel Morse got in 1837 a patent for an "electromagnetic telegraph". The main new idea for the receiving part of this telegraph was the use of the electromagnet for the printing by an embossing lever the combination of dots and dashes—the "Morse-code"—on a moving paper strip. In early receivers, called "Morse-registers", the paper strip was moved by a weight-driven clockwork. The first "Morse-line" from Washington to Baltimore dates from the year 1844. Besides of the USA also most of the countries in Europe adopted the Morse system. Until the midst of the 20th century it stayed in operation for the postal telegram services and in railway systems (Figs. 7.2 and 7.3).

The Morse-code was also used for the transatlantic telegraphy (from 1866 on) and also in the wireless telegraphy of Marconi (from 1897 on).

Fig. 7.2 **a** Siemens magnet inductor telegraph; **b** ABC telegraph of Breguet

Fig. 7.3 **a** German Morse register, about 1848; **b** Austrian Morse register, about 1860

Fig. 7.4 **a** Hughes printing telegraph; **b** Teletype (about 1960)

7.1.4 Print-Telegraphy

The British born American engineer David Hughes invented in 1855 an electromagnetic telegraph of the kind which got later known as "teletype". By a wheel, turning synchronous both, at the transmitter and receiver, printed letters into a moving strip of paper. Hughes telegraphs served in Europe long time for international telegram service and also for the military. The invention of the teletype (about 1930) replaced the Hughes telegraph in such systems (Fig. 7.4). Today the e-mail service of the Internet and also the Short Message Service (SMS) of mobile telephony represents telegraph.

7.2 Electrical Transmission of Speech: Telephony

7.2.1 The Telephone of Reis

The German teacher in Physics Philip Reis presented in 1861 by a lecture in Frankfurt a "telephone" which allowed the electrical transmission of sound and music by a wire. The transmitter consisted of a diaphragm of bird skin on which a contact made of platinum was mounted. The receiver was realized by a coil of isolated wire around a needle of steel ("knitting needle receiver") which was fixed on both ends on a wooden resonance box. Although the telephone of Reis was able to transmit sounds and pieces of music the transmission of speech was very poor (Fig. 7.5).

7.2.2 The Telephone of Bell

Graham Bell of Boston was the first to give a practical solution to the problem of electrical transmission of speech. The telephone of Bell (1876) consists in its final form of a permanent magnet with a coil on one end together with a iron membrane. If sound waves effect the membrane the magnetic flux in the coil is changed which induces an electrical signal (Bell telephone as transmitter). On the other hand, if a electrical signal which is generated by a Bell transmitter is led on the coil of a Bell telephone, by its magnetic forces the membrane oscillates according to the speech (Bell telephone as a receiver) (Fig. 7.6). An important improvement of the Bell telephone was made by Werner von Siemens by the application of a horse-shoe magnet which increased the magnetic flux and improved the sensitivity. David Hughes invented in 1877 the carbon transmitter (the microphone) which replaced the Bell transmitter. The microphone of Hughes used the physical effect of grains of carbon which change its electrical resistance dramatically by a change of pressure. Carbon microphones were used in telephony until recent years. Today they are replaced by microphones which are based on the pressure properties of certain semiconductors.

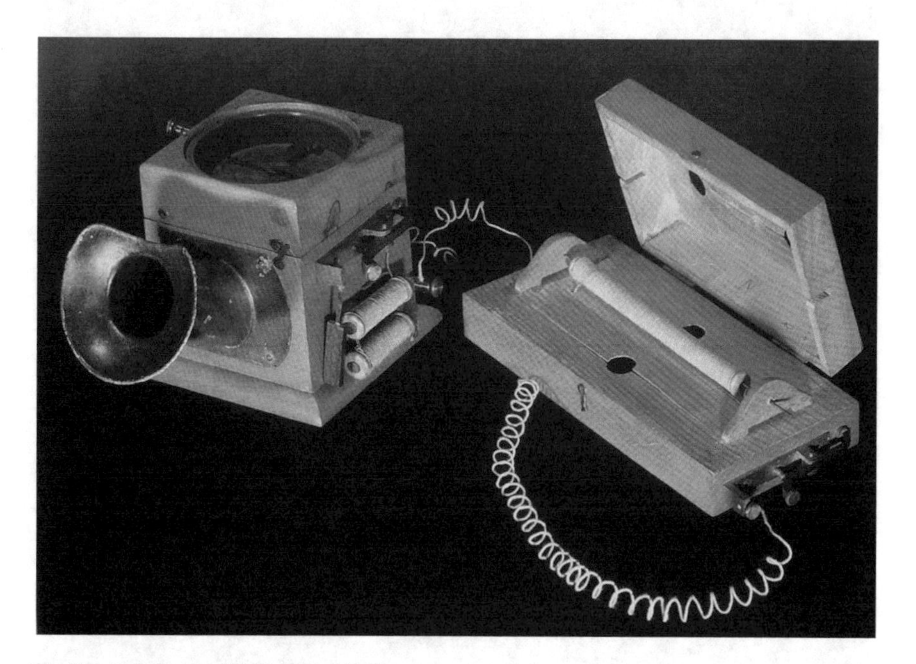

Fig. 7.5 Telephone of Philip Reis (1861)

7.2.3 Automatic Telephone Switching

The great public acceptance of the telephone services brought within a short time in the cities a large number of subscribers such that automation of the task of switching became desirable. Almon B. Strowger of Kansas City, USA, invented in 1889 an electromagnetic switch which made such an automation possible. The Strowger system is based on step by step switching which realizes the connection according

Fig. 7.6 **a** Bell's diaphragm magneto telephone (1875); **b** Bell's butterstamp telephone (1877)

Fig. 7.7 **a** German Strowger switch; **b** Strowger system telephone exchange

to a dialed number. Step by step automatic dialing systems of this kind became in Europe very popular and were in practical until the digital revolution (Fig. 7.7). Besides of switching systems of Strowger type other systems of switching such as systems based on the crossbar switch or the switch of the Ericsson company. Today, in the age of microelectronics telephone switching is realized by networks which are implemented on specific computers.

7.3 Electrical Transmission of Images: Picture Telegraphy

7.3.1 The Pioneering Years

The history of transmission of pictures (facsimile transmission) by electrical means can be traced to the inventions of Bakewell (1848) and Caselli (1855), to mention important inventors. The transmitter of the picture telegraph of Bakewell consists of a drum on which a metallic foil with the drawing (made by a ink which isolates) is mounted. The foil on the turning drum is scanned by a contact line by line and depending on the isolation of the drawing a digital "ON/OFF" signal (electrical current) is produced. The receiver consists of a comparable identical arrangement. However the drum of the receiver is covered by a chemical prepared paper, which gets a colour depending whether the scanning contact produces a current (signal is "ON") or not (signal is "OFF"). The pantelegraph of Caselli uses instead of a drum a pendulum for scanning the picture, otherwise the function is similar to the telegraph of Bakewell. It is reported that the pantelegraph was in 1855 successfully used in France to transmit a picture from Paris to Marseille by a distance of about 800 Km (Fig. 7.8).

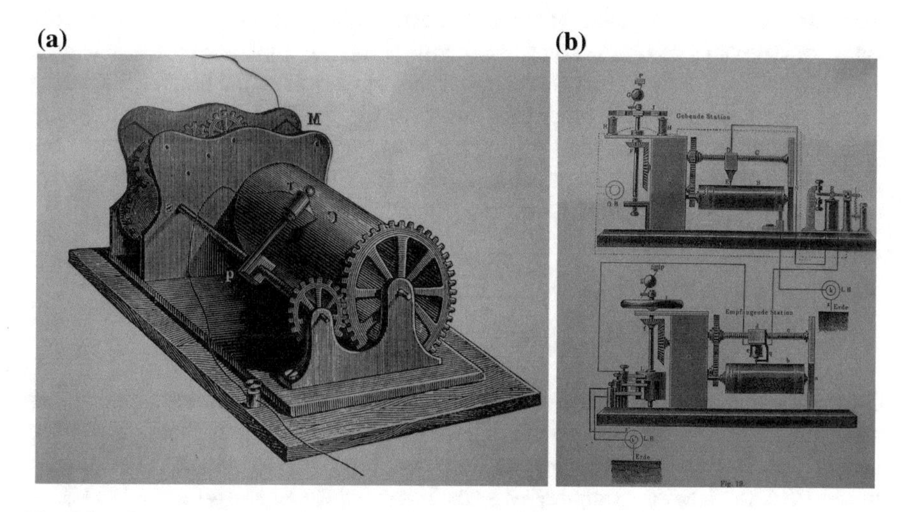

Fig. 7.8 **a** Picture telegraph of Bakewell; **b** Picture telegraph communication system

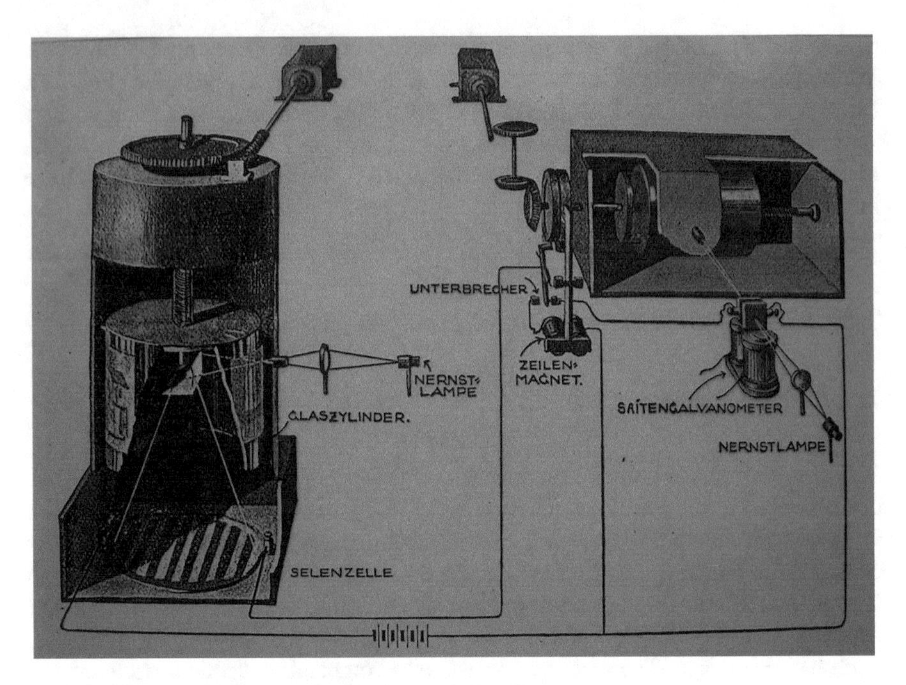

Fig. 7.9 Picture telegraph system according to Arthur Korn

7.3.2 Arthur Korn

Arthur Korn, by his academic education a mathematician, born in Breslau, Germany (today Wroclaw, Poland) felt the strong desire "to see across space and time". The electrical properties of the element Selenium (as result of the research by Hallwachs) together with the development of the light-electric cell by Elster (1888) provided the basis for Korn's system of picture telegraphy. Scanning a picture by the selenium cell allowed the electrical representation of the grey levels of the picture; the string galvanometer, originally developed for electro-cardiography, together with a special light tube which was controlled by a sparc relays made a true reproduction of the picture on a photographic film possible. In 1914 Korn showed successfully the transmission of pictures on a wired loop from Munich to Nürnberg and return (Fig. 7.9).

7.3.3 Development of the Fax-System

On the basis of the systems for picture telegraphy as developed by Arthur Korn and others it was from 1930 on possible to establish practical applications for the general use by organisations such as the press, the police or the weather offices. In the USA the leading company was the *Times Facsimile Corporation* (TFC), in

Fig. 7.10 Belinograph, France ca. 1960 (Museum für Kommunikation, Riquewihr, Alsace)

Germany Siemens & Halske and the Hell company. In France the Belin Company manufactured the "Belinograph" for its applications for the press. An important step forward was achieved by the establishment of international standards for facsimile transmission to make the use of the existing international telephone network possible. In this respect the recommendations of the CCITT (today ITU, the *International Telecommunication Union*) are important. Today, in the age of microelectronics and digital technology, fax machines are integrated with the telephone or the personal computer (Fig. 7.10).

7.4 Mechanical Computing

7.4.1 Handcrafted Machines

The invention and development of instruments which can give support in mathematical computations has a long history. For the computation with ordinary numbers the machine of Schickard (1623) is a milestone (Fig. 7.11). The Schickard machine was able to perform addition and subtraction; multiplication and division were realized by an aggregate of turning drums with tables similar to Neper's sticks. For its construction Schickard got valuable suggestions by the famous astronomer and mathematician Johannes Kepler. The machines of Pascal (ca 1640), Leibniz (1694),

Braun (1727) and Hahn (ca. 1790) are further examples of handcrafted mechanical computing machines which show the high technical skill of the constructors and constitute beautiful pieces of art.

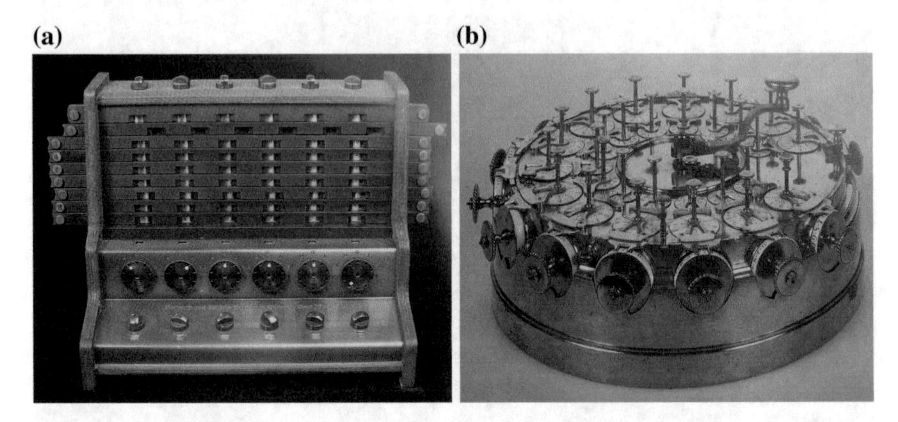

Fig. 7.11 **a** Machine of Schickard (1623); **b** Machine of Müller (1784)

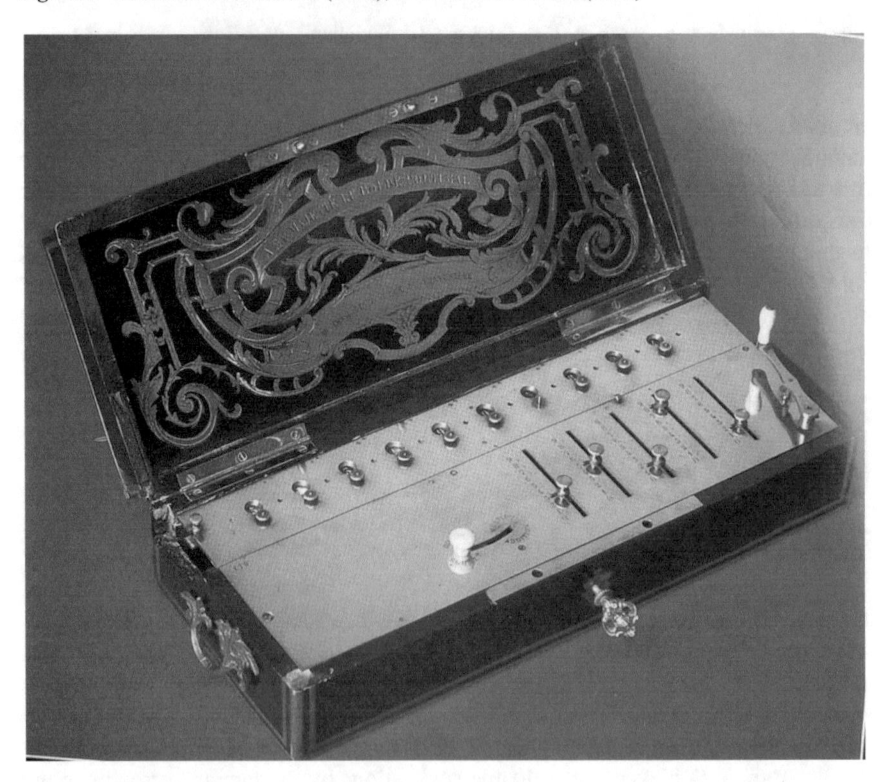

Fig. 7.12 Machine of Thomas de Colmar

7.4.2 Industrial Manufacturing of Machines

Charles Xavier Thomas of Colmar was at the year 1820 the first to start the industrial manufacturing of mechanical computing machines (Fig. 7.12). His "Arithmometer" made use of the principles of the Leibniz machine. Until 1900 about 1500 machines were manufactured in his company. In Germany it was Burkhardt in Glashütte (Saxonia) who started in 1880 the industrial manufacturing of the Thomas-machine being followed by other companies ("Saxonia", "Archimedes" and others). Besides of machines which used the the "Leibniz wheel" as principal construction element also machines which were based on the wheel originally invented by the Italian scientist Giovanni Polenus (1709). Here the companies Odhner (St. Petersburg and Stockholm) and also Brunsviga (Braunschweig, Germany) have to be mentioned being important manufacturers.

7.5 Wireless Telegraphy

7.5.1 The Marconi System

The experimental discovery of the electromagnetic waves by the german physicist Heinrich Hertz (1887) together with the practical and theoretical results as contributed by Oliver Lodge (England), Edouard Branly (France), Alexander Popov (Russia) and Augusto Righi (Italy) to name one of the most important one led the young Italian student Guglielmo Marconi in the years 1896 to the invention of a system for wireless telegraphy (Fig. 7.13). The Marconi Company, founded in England, started immediately to produce stations for wireless telegraphy for the communication between ships and the shore. Already in 1901 Marconi proved that a wireless telegraphic transmission between Europe (Ireland) and America (Newfoundland, Canada) is possible. The first transmitter of Marconi consisted of an induction coil together with a sparc gap as proposed by his teacher Augusto Righi. Later high voltage dynamos were used to get the necessary sparcs for the generation of electromagnetic waves and, to modulate the waves with a tone, rotating sparc gaps were used. The first receivers of Marconi consisted of a coherer as invented by the French scientist professor Branly together with the associated decoherer for interruption and therefore demodulation of the received electromagnetic wave. For registration of the received signal served a usual Morse register. In later Marconi receivers the magnet detector and also the Fleming diode was applied. An important step forward in receiver technology was the introduction of the syntonic system by application of resonance circuits, as developed by the famous British physicist Sir Ambrose Fleming.

Fig. 7.13 Marconi wireless telegraphy ship station (AWA museum, Bloomfield, N.Y.)

7.5.2 *The Telefunken System*

In Germany important research on wireless telegraphy was performed an two different places: In Berlin Professor Slaby, after his participation at early Marconi experiments successfully experimented with wireless telegraphy. He got financial support by the Allgemeine Elektrizitäts Gesellschaft (AEG), one of the leading companies in Germany in the field of electrical engineering. In Straßburg, being at that time a part of Germany, Professor Ferdinand Braun, well known by the invention of the cathode ray tube, got interest in the scientific aspects of wireless telegraphy and investigated the role of resonance to achieve wireless transmission over long distances. His group got support from Siemens & Halske. To concentrate nationally the efforts of the two groups, AEG and Siemens & Halske established in 1903 the company Telefunken. Telefunken and its systems for wireless telegraphy became a strong competitor for the Marconi company. A specific success was the application of the quenched sparc gap, which allowed higher transmitting power and realized at the same time a modulation of the electromagnetic waves with a musical sound. On the part of the receiver the crystal detector, as proposed by Professor Braun, replaced the mechanical demodulators such as tickers and rotary interrupters.

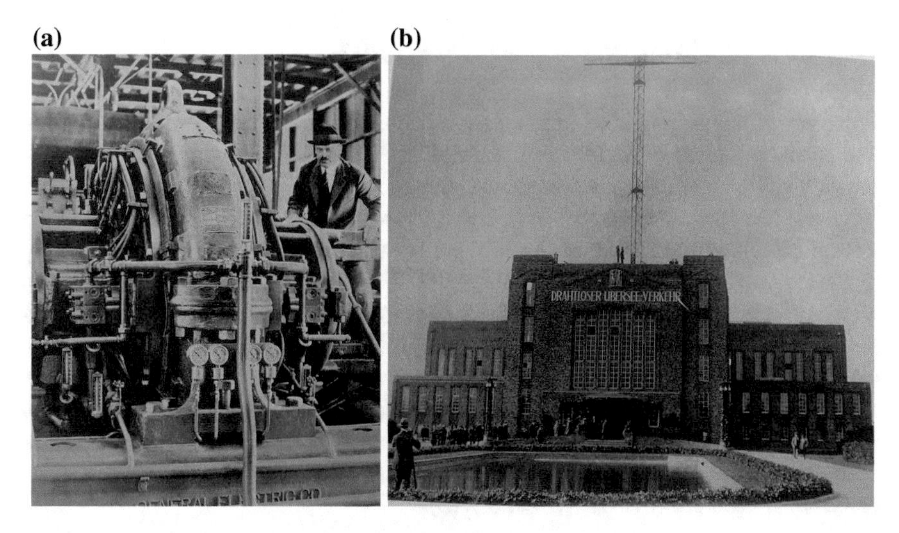

Fig. 7.14 **a** Alexandersson dynamo by GE; **b** Transcontinental station in Nauen, Germany

Fig. 7.15 Telefunken 1.5 kW transmitter with quenched sparc gap

7.5.3 *Long Distance Wireless Telegraphy*

For wireless telegraphy to overseas, here especially to the United States of America and to the colonies of European countries in Africa and Asia high power transmission stations were needed. Marconi erected such stations in Poldhu (Cornwall, England) and in Clifden (Ireland). In Germany the trans-radio station Nauen and the station Königswusterhausen of the Reichspost were important. Big high-frequency dynamos provided the necessary high power. While transmitters for wireless telegraphy of smaller transmission power used wavelength from 200–800 m, the high power transmitters used wavelength above which means extremely low frequencies. In consequence for transmission the propagation along the earth surface was essential. Today such stations for long distance wireless telegraphy have disappeared. However, as a museum the Swedish station SAQ in Grimeton, near the city of Gotheborg, still exists. SAQ uses Alexandersson dynamos from General Electric and stays operational until today (Figs. 7.14 and 7.15).

7.6 Radio Broadcasting

7.6.1 *The Electronic Tube*

The beginning of the electronic age can be set to the year 1906 were important inventions were made. The American physicist and inventor by profession Lee de Forest got a patent for his audion, a sensitive detector for wireless telegraphy. Robert von Lieben from Austria got in that year his patent "Das Kathodenstrahlenrelais" for the friction-less amplification of telephone signals. Both inventions are based on a electrical control of beams of electron in vacuum. The audion of de Forest had only a limited success. However, by the outstanding research of Arnold at the Western Electric Laboratories of AT&T and by Langmuir at General Electric it served as a basis for the development of the high vacuum tube in 1913. Also Robert von Lieben could not get by his patent directly a practical realization. However, in the year 1910 he and his collaborators Eugen Reisz and Siegmund Strauss realized an electrostatic control of their tube by a third electrode, the grid, and were able to show how to build a working telephone repeater (Fig. 7.16). The LRS relays (LRS = Lieben, Reisz, Strauss) which was manufactured by the German industry served in WWI in different devices in communication systems. However, since the function of the LRS relay depended on mercury vapour, it was difficult to get a stable operation. In Germany Telefunken pioneered the development of the high vacuum tube and the LRS relay became obsolete. The high vacuum tube allowed different kind of applications. The most successful one was probably its use in radio broadcasting.

Fig. 7.16 **a** Lieben laboratory in Vienna 1906; **b** Amplifier with Lieben tube (1913)

7.6.2 Radio Broadcasting

After the end of WWI (1918) the idea was born to apply the existing modern technology for wireless communication for the public and to establish broadcasting systems for the purpose of education and entertainment. The station KDKA in Pittsburg, USA, was the first which transmitted such programs in 1920 to its listeners. Radio Broadcasting as it was called was immediately a big success in USA and also in the European countries. Within a few years a large network of station was established and a new industry to manufacture the necessary radio receivers had to be erected. The first radio apparatus needed batteries of galvanic elements for their operation which was expensive (Fig. 7.17). From about 1930 on radio apparatus got a power supply and could be directly connected with the usual electrical power line. Further technological progress in radio engineering was the introduction of the superheterodyne circuit (from about 1925 on) and later the introduction of FM radio (FM= Frequency Modulation) to supplement AM radio (AM = Amplitude Modulation) with improvement in bandwidth and avoidance of statics.

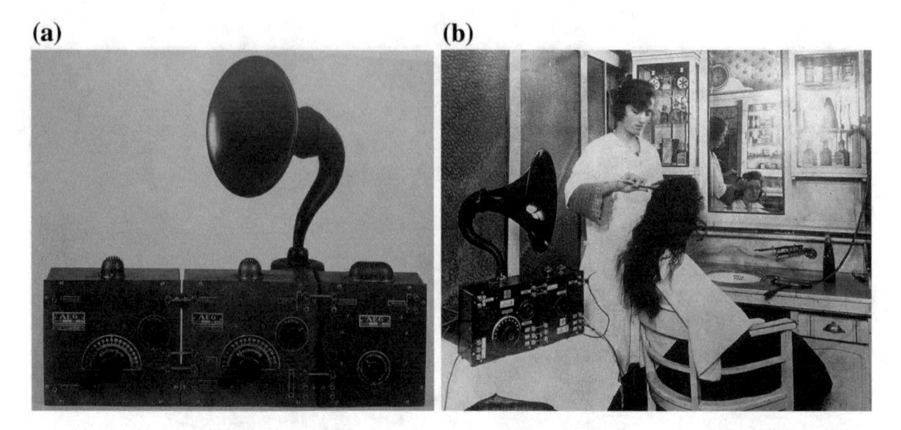

Fig. 7.17 **a** AEG radio (1925); **b** Radio at the hair dressing shop

7.7 Television

7.7.1 Mechanical Television

At the end of the 19th century picture telegraphy had already reached the maturity for practical application. Naturally there appeared also the wish to transmit instantaneous moving pictures by electrical means over distances in space ("television"). It was the young student Paul Nipkow who in 1884 made in Berlin the necessary invention for the realization of television. Nipkow proposed for the scanning of moving pictures (and in consequence also for the reverse operation, the synthesis of the scanned picture) a fast turning disk with a spiral of holes, every single hole being responsible for scanning a row of the picture. A light sensitive cell (at the time of Nipkow a cell of Selenium, later a photocell) transforms the light beam through the holes into a electrical signal. At the receiver the electrical signal drives a fluorescent lamp and the resulting light beam is synthesized by the turning Nipkow disk row by row. It is due the inborn inertia of our eyes that this gives a full picture. From the invention of Paul Nipkow it took more that 30 years until in England the first practical realization of television was made by John Logie Baird. Mechanical television on the basis of Nipkow disks (also mirror screws and mirror wheels were used for scanning) got only limited practical application. In 1935 a British committee decided to stop the development in favour of "electronic television" (Fig. 7.18).

7.7.2 Electronic Television

By the year of 1930 electronic technology had reached a degree of maturity such that the electronic realization of television systems became feasible. A pioneering

Fig. 7.18 **a** Building a Nipkow television receiver; **b** Televisor of John Logie Baird

contribution in this respect was the application of the cathode ray tube to realize a scanner (today called a "flying spot scanner") by the german inventor Manfred von Ardenne in 1930. This allowed for the first time the electronic scanning of motion pictures. The electronic realization of the camera operation was proposed by the American inventors Philo Farnsworth by the patent for his Image Dissector in 1927 and 1933 and by Vladimir Zworykin for the patent to his Iconoscope. Both camera tubes were in 1935 in operational use and were with success applied on occasion of the Olympic Games in Berlin 1936 (Fig. 7.19). For the electronic realization of television pictures in the receivers the cathode ray tube as developed in Germany by the Leypold company under assistance of Manfred von Ardenne. The outbreak of WWII delayed the introduction of public television systems in the USA, England and Germany, the leading countries in television technology. By end of WWII new efforts were taken and the known standards, NTSC from USA, SECAM from France and PAL from Germany were defined. Today the existing semiconductor electronics together with

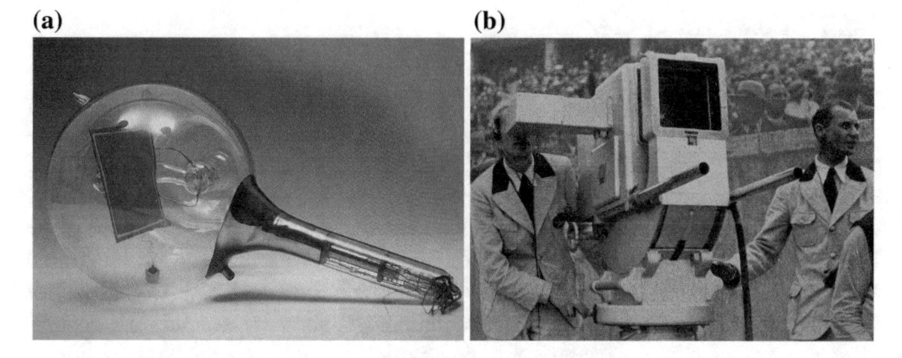

Fig. 7.19 **a** Ikonoscope of Zworykin (1923); **b** Ikonoscope camera at the Olympic games 1936

the high intergration of microelectonics allow new solutions to realize television systems ("digital television"). As an example for a high integrated micro electro mechanical system (MEMS) we mention the mirror matrix of Texas Instruments which realizes pictures in the size of 1024 times 1024 pixels and which is used for large screen projection.

7.8 Electromagnetic Computing: Relay Computer

7.8.1 Introduction

A relay R is a electromagnetic device with a moving lever which can assume two states: state "ON" in the case that the lever has been moved by the magnetic field of the coil; state "OFF" otherwise. The lever of R moves electrical switches r and r^* of two kind, one if it being closed in state "ON" otherwise open, the other working just in the opposite, being closed in case of state "OFF" otherwise open. With two Relay R and S it is possible to realize electrically the mathematical structure of a switching algebra with Boolean variables r and s in the following way: $r + s$ corresponds to a parallel circuit of the switches r and s, $r \cdot s$ corresponds to a serial connection of the switch r with the switch s. A switch r^* realizes the negation of the variable r. It follows that an aggregate of n relays R_1, R_2, ..., R_n is able two realize any binary operation defined on the set of binary words of length n. To prove that it is possible to build digital computing devices by relays we have to show that also a digital memory can be realized. This can be done by the concept of a register of relay flip flop circuits as memory cells. Each such flip flop can be realized by two relays Q and Q^*. This shows that with relays it is possible to realize by electrical means mathematical structures which are finite state machines, which is, from a mathematical point of view, sufficient to build a computing machine.

7.8.2 Relay Computer

The introductory chapter pointed out that relays, which are well known from their common use in telephone switching systems, can be used to build a electromagnetic computer. This idea was followed at different places and "relay computers" of dramatically increased computing power, as compared to the existing mechanical computing devices, could be realized. In the USA the Model I-VI (1940–1949) of the Bell Laboratories, the model Mark II (constructed by Professor Howard H. Aiken) at Harvard University and the PSRC (=Pluggable Sequence Relay Computer) of IBM can be considered as pioneering installations. In Germany Konrad Zuse accomplished his outstanding contribution in computer technology by the construction of the machines Z1 (1936), Z3 (1941) and Z4 (1944). Unfortunately for Z1 and Z3 no original designs have been preserved (although a replica of Z1 can be seen at the most interesting exhibition of ZUSE computers an the *Deutsches Technik Museum Berlin*). The machine Z4 could survive WWII and was later for many years in service at the Institute of Mathematics at ETH Zürich. Today the Z4 is on display at the *Deutsches Museum* in Munich. The ZUSE company continued after WWII the production of relay computers. The model Z11 was especially successful for applications in optical and geodetic computations (Figs. 7.20, 7.21).

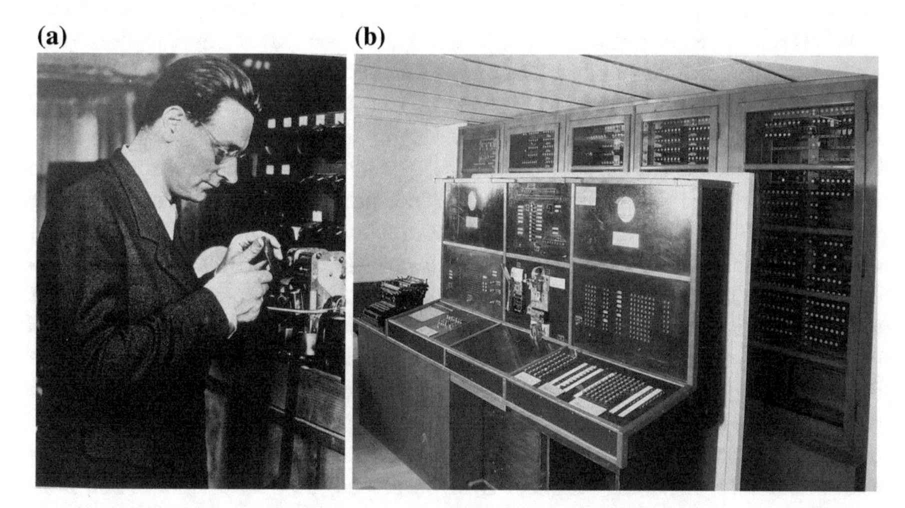

Fig. 7.20 **a** Konrad Zuse (1910–1998); **b** Zuse Z4 relay computer (*Deutsches Museum* Munich)

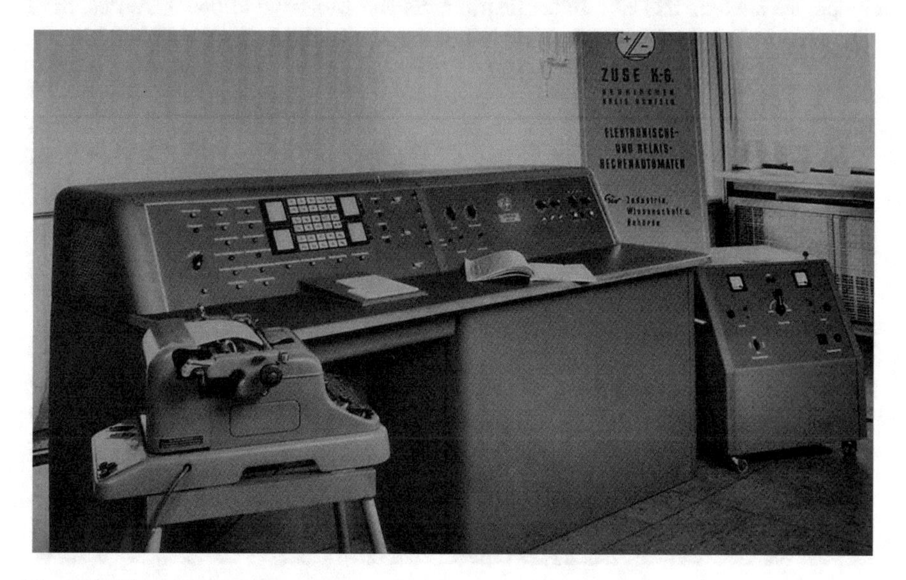

Fig. 7.21 The ZUSE Z11 Computer

7.9 Electronic Computers and Electronic Calculators

7.9.1 Technological Prerequisites

Relay computer are necessarily limited in the speed of their operation. The mechanical moving parts have a certain inertia which can not be avoided. To increase the speed of computing it was necessary to realize the switches and the memory by electronic devices. Electronic tubes, which found in different military high speed systems of WWII such as Radar systems a successful application (for example in high speed counters), proved to be suitable. The today famous first electronic computer which used ordinary radio tubes for its realization (18.000 in number!) was the ENIAC machine (ENIAC = Electronic Numerical Integrator and Computer) of 1946 at the University of Pennsylvania, Philadelphia, USA. Other examples of "electronic tube computers" followed immediately and also industrial production got started in the years after. We mention as examples the production of the model 650 from IBM and the model Z22 from the ZUSE KG. However in that time the market for selling electronic computers was still small, besides the prices were very high. It was the technological revolution initiated by the invention of the transistor (1948) and the invention of the integrated circuit which changed the situation and allowed the design and implementation of a new generation of computers. New companies such as for example the Digital Equipment Corporation (DEC) which manufactured successfully computers for real time applications for process control (e.g. the models PDP-8 and PDP-11) were founded. IBM developed the famous model 360 family with a new architectural design. The development of the microprocessor by US companies such as INTEL, Texas Instruments and Motorola allowed the design of "home computers" which eventually led to the personal computer and notebook of

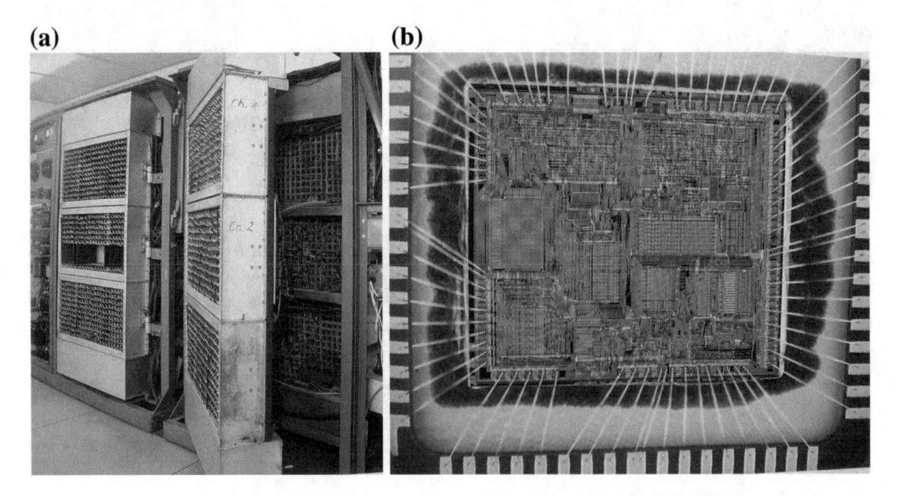

Fig. 7.22 **a** IBM 650 computer with electronic tubes; **b** Intel R 80186 microprocessor chip

our time. The development of software for the operating and application of personal computers gave a chance of the founding of new companies. One of it was the founding of MICROSOFT, today one of the leading companies in software development (Fig. 7.22).

7.10 Conclusion and Outlook

The paper covers the overall development of electrical information technology and the invention and manufacturing of the associated equipment and systems which is needed to enable the communication and control between men and machines. The discovery of the physical phenomena of electricity as observed in nature, both in space and in the world of materials, was without any doubts one of the most important one. It opened the invention and development of technical means for the transmission and processing of electrical signals which carry information. The history of the development of information technology shows us, that new results in physics are accompanied by the invention and development of new engineering systems. The goals which are represented by the requirements which a engineering system has to meet can be considered as constant if we allow also futuristic goals. The feasibility of goals depends however always on the state of the currently available scientific knowledge and on the existing methods and tools for engineering design and implementation. The knowledge of the history may help to give support in getting this means.

References

1. Albert Abramson: *The History of Television, 1880 to 1941*, Jefferson NC: Mc Farland & Company, 1987.
2. *From Semaphore to Satellite*. Published by the International Telecommunication Union, Geneva 1965.
3. Herman H. Goldstine: *The Computer from Pascal to von Neumann*, Princeton, NJ: Princeton University Press, 1972.
4. Tapan K. Sarkar; Robert J. Mailloux; Arthur A. Oliner; Magdalena Salazar-Palma; Dipak L. Sengupta: *History of Wireless*, Hoboken, NJ: John Wiley & Sons, 2006.

Chapter 8
On the Ability of Automatic Generation Control to Manage Critical Situations in Power Systems with Participation of Wind Power Plants Parks

Suad S. Halilčević and Claudio Moraga

8.1 Introduction

Synchronization of machines and the parallel switching of power systems is always a very important and sensitive step to change the conditions of a power systems work. At the time of closing the parallel switch (circuit breaker) a minimum of energy transfer is desired [1]. To have that condition realized, one must satisfy the conditions in which it is possible to get the machines run in parallel. Whether the synchronism conditions are fulfilled or not is a process that can be done by either the electromechanical or electronic synchronizing equipment (*SE*).

There are two ways of paralleling power systems: that based on the attended synchroscope-furnished stations, and that designed as a remote synchronism indicator, the information registered by the latter, forwarded to the operator through special communication channels. Today, however, with large interconnected power systems and new competitive environment for electricity, paralleling is very frequently a task for the system's operator and demands a great attention.

Nowadays, *SE* is set into operation by the system operator. However, utilization of real-time rating systems and integration with existing utility communication networks (*SCADA*) gives chances for automation of starting *SE*. Having this in mind, sophisticated *SE* needs to be responsive to that kind of signals or information that will excite it and which will act toward circuit breaker installed to connect the power systems (or generating unit to a power system).

S.S. Halilčević (✉)
Faculty of Electrical Engineering, University of Tuzla, Tuzla, Bosnia and Herzegovina
e-mail: suad.halilcevic@fet.ba

C. Moraga
Technical University of Dortmund, Dortmund, Germany

R. Seising and H. Allende-Cid (eds.), *Claudio Moraga: A Passion for Multi-Valued Logic and Soft Computing*, Studies in Fuzziness and Soft Computing 349, DOI 10.1007/978-3-319-48317-7_8

The crisis such as the one arising from the system's own generator outage will only be deepened should the operator have to waste time in considering whether or not to resort to some of the ancillary services (exploiting the available spinning reserve or introducing the appropriate quantity of the generation ready-reserve power). Though, as a matter of fact, to have access to such a resort is what heavily outages must be followed by; otherwise, nothing would save a great number of related consumers from having a black-out time. Therefore, without the continuous control of the related parameters—especially those of an off-on stage of generating units (or generating areas), systems load (power demand), and available power of the numerous wind power plants introduced into the power systems—neither the opportune paralleling nor the systems integrity preservation would be possible.

Response rate (MW/min.) requires maintaining an active power margin so that the generators have the thermal stored energy and generators capability to actually respond when the turbine's governor opens the valve. The quantity of response is important, and the appropriate policy needs to define requirements (which generators will maintain a MW-margin, what is the minimum response rate requirement). In addition, the owners of generators need to maintain load angles and a MVAr-margin of their generators within allowable limits. This is necessary for contingencies in which the generators, in these crisis situations, should keep their transient stability and provide voltage support. That will be successful if an appropriate quantity and also a spatial distribution of active and reactive power reserve have been maintained.

In the past two decades the power systems has hosted a huge number of wind power plants. The inertia and frequency responses of wind power plants with different wind turbine technologies can contribute to the inertial and primary frequency response during the frequency drop [2, 3]. There are numerous studies that document frequency response implications in cases with high levels of wind generation. It has been found out that reduction in system inertia due to higher levels of renewable generation will not have a significant impact on frequency response when compared with governor action of synchronous generators driven by steam and/or hydro-power. The fast transient frequency support using controlled inertial response from wind power will help increase the underfrequency load-shedding margin and avoid load shedding due to low frequencies. It has been demonstrated here that many systems and regional transmission operators in different countries began recognizing the value of inertial response by wind power and its importance for system reliability [4–6]. However, there are no studies on recognizing available wind power in the critical situations where the system operator should know the number of MWs that will stay at his disposal for overcoming the frequency drop and possible frequency and transient instability. That know-how, presented in this paper, is based on the logical decision-making and the degree of satisfaction of the current WPPP available power φ_W by particular $WPPP$ available power ranges W_i. In that way, the stochastic nature of the wind and, accordingly, the power available from WPPP, can be identified, and so offer a basis on which the needed RR can be calculated and eventually have the SE initialized.

The range of the load shifts that a power system may happen to undergo is very wide (load duration diagram). On the other side, the load duration diagram can be divided into few classes, each having a mean value of the load quantity, and the same duration. All the classes can further be attributed by the *RR* amounts that—if applied—will achieve that the power system generation meets consumption demands. There are several methods by which the information on the emergency power demands may be obtained. One of them—based on the chart of the *RR* generators in function of the outage affected generators and load classes—enables one to find out how much *RR* to apply, or—if the reserve should prove insufficient—how much power to buy at the power market [7]. In any case, such an operation is to be carried out by a skillful operator.

The need for an operating reserve is emphasized in competitive environment of power system work. That reserve is important for frequency response, security (load angles) and adequacy (generation meets consumption); but to keep the appropriate synchronous generation margins at a satisfying level, the chosen generators that are in operation have to maintain the difference between nominal and actual engaged power (spinning reserve). Among inertial power from wind power plants, the WPPP can provide the power in the form of spinning reserve, too. Due to stochastic nature of the wind, the electric power from wind power plants is also stochastic. That is why the need for RR to recover the emerged power deficit, considering $WPPP$, should be estimated on the base of a statistic approach, which includes a Rayleigh distribution [8]. The Rayleigh distribution is observed when the magnitude of a variable is related to its directional component such as it is a case with wind speed. In that case, there is an assumption that two components that describe the variable (wind speed and direction of wind) are not correlated, normally distributed with an equal variance and a zero mean.

The available spinning reserve of the conventional generators and available output electric power from $WPPP$ in the circumstances of the current power demand, are the starting point for a calculation of the needed *RR* and determination of the need to initialize the SE. The appropriate quantity of the *RR* also needs to be maintained. *R*R (or contingency reserve-supplemental) consists of the fast start-up generating units that can be in operation within 10 min at most. Having *RR* generators available, the process of paralleling should be made first.

The aim of this paper is describe a fast way of *SE* initiation for paralleling *RR* generators. Namely, the *SE* start is automated in the process of paralleling *RR* generator to a power system. By this innovative method the *RR* power can be installed in a very short time reducing customers loss of power. This method enables an immediate emergency response of the *RR* generator to the power system. Such an action requires an efficient mechanism of HNM. Neural computing is appropriate because, if a neural network (*NN*) is properly trained, it can undertake an adequate action in all situations including those that neither the operator nor the neural network itself has ever faced [9, 10]. The *HNM* designed can be understood as a part of the automatic generation control capable to manage critical situations.

8.2 The Hybrid Neural Model

The hybrid neural model—developed so as to recognize the information signals able to excite (start) the *SE*—contains two kinds of neurons: those with a sigmoid activation function and a perceptron [11]. The latter enables dividing the possible stage space into two areas characterizing the need for *RR* or sufficient power available. Neurons with sigmoid activation function are used to build a feedforward, three-layer neural network that is trained (learned) by a supervised, backpropagation learning rule based on the gradient descent algorithm. The second part of the adopted *HNM* is a perceptron that is suited for problems in pattern classification due to its hard limiting activation function.

The back-propagation learning rule is used to continually adjust the weights and biases of *NN* in the direction of the steepest descent with respect to the error:

$$\begin{aligned} W_{(t+1)} &= W_{(t)} + dW \\ B_{(t+1)} &= B_{(t)} + dB \end{aligned} \tag{8.1}$$

where:

$W_{(t+1)}$—the updated weight matrix (weights from the $(t + 1\text{st})$- phase of the *NN* training procedure),
$W_{(t)}$—the weight matrix from the t-phase of the *NN* training procedure,
$B_{(t+1)}, B_{(t)}$—the bias vector from the $(t + 1\text{st})$-phase and t-phase of the *NN* training procedure, respectively,
dW, dB—the weight change matrix and the bias change vector, respectively.

In addition, an improved backpropagation learning rule is used which includes elements for decreasing backpropagation's sensitivity to details in the error surface and training time. These elements are momentum m and adaptive learning rate *alr*, respectively. In that way, the weight change matrix can be expressed as:

$$\Delta W_{i,j}(t+1) = m\Delta W_{(i,j)}(t) + (1-m)\cdot alr \cdot \Delta_{(i)}(t)\cdot P_{(j)} \tag{8.2}$$

where:

$\Delta_{(i)}$—the layer's delta vector, and
$P(j)$—the layer's input vector.

The backpropagation learning rule is used to adjust the weights and biases of *NN* in order to minimize the sum squared error of the corresponding output signals. In this way, the values of the network weights and biases are continually changed until a previously defined acceptable error is reached. Delta vectors Δ (as a difference between the actual and the desired output's value) are calculated for the network's output layer, and then backpropagated through the network until delta vectors are available for a given hidden layer. In order to speed up the learning procedure, a batch presentation of input data (vectors) and delta vectors is used.

Since the *NN* outputs may be too noisy to make a decision in favor of either making (1), or not making (0) a demand to the *SE*, the output is forwarded to the perceptron which is based on the hard-limit transfer function (bias *b* for the presented case being 0.5). Therefore, the perceptron neuron's output is:

$$O = hard\,\lim(O_{NN} + b). \tag{8.3}$$

where

O_{NN}—output of the feed-forward *NN*,
b—bias that enables shifting of the *NN*'s output to the left by the accepted amount of *b*.

In other words, the applied hard limit transfer function enables the perceptron to classify (divide) the output signal of the feedforward *NN* into two areas:

Area 1 (output = 1): activation of *SE*,
Area 2 (output = 0): no signal (*SE* does not need to operate).

The input vector to *HNM* (Fig. 8.1) consists of two parts: one concerning the generator (or generating area) that has suffered outage, and another one concerning the load (load class) at which the outage has taken place. The outage-affected generator is marked with 1; the sounds ones, with 0.

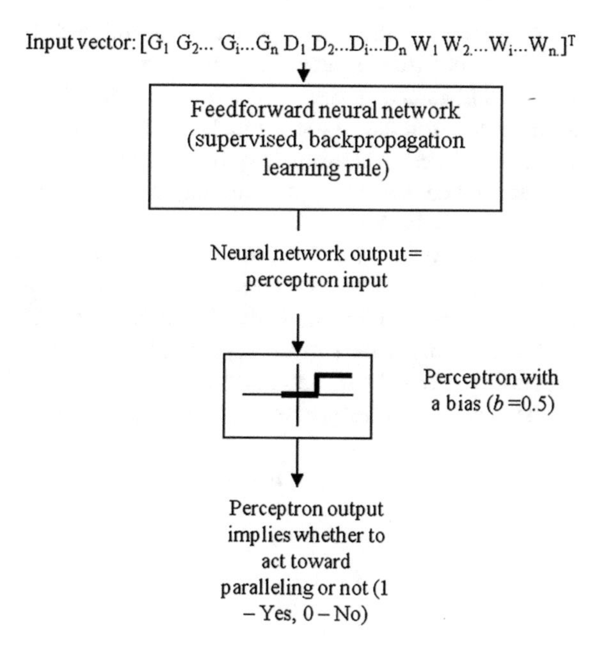

Fig. 8.1 Hybrid neural network model

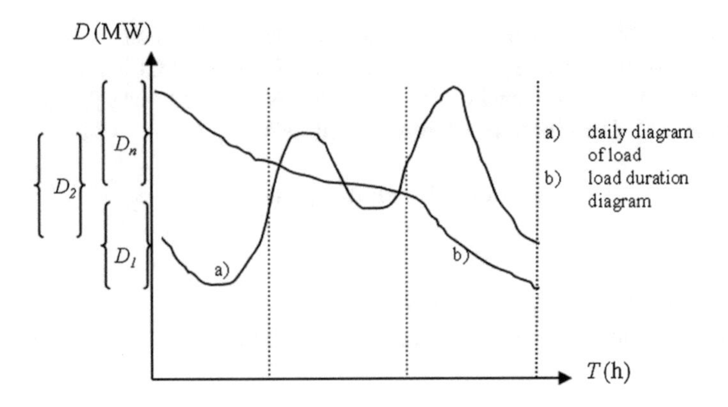

Fig. 8.2 Daily diagram of load (**a**) and load duration curve (**b**)

8.2.1 The Logical Decision-Making Strategy

The second part of the input vector comprises the indicators of the matching of the current load (power demand) regime with the defined load ranges. The load duration curve (daily diagram of load Fig. 8.2) is divided into the load classes: $D_1 = (D_{1max} - D_{1min}), \ldots, D_n = (D_{nmax} - D_{nmin})$. These load classes are defined in accordance with the load shifting in a positive and negative direction with respect to their mean values. Unpredictability of the load shift is estimated by classical methods of estimation based on the Central Limit Theorem, [12].

As a result of the stochastic nature of the load, overlaps of the power demand ranges defined for the particular time intervals have to be considered. For each of the chosen time intervals (one or two hours or other time intervals) an estimation of the load ranges with the "three-sigma" rule will be made. This rule provides a 99, 7 % confidence interval of estimation. In order to find out to which power demand range belongs the current power demand regime of the power system, a decision-making strategy has to be designed.

The logical decision-making uses the operating condition—the current power demand regime and defined power demand ranges—in accordance with the normal distribution of load (power demand) uncertainties and the rules given below.

Let D_i represent the i-th power demand range and φ_{Di}, the degree to which the current power demand may be satisfied by the i-th power demand range. The following decision rules may be used for a given current power demand D:

IF

$$\varphi_{D1}(D) > max\left(\varphi_{D2}(D), \varphi_{D3}(D), \ldots, \varphi_{Dn}(D)\right), \tag{8.4}$$

THEN

$$\begin{aligned} D_1 &= 1 \\ D_2 &= 0 \\ &\ldots\ldots \\ D_n &= 0 \end{aligned} \tag{8.5}$$

$\forall 1 \leq i \leq n$; n-number of power demand ranges.

IF

$$\varphi_{Di}(D) > max\left(\varphi_{D1}(D), \ldots, \varphi_{Di-1}(D), \varphi_{Di+1}(D), \ldots, \varphi_{Dn}(D)\right), \tag{8.6}$$

THEN

$$\begin{aligned} D_1 &= 0 \\ &\ldots\ldots \\ D_{i-1} &= 0 \\ D_i &= 1 \\ D_{i+1} &= 0 \\ &\ldots\ldots \\ D_n &= 0 \end{aligned} \tag{8.7}$$

IF

$$\varphi_{Dn}(D) > max\left(\varphi_{D1}(D), \ldots, \varphi_{Dn-1}(D)\right), \tag{8.8}$$

THEN

$$\begin{aligned} D_1 &= 0 \\ &\ldots\ldots \\ D_{n-1} &= 0 \\ D_n &= 1 \end{aligned} \tag{8.9}$$

The left side of the implications presented in (8.9) are defined as illustrated in Fig. 8.3 and determined in (8.1). A degree value of 1 indicates that the current power demand corresponds to the appropriate power demand range; while 0, that the current power demand cannot be satisfied by the corresponding power demand range.

Let us recall that RR is defined on the base of estimation under "three sigma" rule with respect to the upper limit of the confidence interval of power demand and lower limit of the confidence interval of electric power available from $WPPP$:

$$RR_{Gx} = \sum_{i_D=1}^{N_D}(D_i^0 + 3\sigma_{D,i}) - \sum_{i_G=1}^{N_G-1} P_{Gi} - \sum_{i_W=1}^{N_W}(W_i^0 - 3\sigma_{W,i}) \tag{8.10}$$

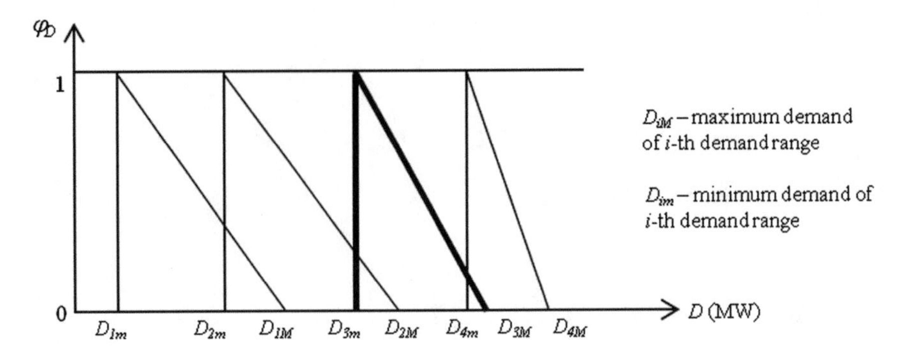

Fig. 8.3 The defined power demand ranges

where:

$(RR)_{Gx}$—capacity of *RR* needed to compensate for the outage of the generating unit "x",
D_i^0—mean value of power demand (MW) for the considered time interval; i.e., the mean value of particular power demand range,
W_i^0—mean value of available wind power (MW) for the considered time interval; i.e., the mean value of particular wind power range,
$\sigma_{D,i}$—standard deviation for the i-th power demand range in the considered time interval (MW),
$\sigma_{W,i}$—standard deviation for the i-th wind power range in the considered time interval (MW)
N_D—number of consumers' buses,
N_G—number of generating units,
N_W—number of wind power plants (or WPPPs),
P_{gi}—nominal active power of the i-th generating unit (MW),
$D_{i,min}$, $D_{i,max}$—minimum and maximum level of the i-th power demand range, respectively (MW) (these levels mach the $-3\sigma_{D,i}$, i.e. $+3\sigma_{D,i}$),
$W_{i,min}$, $W_{i,max}$—minimum and maximum level of the i-th power range of WPPP, respectively (MW) (these levels mach the $-3\sigma_{W,i}$, i.e. $+3\sigma_{W,i}$).

If one takes into consideration the security side of the evaluation of the *RR* power for outage of a generating unit G_x, the degree of satisfaction of the current demand regime φ_D by particular demand range D_i may be defined as follows:

$$\varphi_D = \begin{cases} 1 & \text{if } D = D_{i,min} \\ \frac{D_{i,max}-D}{D_{i,max}-D_{i,min}} & \text{if } D_{i,min} < D < D_{i,max} \\ 0 & \text{else } D \geq D_{i,max} \end{cases} \quad (8.11)$$

where D is the current power demand.

Example Let the current power demand of a power system be 1,200 MW. On the basis of the load duration diagram, four power demand ranges (for four time intervals of six hours each) are chosen: 500–1,000 MW; 800–1,450 MW; 950–1,700 MW and 1,500–2,000 MW. Each demand range represents a 3-sigma confidence interval of power demand estimation. Taking into consideration the proposed decision rules, the demand range of 1,200 MW corresponds to the third power demand range or more precisely, to the interval (D_{3m}, D_{3M}) (shown in bold in Fig. 8.3). This implies that D_3 as an input data for the second part of the input vector is equal to 1 (all other D_i are equal to 0). In addition, $\left(D_3^0 + 3\sigma_{D3}\right)$ is used as a relevant data in the calculation of the needed RR. In [13] the input vector consisted of two parts has been applied to *HNN*. Here the input vector is expanded by indicators of the current electric power available from *WPPP* with the defined generation ranges, Fig. 8.4.

If one takes into consideration the security side of the evaluation of the *RR* power for outage of a generating unit G_x, the degree of satisfaction of the current *WPPP* output electric power φ_W by particular power range W_i may be defined in a similar way as in (8.11):

$$\varphi_W = \begin{cases} 1 & \text{if } W = W_{i,max} \\ \frac{W_{i,max} - W}{W_{i,max} - W_{i,min}} & \text{if } W_{i,min} < W < W_{i,max} \\ 0 & \text{else } W \leq W_{i,min} \end{cases} \tag{8.12}$$

Example Let the current available power from one wind power plants park (WPPP) (A wind-farm or wind park is a group of wind power plants in the same location used to produce electricity) be 8.6 MW. Four available power ranges of WPPP are chosen: 0.0–4.0 MW; 2.0–7.5 MW; 6.0–11.0 MW and 7.5–14 MW (for four six-hour intervals). Each power range represents a 3-sigma confidence interval of the available power estimation. Taking into consideration the proposed decision rules, the available power of 8.6 MW corresponds to the fourth power range with degree of satisfaction of 0.83 (Fig. 8.4). This implies that W_4 as an input data for the third part of the input vector, is equal to 1 (all other W_i are equal to 0). In addition, $(W_4^0 - 3\sigma_{W_4})$ is used as a relevant data in the calculation of the needed *RR* in (8.10). The input

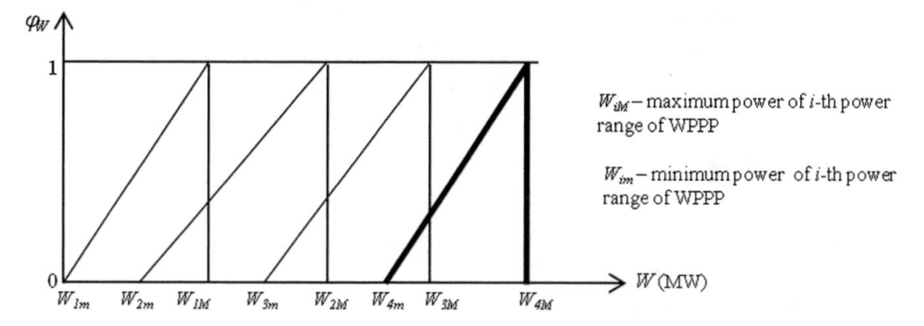

Fig. 8.4 The defined electric power ranges available from WPPP

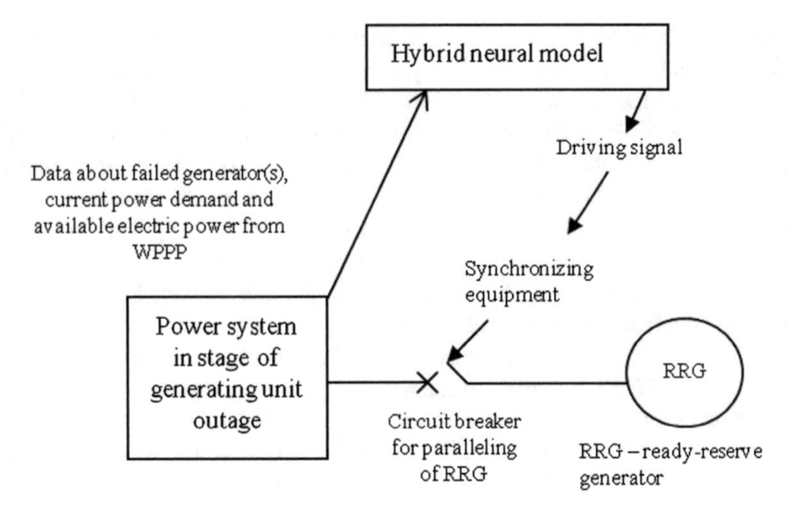

Fig. 8.5 Model of the studied situation

data for the third part of the input vector are all equal to zero when the WPPP is disconnected from power system.

If a neural network is used to activate the SE, the inputs to the neural network should comprise information about the generator state (or state of the generating area) received through the communication channel (1—outage, 0—operating state), data about the present demand class, and current available electric power from *WPPP* provided by the above rules.

An input vector with the following structure drives the *HNM*:

$$I = [G_1 G_2 \dots G_i \dots G_n D_1 D_2 \dots D_i \dots D_n W_1 W_2 \dots W_i \dots W_n]^T \tag{8.13}$$

with G_i, D_i, and W_i taking values 0 or 1, depending on the current working regime.

After a proper training, the hybrid neural network gives valid information, i.e. the signal for either activating (marked by 1) or not activating *SE* (marked by 0).

The acting scheme of HNM for automation of the power system *SE* exciting is presented in Fig. 8.5.

8.3 Test Model

The results with adopted neural network model have been verified with those reported in [14],—Fig. 8.6 and Table 8.1.

The difference between results reported in [14] and those found here is that we have one WPPP nominal power of 14 (MW). The feed-forward neural network is trained for several cases of generator outages, system demand and available electric

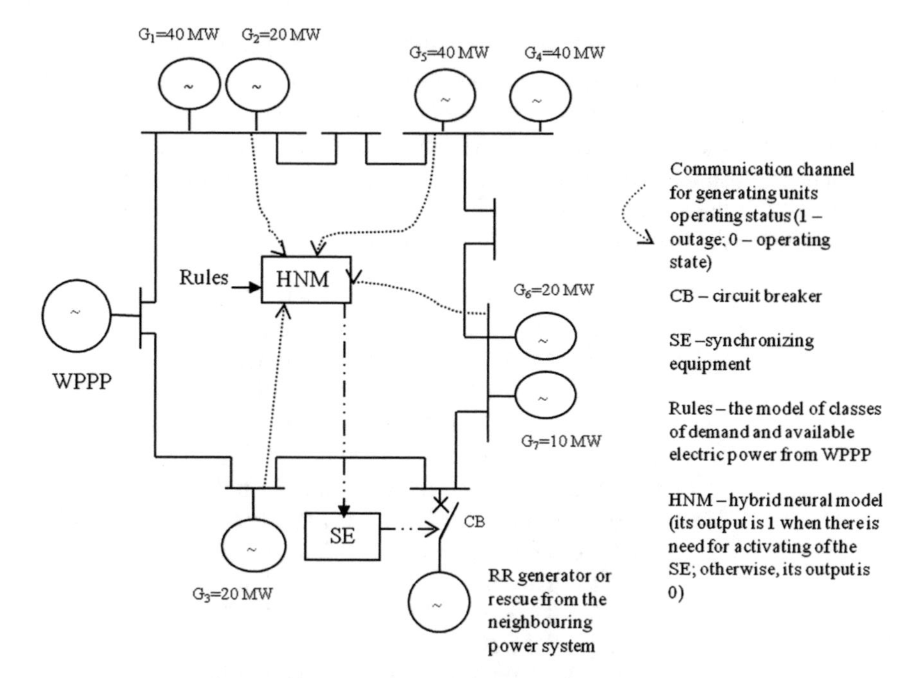

Fig. 8.6 Test model for checking the *HNM* performance

Table 8.1 One part of the training data obtained by using of (8.10)

Outage	Upper limit of the power demand range *D* (MW)				Upper limit of the *WPPP* range *W* (MW)				1− *SE* should act, 0− *SE* should not act
	155	165	175	185	4	7,5	11	14	
$G_2 + G_3$	0	1	0	0	1	0	0	0	1
G_2	0	0	1	0	0	0	1	0	1
G_3	0	0	0	1	0	1	0	0	1
G_4	0	1	0	0	0	0	1	0	0
G_1	0	0	1	0	0	1	0	0	0
$G_1 + G_2$	0	0	0	1	0	0	0	1	1
G_4	0	0	1	0	0	0	0	1	1
G_5	1	0	0	0	0	0	1	0	0
G_6	0	1	0	0	1	0	0	0	1
G_7	0	0	0	1	0	1	0	0	0
$G_1 + G_4$	1	0	0	0	0	0	0	1	1
WPPP	0	1	0	0	0	0	1	0	0
WPPP	0	0	0	1	0	0	0	1	1
$WPPP + G_5$	0	0	1	0	0	1	0	0	1

power from WPPP (Mathlab's Neural Network Toolbox—1998). The neural network has three layers. The input layer has fifteen neurons, the hidden layer twenty-seven, and the output layer one neuron. (This architecture was obtained experimentally, however an evolutionary design could have been used [15]). Test cases not applied in the training phase were used to evaluate the generalizing performance of the *HNM*.

The overall input matrix for a batching operation, that is, as required for the learning process of the adopted *HNM* has the following structure:

$$I = \begin{matrix} G_1 & 0\,0\,0\,0\,1\,1\,1\,0\,0\,0\,0\,0\,0\,0\,1\,0\,0\,0\,0\,0 \\ G_2 & 1\,1\,0\,0\,0\,1\,0\,0\,0\,0\,1\,0\,0\,0\,0\,0\,0\,0\,1\,0 \\ G_3 & 1\,0\,1\,0\,0\,0\,0\,0\,0\,0\,0\,0\,0\,0\,0\,1\,0\,0\,0\,0 \\ G_4 & 0\,0\,0\,1\,0\,0\,1\,1\,1\,0\,0\,0\,0\,0\,0\,0\,0\,0\,0\,0 \\ G_5 & 0\,0\,0\,0\,0\,0\,0\,0\,0\,1\,0\,1\,0\,0\,0\,0\,0\,0\,0\,1 \\ G_6 & 0\,0\,0\,0\,0\,0\,0\,0\,0\,0\,1\,0\,1\,0\,0\,0\,1\,0\,0\,0 \\ G_7 & 0\,0\,0\,0\,0\,1\,0\,0\,0\,0\,0\,0\,0\,1\,0\,0\,0\,1\,0\,0 \\ D_1 & 0\,0\,0\,0\,0\,0\,1\,0\,0\,0\,0\,1\,0\,0\,1\,0\,0\,1\,0\,0 \\ D_2 & 1\,0\,0\,1\,0\,0\,0\,0\,0\,0\,1\,0\,1\,0\,0\,0\,0\,0\,0\,1 \\ D_3 & 0\,1\,0\,0\,1\,0\,0\,1\,0\,0\,0\,0\,0\,0\,0\,0\,0\,0\,0\,0 \\ D_4 & 0\,0\,1\,0\,0\,1\,0\,0\,1\,1\,0\,0\,0\,1\,0\,1\,1\,0\,1\,0 \\ W_1 & 0\,0\,0\,0\,0\,0\,0\,0\,0\,0\,0\,0\,0\,0\,0\,0\,0\,0\,0\,0 \\ W_2 & 1\,0\,0\,0\,0\,0\,0\,0\,0\,0\,0\,0\,0\,0\,0\,0\,0\,0\,0\,0 \\ W_3 & 0\,0\,0\,0\,0\,0\,0\,0\,0\,0\,0\,0\,0\,0\,0\,0\,0\,0\,0\,0 \\ W_4 & 0\,0\,0\,0\,0\,0\,0\,0\,0\,0\,0\,0\,0\,0\,0\,0\,0\,0\,0\,0 \end{matrix} \tag{8.14}$$

The signals produced by *HNM* for unseen cases are given in Table 8.2.

Table 8.2 Test results

Outage	Upper limit of the power demand range *D* (MW)				Upper limit of the *WPPP* range *W* (MW)				1– SE should act, 0– *SE* should not act
	155	165	175	185	4	7,5	11	14	
$G_2 + G_3$	0	1	0	0	0	0	1	0	1
G_2	0	1	0	0	0	0	1	0	0
G_3	0	0	0	1	0	0	0	1	1
G_4	1	0	0	0	1	0	0	0	0
G_1	0	1	0	0	0	0	1	0	1
$G_1 + G_2$	0	1	0	0	1	0	0	0	1
G_4	0	0	0	1	1	0	0	0	1
G_5	1	0	0	0	0	0	0	1	0
G_6	0	0	1	0	0	0	1	0	0
G_7	0	0	1	0	1	0	0	0	0
$G_1 + G_4$	0	1	0	0	0	0	0	0	1
WPPP	1	0	0	0	0	0	1	0	0

By simulating a power system work, where *HNM* is served for an automatic sending of a driving signal for the *SE* initiation has been illustrated. The response of the *HNM* has been correct in all cases. This automatic work takes just a couple of moments, and one can say that the automatic production of an activating signal for the *SE* is done in a real time.

Furthermore, we made additional time domain simulations to determine frequency as indication of a momentary imbalance in the power system and responses of the *RR* generator in the case of its manual introduction to operation and by means of *HNM* through *SE*. We present two characteristic cases:

Case I.
Generator G_6 outage at the power system demand of 175 MW. The loading of the generator G_6 at the moment of outage is 18 MW. Taking into account governor speed load characteristic of 5 % the total composite self-regulation of the generators for described case is 75 MW/Hz. The self-regulation of the consumers, taking into account the factor of consumer self-regulation of 4 (%/Hz), is 7 MW/Hz. The overall power system self-regulation is the sum of the above mentioned self-regulation factors and has value of the 82 MW/Hz. In that way, the frequency change is 18 (MW)/82 (MW/Hz) = 0.219 Hz (Fig. 8.7). Comparing with results in [13] it can be concluded that the frequency drop is lower in this case due to benefit of WPPP introduction into the power system, and as result of that, more spinning reserve on the synchronized classical generators.

The output of HNM is "0" which means the stand-by of the *SE*. The power deficiency that is caused by the failed generator is recovered by the spinning reserve of the sound generators (including available electric power from *WPPP*). The power system has been returned to the normal operating state (50 Hz) after a transient process determined by the response rate of the available spinning reserve.

Case II:
Generator G_1 outage at the power system demand of 185 MW. The loading of the generator G_1 at the moment of outage is 37 MW. Taking into account governor

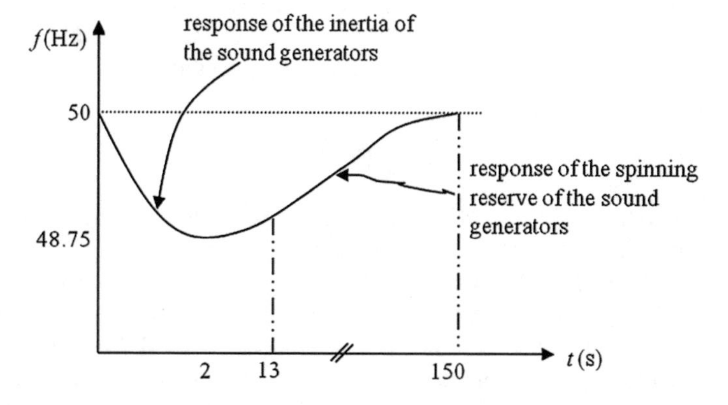

Fig. 8.7 Frequency change for the first case

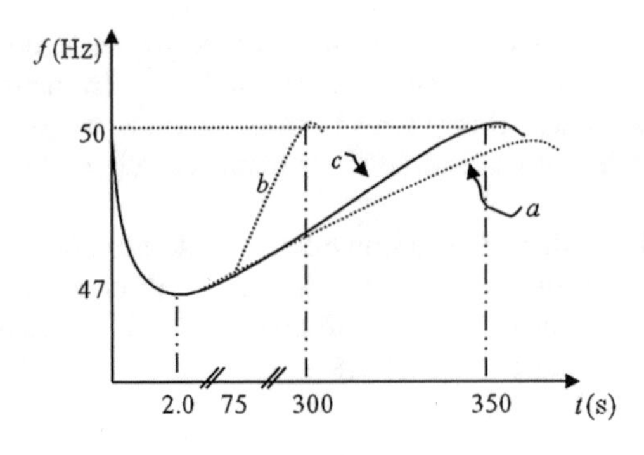

Fig. 8.8 Frequency change for the second case

speed load characteristic of 5% the total composite self-regulation of the generators for the described case is 66 MW/Hz. The self-regulation of the consumers, taking into account the factor of consumer self-regulation of 4 (%/Hz), is 7.4 MW/Hz. The overall power system self-regulation is the sum of the above mentioned self-regulation factors and has value of the 73.4 MW/Hz. In that way the frequency change is 37 (MW)/71.4 (MW/Hz) = 0.518 Hz (Fig. 8.8). As in the first case, the frequency drop is lower with respect to the case described in [13] due to more installed power (in the view of *WPPP*); however, it is not enough to recover the emerged power deficit caused by outage of generator G_1.

The output of the *HNM* is "1" which requires activation of *SE*. The power deficiency that is caused by the failed generator cannot be recovered by the spinning reserve of the sound generators and available power from *WPPP*. The power system has been returned to the normal operating state (50 Hz) after a transient process determined by the response rate of the available spinning and supplemental reserve (ready reserve generator).

The effects of generator outages and putting of the RR generator into operation manually (curve "*c*" in Fig. 14.8) and by means of *HNM* (curve "*b*" in Fig. 8.8) may be approximately modeled by examining Hz/sec (the Euler numerical method "step-by-step" is applied). Deceleration and calculating the change in system frequency is done for the described system by using the following relationship:

$$\frac{\Delta_f}{\Delta_t} = \frac{P_{dec}}{2H} \cdot f_{rated} \tag{8.15}$$

with f_{rated} in Hz, P_{dec} in p.u. MW, and H (the total system inertia) in p.u. MWs (in the first case it is amounted to 4 p.u. MWs and in the second case it is amounted to 3 p.u. MWs). We assume that all remaining plants have inertia constant (on their own base) of 2 s. Substituting the appropriate data into (8.15) the frequency deceleration for the first and the second case are 1.25 Hz/s (Fig. 8.7) and 3.0 Hz/s (Fig. 8.8), respectively. In the previous work [13] without introduction of WPPP into power system these changes have had values of 1.47 Hz/s and 3.78 Hz/s, respectively. The benefit of the *WPPP* to the power system management in the case of load changes is obvious and it can be and has to be treated as a positive contributor in the process of keeping the power system frequency stable.

If *SE* is governed by *HNM*, the power system takes 300 s to restore a normal operation. If not, it takes 350 s. The reason for the promptness of the former case is the ability of *HNM* to recognize the crisis symptoms immediately, and to consequently suggest introduction of the *RR* generator through *SE*.

8.4 Conclusion

The work presents *HNM* as a means of an on-line outage crisis counteraction. The proposal is particularly helpful in making a decision on whether or not to have the outage affected power system paralleled with the RR generator. Based on the logical decision-making and neural calculation procedures, the *HNM* is a prompt and a precise starter of the SE, thus making it possible for the outage-suffering system to have a power compensating resort in a very short time. In this way, the high level of reliability can be kept. By leaving the decision on paralleling to the *HNM*-based automatic starter, the operator can devote himself to taking care of other aspects of the crisis.

In the competitive environment of the area of electricity, opportune counter-emergency action is certainly a step towards having reliable energy management services.

Future research in the field of a crisis regime of the power system work will probably include a decision-making automation in the load shedding process in addition to the presented model.

References

1. R. H. Miller, J. H. Malinowski: *Power System Operation*, third edition, McGraw-Hill, Inc., 1994.
2. E. Muljadi, V. Gevorgian, M. Singh, and S. Santoso: Understanding Inertial and Frequency Response of Wind Power Plants, IEEE Symposium on Power Electronics and Machines in Wind Applications Denver, Colorado, July 16–18, 2012.
3. Caixia Wang, James Mccalley: Impact of wind power on control performance standards, *International Journal of Electrical Power & Energy Systems* 47(1), May 2013, pp. 225–234.
4. N. Miller, M. Shao, and S. Venataraman: California ISO: Frequency response study, Final draft, Nov 9, 2011. Available at http://www.uwig.org/Report-FrequencyResponseStudy.pdf
5. P. W. Christensen and G. T. Tarnowski: Inertia of wind power plants State-of-the-art review, presented at 10th International Workshop on Large-Scale of Wind Power, Aarhus, Denmark, Oct 25–26, 2011.
6. S. Sharma, S. H. Huang, and N. D. R. Sarma: System inertial frequency response estimation and impact of renewable resources in ERCOT interconnection, in: IEEE Power and Energy Society General Meeting Proc., July 24–29, 2011, pp. 16.

7. S. S. Halilčević: *Determining of the generation ready-reserve capacity*, Ph.D. Thesis, Tuzla University, Bosnia and Herzegovina, March 1998.
8. Christian Walck: *Hand-book on Statistical Distributions for experimentalists, Particle Physics Group Fysikum*, University of Stockholm, Internal Report SUF PFY/96 01Stockholm, 1996, 1st revision 1998, last modification 2007.
9. K. Hornik, M. Stinchcombe, H. White: Multilayer Feed-forward Networks are Universal Approximators, *Neural Networks* 2, 359–366, 1989.
10. C. Moraga: Properties of Parametric Feed-forward Networks, Proceedings XXIII Conferencia Latinoamericana de Informatica, Valparaiso, Chile, 861–870, 1997.
11. M. Caudill, C. Butler: *Understanding Neural Networks Computer Explorations*, Vol. 1 and 2, Cambridge, MA, MIT Press, 1992.
12. R. E. Walpole, R. H. Myers: *Probability and Statistics for Engineers and Scientists*, fourth edition, Maxwell MacMillan Publishing Company Pte. Ltd. 1990.
13. S. S. Halilčević, C. Moraga: On the Ability of Automatic Generation Control to Manage Critical Situations, *International Journal of Power and Energy Systems*, Issue 2, Vol. 24/2004, 203–3353, ACTA Press, Canada.
14. S. S. Halilčević, F. Gubina: An On-Line Determination of the Ready Reserve Power, *IEEE Trans. On Power Systems*, Nov. 1999, Vol.14, Number 04, p. 1514.
15. R. Heider, C. Moraga: Evolutionary Synthesis of Neural Networks Based on Graph Grammars, Proc. Int. Panel Conference on Soft and Intelligent Computing, 119–126, Budapest, ISBN 963 420 510 0, 1996.

Chapter 9
The Reed-Muller-Fourier Transform—Computing Methods and Factorizations

Radomir S. Stanković

9.1 Introduction

In late 1960s and early 1970s, there was apparent interest in discrete dyadic analysis based on the discrete Walsh functions, as can be seen from the organization of a series of international workshops dedicated exclusively to this subject. For more information see a discussion of that in [39]. From the abstract harmonic analysis point of view, these functions are kernels of the discrete Walsh transform which can be viewed as the Fourier transforms on finite dyadic groups consisting of a set of binary n-tuples equipped with the addition modulo 2, the operation that is usually called EXOR in switching theory.

Dr. James Edmund Gibbs from the National Physical Laboratory, Teddington, Middlesex, UK, was deeply involved in these research activities and his work in this area led to the definition of the so-called logical derivative, or the discrete Gibbs derivative, as an operator enabling the differentiation of piecewise constant functions, such as the Walsh functions [13, 14].

By looking for a counterpart of the Walsh (Fourier) analysis in the Boolean domain, J.E. Gibbs defined a transform that he called the Instant Fourier transform [15]. The name of the transform comes from the possibility to compute the related spectrum instantaneously through a network consisting of AND and EXOR logic gates. A study of the relationships between the discrete Walsh series and the Reed-Muller expressions of Boolean functions was carried out several years after [30]. In the frame of this research, it was recognized that the set of basis functions in terms of which the Instant Fourier transform is defined is identical with the basis functions used in the definition of the Reed-Muller (RM) expressions. The spectral interpretation of the Reed-Muller expressions, where the coefficients in the RM-expressions are viewed as spectral coefficients of a Fourier-like transform over the finite Galois

R.S. Stanković (✉)
Faculty of Electronics, Department of Computer Science, University of Nis, Niš, Serbia
e-mail: radomir.stankovic@gmail.com

R. Seising and H. Allende-Cid (eds.), *Claudio Moraga: A Passion for Multi-Valued Logic and Soft Computing*, Studies in Fuzziness and Soft Computing 349, DOI 10.1007/978-3-319-48317-7_9

(GF) field $GF(2)$, i.e., with computations modulo 2, or equivalently in terms of logical operations AND and EXOR, was championed by Ph.W. Besslich [2–5], and the same approach was accepted by some other authors [18, 19].

The main difference between these two concepts, the Reed-Muller (RM) transform and the Instant Fourier transform, is in the underlying algebraic structures, the Boolean ring for the Reed-Muller expressions, and the Gibbs algebra defined in terms of the Gibbs multiplication for the Instant Fourier transform [15]. The main idea behind the definition of this new algebraic structure was to derive properties of the Instant Fourier transform (equivalently, the Reed-Muller transform) corresponding to the properties of the classical Fourier transform on the real line.

A generalization of the Gibbs algebra to multiple-valued functions presented in [31] provided a way to define a transform, called the Reed-Muller-Fourier (RMF) transform, that can be viewed as a generalization of the Reed-Muller transform for binary functions to multiple-valued functions. Professor Claudio Moraga immediately realized that the RMF-transform is an interesting concept and already in 1993 started working on the development of a related theory by putting a lot of effort towards the refinement of its definition and the study of its properties for different classes of multiple-valued functions [24, 42]. That was the motivation for the selection of the Reed-Muller-Fourier (RMF) transform as the subject in this book dedicated to Professor Moraga on the occasion of his 80th birthday. In particular, we will discuss the different methods used to compute the Reed-Muller-Fourier spectra of multiple-valued functions efficiently in time and space. The definition of the so-called algorithms with constant geometry for the Reed-Muller-Fourier transform as well as various factorizations of the RMF-matrix derived from the factorization of the Pascal matrices can be consider as new contributions to the field.

9.2 Reed-Muller-Fourier Transform

In this section, we present the basic definitions of the Reed-Muller-Fourier transform.

9.2.1 The Gibbs Algebra

Denote by G a group of n-ary p-valued sequences $x = (x_1, \ldots, x_n)$ with the group operation defined as componentwise addition modulo p. Thus, for all $x = (x_1, \ldots, x_n)$, $y = (y_1, \ldots, y_n) \in G$,

$$\begin{aligned} x \oplus y &= (x_1, \ldots x_n) \oplus (y_1, \ldots, y_n) \\ &= ((x_1 \oplus y_1), \ldots, (x_n \oplus y_n)) \bmod p. \end{aligned}$$

Denote by Z_q the set of first q non-negative integers. For each $x \in G$, the p-adic contraction is defined as a mapping $\sigma : G \rightarrow Z_q$ given by

$$\sigma(x) = \sum_{i=1}^{n} x_i p^{n-i}.$$

We denote by $P(G)$ the set of all functions $f : G \to Z_q$. In $P(G)$, we define the addition as modulo p addition,

$$(f \oplus g)(x) = f(x) \oplus g(x), \forall x \in G,$$

and multiplication as a convolutionwise (Gibbs) multiplication [15]

$$(fg)(0) = 0$$
$$(fg)(x) = \sum_{s=0}^{\sigma(x)-1} f(\sigma(x) - 1 - s)g(s), \forall x \in G,\ x \neq 0.$$

Denote by W a particular function in $P(G)$ such that

$$W(x) = p - 1, \quad \forall x \in G,$$

and by S the set of first q positive integer powers of W, i.e., $S = \{W^1, \ldots, W^q\}$ with exponentiation in terms of the Gibbs multiplication defined above. The set S is a basis in $P(G)$ with respect to which the Reed-Muller-Fourier (RMF) transform is defined [31]

$$f = \sum_{i=0}^{q-1} c_i W^{i+1} \bmod p,$$

where $c_i \in Z_p$.

As is shown in [31], if a p-valued variable x_i is considered as a particular function in $P(G)$, $f(x_1, \ldots, x_n) = x_i$, then the RMF-transform matrix can be expressed in terms of the variable x_n as

$$f(x_1, \ldots, x_n) = \sum_{i=0}^{q-1} c_i \phi_{i+1}(x_n) \bmod p, c_i \in \{0, \ldots, p-1\},$$

where

$$\phi_i(x_n) = \begin{cases} (p-1) \cdot x_n^{\lceil i/2 \rceil}, & i\text{-odd}, \\ x_n^{i/2}, & i\text{-even}, \end{cases}$$

where $\lceil a \rceil$ is the integer part of a, and $x_n^r = x_n \cdot x_n \ldots x_n$ r times with the multiplication defined as convolutionwise (Gibbs) multiplication in $P(G)$.

This definition corresponds to the interpretation of the RMF-expressions as an analogue of the Fourier series expressions with functions $\phi_i(x_n)$ viewed as counterparts of the exponential functions. The following alternative definition allows us to consider the RMF-expressions as polynomial expressions. In this way, the RMF-expressions can be viewed as a generalization of the Reed-Muller expressions for binary logic functions to multiple-valued functions. At the same time, being computed modulo p, with no restrictions to p prime, the RMF-expressions can be viewed as a counterpart of GF-expressions for multiple-valued functions [34].

Definition 9.2.1 (*Reed-Muller-Fourier expressions*) Any p-valued n-variable function $f(x_1, \ldots, x_n)$ can be expanded in powers of variables x_i, $i = 1, \ldots, n$ as

$$f(x_1, \ldots, x_n) = (-1)^n \sum_{a \in V^n} q(a) x_1^{*a_1} \ldots x_n^{*a_n},$$

where V^n is the set of all p-valued n-tuples, $q(a) \in \{0, 1, 2, \ldots p-1\}$, and the exponentiation is defined as $x^{*0} = -1$ modulo p, and for $i > 0$, x^{*i} is determined in terms of the convolutionwise (Gibbs) multiplication defined above.

When discussed in the Gibbs algebra, properties of the RMF-transform correspond more likely to the properties of the classical Fourier transform than the properties of the Galois field (GF) transforms [34]. We point out the following two chief features among those expressing the differences between the RMF-transforms and the GF-transforms, since these features influence performances of the computation methods discussed below.

1. The operations used in the definition of the RMF-transform are modulo p operations, also in the case of a non-prime p. Therefore, in computations, these operations can be performed faster than the field operations used in GF-transforms. Note that modulo operations are provided in contemporary programming environments such as CUDA for computing over Graphics Processing Units (GPUs) [26].
2. The RMF-transform matrix is a triangular matrix, but expresses a Kronecker product structure in the same way as the GF-transform matrices. In this respect, the triangular structure of this transform matrix corresponds directly to the RM-transform matrix, which is the RMF-transform for $p = 2$. Recall that the GF-transform matrices are not triangular and have a larger number of non-zero elements than the RMF-transform matrices.

Although the entire theory of RMF-transforms is valid for any p, for simplicity, the discussion of computation methods will be given on the example of functions for $p = 3$ and, therefore, we provide the corresponding case examples of the Reed-Muller-Fourier transform. Examples for $p = 4$ elaborated in detail are presented in [34].

Table 9.1 Addition and multiplication modulo 3

$\oplus$	0	1	2
0	0	1	2
1	1	2	0
2	2	0	1

$\cdot$	0	1	2
0	0	0	0
1	0	1	2
2	0	2	1

Table 9.2 3EXP and 3AND

$*$	0	1	2
0	2	0	0
1	2	1	0
2	2	2	2

$\odot$	0	1	2
0	0	0	0
1	0	2	1
2	0	1	2

9.2.2 *RMF Expressions for* $p = 3$

Consider the ring of integers modulo 3 defined in terms of addition and multiplication modulo 3. For a uniform presentation of all the operations that will be used, we present these operations in Table 9.1. In order to generate the product terms of three-valued variables corresponding to those appearing in the RM-expressions for switching functions and GF-expressions for multiple-valued functions, we define in Table 9.2 the 3AND multiplication and 3EXP exponentiation, denoted by $\odot$ and $*$, respectively. Note that the 3AND table is actually the multiplication modulo 3 table multiplied by 2, and $2 = -1$ modulo 3.

We generate a set of 3^n product terms given in the matrix notation by

$$\mathbf{X}_3(n) = \bigotimes_{i=1}^{n} \left[x_i^{*0} \; x_i^{*1} \; x_i^{*2} \right] = \bigotimes_{i=1}^{n} \left[2 \; x_i \; x_i^{*2} \right],$$

with 3AND and 3EXP applied to the three-valued variables. In matrix notation, the basis functions are expressed as columns of the matrix

$$\mathbf{X}_3(1) = \begin{bmatrix} 2 & 0 & 0 \\ 2 & 1 & 0 \\ 2 & 2 & 2 \end{bmatrix}.$$

Definition 9.2.2 Each n-variable 3-valued logic function given by the truth-vector $\mathbf{F}(n) = [f(0), \ldots, f(3^n - 1)]^T$ can be represented as a Reed-Muller-Fourier (RMF) polynomial given by

$$f(x_1, \ldots x_n) = (-1)^n \mathbf{X}_3(n) \mathbf{S}_{f,3}(n),$$

with calculations modulo 3, where the vector of RMF-coefficients $\mathbf{S}_{f,3}(n) = [a_0, \ldots, a_{3^n-1}]^T$ is determined by the matrix relation

$$\mathbf{S}_{f,3}(n) = \mathbf{R}_3(n)\mathbf{F}(n),$$

$$\mathbf{R}_3(n) = \bigotimes_{i=1}^{n} \mathbf{R}_3(1),$$

where

$$\mathbf{R}_3(1) = \mathbf{X}_3^{-1}(1) = 2\begin{bmatrix} 1\,0\,0 \\ 1\,2\,0 \\ 1\,1\,1 \end{bmatrix}.$$

Note that $\mathbf{X}_3^{-1}(1)$ is its own inverse.

Example 9.2.1 For $p = 3$ and $n = 2$, the RMF-transform matrix is

$$\mathbf{R}_3(2) = 2\begin{bmatrix} 1\,0\,0 \\ 1\,2\,0 \\ 1\,1\,1 \end{bmatrix} \otimes 2\begin{bmatrix} 1\,0\,0 \\ 1\,2\,0 \\ 1\,1\,1 \end{bmatrix}$$

$$= \begin{bmatrix} 1\,0\,0\,0\,0\,0\,0\,0\,0 \\ 1\,2\,0\,0\,0\,0\,0\,0\,0 \\ 1\,1\,1\,0\,0\,0\,0\,0\,0 \\ 1\,0\,0\,2\,0\,0\,0\,0\,0 \\ 1\,2\,0\,2\,1\,0\,0\,0\,0 \\ 1\,1\,1\,2\,2\,2\,0\,0\,0 \\ 1\,0\,0\,1\,0\,0\,1\,0\,0 \\ 1\,2\,0\,1\,2\,0\,1\,2\,0 \\ 1\,1\,1\,1\,1\,1\,1\,1\,1 \end{bmatrix} = \mathbf{X}_3(2).$$

The basis functions used to define the RMF-expressions for $p = 3$ are defined as

$$2\mathbf{X}_3(2) = [2, x_2, x_2^{*2}, x_1, x_1 \odot x_2, x_1 \odot x_2^{*2}, x_1^{*2}, x_1^{*2} \odot x_2, x_1^{*2} \odot x_2^{*2}].$$

Example 9.2.2 For the ternary function $f(x_1, x_2) = x_1 \oplus x_2$, specified by the function vector $\mathbf{F} = [0, 1, 2, 1, 2, 0, 2, 0, 1]^T$, the RMF-spectrum is $\mathbf{S}_{f,3} = [0, 2, 0, 2, 0, 0, 0, 0, 0]^T$.

Due to its Kronecker product structure, the RMF-matrix $\mathbf{R}_3(n)$ can be defined recursively as

$$\mathbf{R}_3(n) = 2 \begin{bmatrix} \mathbf{R}_3(n-1) & \mathbf{0}(n-1) & \mathbf{0}(n-1) \\ \mathbf{R}_3(n-1) & 2\mathbf{R}_3(n-1) & \mathbf{0}(n-1) \\ \mathbf{R}_3(n-1) & \mathbf{R}_3(n-1) & \mathbf{R}_3(n-1) \end{bmatrix}, \quad (9.1)$$

where $\mathbf{0}(n-1)$ is the $((n-1)\times(n-1))$ matrix all of whose elements are equal to 0.

9.3 Methods for Computing the RMF-Transform

The definition of RMF-expressions and the corresponding arithmetic counterparts can be uniformly extended to functions for p non-prime.

In applications, efficient computing of RMF-spectra is essential for the feasibility of algorithms based on GF-coefficients. Therefore, in the following sections we discuss different methods to compute the RMF-coefficients over vectors and decision diagrams as underlying data structures.

9.3.1 Cooley-Tukey Algorithms for GF-Expressions

Elements of the set $P(X)$ (the basis functions in RMF-expressions) can be generated as elements of a matrix defined by the Kronecker product of basic matrices corresponding to all variables $\mathbf{X}_i(1) = \left[x_i^0 \; x_i^1 \; \ldots \; x_i^{p-1} \right]$, $i = 1, \ldots, n$. Due to the properties of the Kronecker product, the matrix $\mathbf{R}_p$, inverse to $\mathbf{X}_p$, can also be generated as the Kronecker product of basic matrices $\mathbf{R}_p(1)$ inverse to $\mathbf{X}_i(1)$. Thus,

$$\mathbf{R}_p(n) = \bigotimes_{i=1}^{n} \mathbf{R}_p(1). \quad (9.2)$$

Due to this Kronecker product representation, the RMF-transform matrix can be factorized as

$$\mathbf{R}_p(n) = \mathbf{C}_1\mathbf{C}_2 \ldots \mathbf{C}_n, \quad (9.3)$$

where factor matrices $\mathbf{C}_i$ are defined as

$$\mathbf{C}_i = \bigotimes_{k=1}^{n} \mathbf{R}_k,$$

with

$$\mathbf{R}_k = \begin{cases} \mathbf{R}_i(1), & k = i, \\ \mathbf{I}_i, & \text{otherwise,} \end{cases}$$

Fig. 9.1 Basic RMF-transform matrices for $p = 3$ and $p = 4$

$p = 3$	$p = 4$
$\mathbf{R}_3(1) = \begin{bmatrix} 1 & 0 & 0 \\ 1 & 2 & 0 \\ 1 & 1 & 1 \end{bmatrix}$	$\mathbf{R}_4(1) = \begin{bmatrix} 1 & 0 & 0 & 0 \\ 1 & 3 & 0 & 0 \\ 1 & 2 & 1 & 0 \\ 1 & 1 & 3 & 3 \end{bmatrix}$

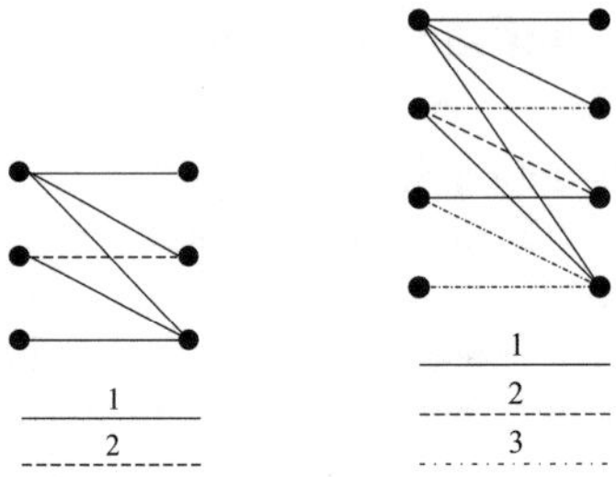

where $\mathbf{R}_i(1) = \mathbf{R}_p(1)$ and $\mathbf{I}_i$ is the $(p \times p)$ identity matrix. See for instance [33].

The computations specified by $\mathbf{R}_p(1)$ can be represented by a flow-graph called the butterfly by analogy to the corresponding FFT-like algorithm.

Example 9.3.1 Figure 9.1 shows the basic *RMF*-matrices for $p = 3$ and $p = 4$ and the corresponding butterfly operations.

Matrices $\mathbf{C}_i$ are obviously sparse, since except for $k = i$, all other factors of the Kronecker product are identity matrices $\mathbf{I}_{g_i}$, and even $\mathbf{R}_3(1)$ is by itself a triangular matrix. It follows that, as in the case of FFT, this factorization leads to a fast computation algorithm consisting of butterflies performing the computations specified by $\mathbf{R}_3(1)$.

The algorithms derived from this factorization belong to the class of Good-Thomas FFT algorithms, since the factorization of $\mathbf{R}_p$ in (9.3) is usually reported as the Good-Thomas factorization by referring to the work of Thomas [45], and Good [16]. In the case of FFT, the Good-Thomas algorithms are tightly related to the Cooley-Tukey algorithms, and differ from them in the correspondence between the single and multiple dimensional indexing [17] and the requirement that factors in the factorization of the transform length have to be mutually prime. The difference with respect to dealing with twiddle factors in these two classes of FFT algorithms is important in the case of DFT. However, it is irrelevant in the case of multiple-valued functions due to computations in finite fields.

Example 9.3.2 For $p = 3$ and $n = 3$, the *RMF*-transform matrix is defined as

$$\mathbf{R}_3(3) = \mathbf{R}_3(1) \otimes \mathbf{R}_3(1) \otimes \mathbf{R}_3(1).$$

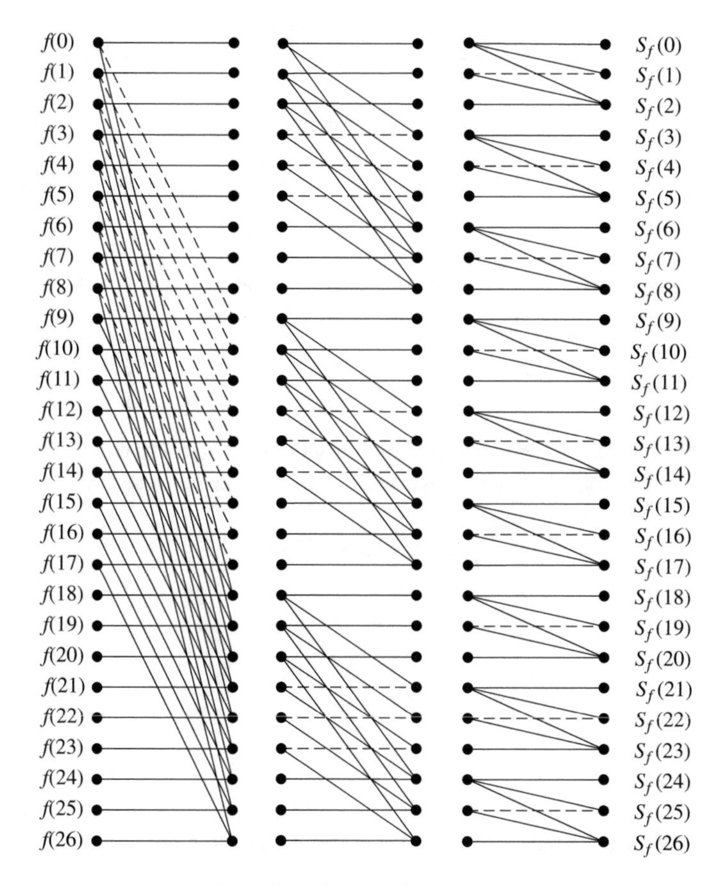

Fig. 9.2 Flow-graph of the Cooley-Tukey algorithm for the RMF-transform for $p = 3$ and $n = 3$

Then,

$$\mathbf{R}_3(3) = \mathbf{C}_1\mathbf{C}_2\mathbf{C}_3,$$

with

$$\begin{aligned}
\mathbf{C}_1 &= \mathbf{R}_3(1) \otimes \mathbf{I}_1 \otimes \mathbf{I}_1, \\
\mathbf{C}_2 &= \mathbf{I}_1 \otimes \mathbf{R}_3(1) \otimes \mathbf{I}_1, \\
\mathbf{C}_3 &= \mathbf{I}_1 \otimes \mathbf{I}_1 \otimes \mathbf{R}_3(1),
\end{aligned}$$

where $\mathbf{I}_1$ is the (3×3) identity matrix.

Figure 9.2 shows the flow-graphs of the Cooley-Tukey algorithm for $p = 3$ and $n = 3$ based on this factorization of $\mathbf{R}_3(3)$.

9.3.2 Constant Geometry Algorithms for RMF-Transform

In this section, we derive the so-called constant geometry algorithms for computing coefficients in *RMF*-expressions.

We consider the rows of the basic *RMF*-matrix in $GF(p)$ as vectors of length p

$$\mathbf{R}_p(1) = (\mathbf{X}_p(1))^{-1} = \begin{bmatrix} q_{0,0} & q_{0,1} & \cdots & q_{0,p-1} \\ q_{1,0} & q_{1,1} & \cdots & q_{1,p-1} \\ \vdots & \vdots & \vdots & \vdots \\ q_{p-1,0} & q_{p-1,1} & \cdots & q_{p-1,p-1} \end{bmatrix} = \begin{bmatrix} \mathbf{q}_0 \\ \mathbf{q}_1 \\ \vdots \\ \mathbf{q}_{p-1} \end{bmatrix}.$$

We define a row zero matrix with p elements which are equal to 0, $\mathbf{0} = [0, 0, \ldots 0]$.

With this notation, the *RMF*-transform matrix can be factorized as

$$\mathbf{R}_p(n) = (\mathbf{X}_p(n))^{-1} = \mathbf{Q}^n$$

where

$$\mathbf{Q} = \begin{bmatrix} \mathbf{Q}_0 \\ \mathbf{Q}_1 \\ \vdots \\ \mathbf{Q}_{p-1} \end{bmatrix},$$

where $\mathbf{Q}_i$ are $(p \times p)$ diagonal matrices whose non-zero elements are vectors of p elements

$$\mathbf{Q}_i = diag(\mathbf{q}_i, \mathbf{q}_i, \ldots, \mathbf{q}_i).$$

Thus, the $(p \times p)$ matrix $\mathbf{Q}_i$ converts into a $(p \times p^n)$ matrix with elements in $GF(p)$. Since in $\mathbf{Q}$ there are p matrices $\mathbf{Q}_i$, the matrix $\mathbf{R}_p(n)$ is obtained.

Example 9.3.3 For $\mathbf{R}_3(1)$, it is

$$\begin{aligned} \mathbf{q}_0 &= [1, 0, 0], \\ \mathbf{q}_1 &= [1, 2, 0], \\ \mathbf{q}_2 &= [1, 1, 1], \\ \mathbf{0} &= [0, 0, 0]. \end{aligned}$$

Therefore, for $n = 3$,

$$\mathbf{Q} = \begin{bmatrix} \mathbf{Q}_0 \\ \mathbf{Q}_1 \\ \mathbf{Q}_2 \end{bmatrix} = \begin{bmatrix}
\mathbf{q}_0 & \mathbf{0} & \mathbf{0} & \mathbf{0} & \mathbf{0} & \mathbf{0} & \mathbf{0} & \mathbf{0} & \mathbf{0} \\
\mathbf{0} & \mathbf{q}_0 & \mathbf{0} & \mathbf{0} & \mathbf{0} & \mathbf{0} & \mathbf{0} & \mathbf{0} & \mathbf{0} \\
\mathbf{0} & \mathbf{0} & \mathbf{q}_0 & \mathbf{0} & \mathbf{0} & \mathbf{0} & \mathbf{0} & \mathbf{0} & \mathbf{0} \\
\mathbf{0} & \mathbf{0} & \mathbf{0} & \mathbf{q}_0 & \mathbf{0} & \mathbf{0} & \mathbf{0} & \mathbf{0} & \mathbf{0} \\
\mathbf{0} & \mathbf{0} & \mathbf{0} & \mathbf{0} & \mathbf{q}_0 & \mathbf{0} & \mathbf{0} & \mathbf{0} & \mathbf{0} \\
\mathbf{0} & \mathbf{0} & \mathbf{0} & \mathbf{0} & \mathbf{0} & \mathbf{q}_0 & \mathbf{0} & \mathbf{0} & \mathbf{0} \\
\mathbf{0} & \mathbf{0} & \mathbf{0} & \mathbf{0} & \mathbf{0} & \mathbf{0} & \mathbf{q}_0 & \mathbf{0} & \mathbf{0} \\
\mathbf{0} & \mathbf{0} & \mathbf{0} & \mathbf{0} & \mathbf{0} & \mathbf{0} & \mathbf{0} & \mathbf{q}_0 & \mathbf{0} \\
\mathbf{0} & \mathbf{0} & \mathbf{0} & \mathbf{0} & \mathbf{0} & \mathbf{0} & \mathbf{0} & \mathbf{0} & \mathbf{q}_0 \\
\mathbf{q}_1 & \mathbf{0} & \mathbf{0} & \mathbf{0} & \mathbf{0} & \mathbf{0} & \mathbf{0} & \mathbf{0} & \mathbf{0} \\
\mathbf{0} & \mathbf{q}_1 & \mathbf{0} & \mathbf{0} & \mathbf{0} & \mathbf{0} & \mathbf{0} & \mathbf{0} & \mathbf{0} \\
\mathbf{0} & \mathbf{0} & \mathbf{q}_1 & \mathbf{0} & \mathbf{0} & \mathbf{0} & \mathbf{0} & \mathbf{0} & \mathbf{0} \\
\mathbf{0} & \mathbf{0} & \mathbf{0} & \mathbf{q}_1 & \mathbf{0} & \mathbf{0} & \mathbf{0} & \mathbf{0} & \mathbf{0} \\
\mathbf{0} & \mathbf{0} & \mathbf{0} & \mathbf{0} & \mathbf{q}_1 & \mathbf{0} & \mathbf{0} & \mathbf{0} & \mathbf{0} \\
\mathbf{0} & \mathbf{0} & \mathbf{0} & \mathbf{0} & \mathbf{0} & \mathbf{q}_1 & \mathbf{0} & \mathbf{0} & \mathbf{0} \\
\mathbf{0} & \mathbf{0} & \mathbf{0} & \mathbf{0} & \mathbf{0} & \mathbf{0} & \mathbf{q}_1 & \mathbf{0} & \mathbf{0} \\
\mathbf{0} & \mathbf{0} & \mathbf{0} & \mathbf{0} & \mathbf{0} & \mathbf{0} & \mathbf{0} & \mathbf{q}_1 & \mathbf{0} \\
\mathbf{0} & \mathbf{0} & \mathbf{0} & \mathbf{0} & \mathbf{0} & \mathbf{0} & \mathbf{0} & \mathbf{0} & \mathbf{q}_1 \\
\mathbf{q}_2 & \mathbf{0} & \mathbf{0} & \mathbf{0} & \mathbf{0} & \mathbf{0} & \mathbf{0} & \mathbf{0} & \mathbf{0} \\
\mathbf{0} & \mathbf{q}_2 & \mathbf{0} & \mathbf{0} & \mathbf{0} & \mathbf{0} & \mathbf{0} & \mathbf{0} & \mathbf{0} \\
\mathbf{0} & \mathbf{0} & \mathbf{q}_2 & \mathbf{0} & \mathbf{0} & \mathbf{0} & \mathbf{0} & \mathbf{0} & \mathbf{0} \\
\mathbf{0} & \mathbf{0} & \mathbf{0} & \mathbf{q}_2 & \mathbf{0} & \mathbf{0} & \mathbf{0} & \mathbf{0} & \mathbf{0} \\
\mathbf{0} & \mathbf{0} & \mathbf{0} & \mathbf{0} & \mathbf{q}_2 & \mathbf{0} & \mathbf{0} & \mathbf{0} & \mathbf{0} \\
\mathbf{0} & \mathbf{0} & \mathbf{0} & \mathbf{0} & \mathbf{0} & \mathbf{q}_2 & \mathbf{0} & \mathbf{0} & \mathbf{0} \\
\mathbf{0} & \mathbf{0} & \mathbf{0} & \mathbf{0} & \mathbf{0} & \mathbf{0} & \mathbf{q}_2 & \mathbf{0} & \mathbf{0} \\
\mathbf{0} & \mathbf{0} & \mathbf{0} & \mathbf{0} & \mathbf{0} & \mathbf{0} & \mathbf{0} & \mathbf{q}_2 & \mathbf{0} \\
\mathbf{0} & \mathbf{0} & \mathbf{0} & \mathbf{0} & \mathbf{0} & \mathbf{0} & \mathbf{0} & \mathbf{0} & \mathbf{q}_2
\end{bmatrix}.$$

Then,

$$\mathbf{R}_3(3) = \mathbf{Q}^3.$$

Example 9.3.4 Figure 9.3 shows the flow-graph of the constant geometry algorithm for $p = 3$ and $n = 3$.

9.3.3 The Difference Between Algorithms

In this section we briefly summarize the main features of Cooley-Tukey algorithms and constant geometry algorithms.

In both algorithms the computations are determined by the basic transform matrices $\mathbf{R}_p(1)$ and organized as butterflies performed in parallel within each step, with steps performed sequentially. The number of steps is equal to the number of variables

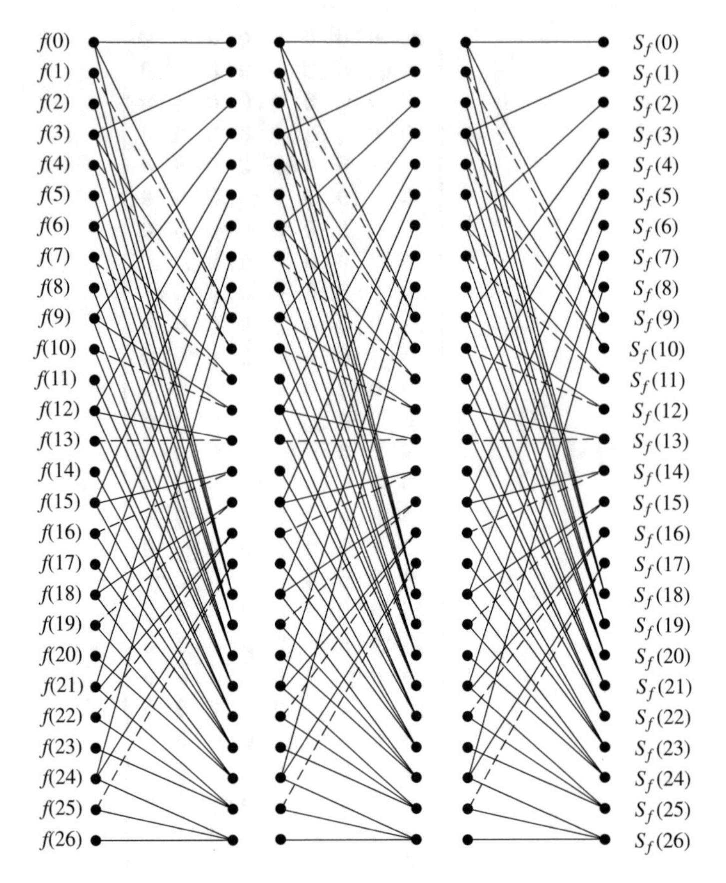

Fig. 9.3 Flow-graph of the constant geometry algorithm for $p = 3$ and $n = 3$

in the function whose spectrum is computed, since in each step the RMF-transform with respect to a variable is performed.

In Cooley-Tukey algorithms and algorithms based on the Good-Thomas factorization, the factor matrices C_i are mutually different, which causes that in each step the address arithmetic is different, meaning that in each step we have to calculate the addresses of memory locations from which the data are read (fetched). This computing of addresses for input data repeated in each step is a kind of drawback of the algorithm. The good feature is that the output data, the results of the computation in the step, are saved in the same locations from which the input data are read. It follows that computations can be performed in-place, meaning that the memory requirements are minimal and equal to those to store the function vector to be processed. As noticed above, the computations performed in each step of the algorithm consists of p^{n-1} sets of identical computations. These are computations defined by the basic RMF-transform matrix $\mathbf{R}_p(1)$. When graphically represented, the computations are called butterflies by analogy with similar computations in FFT.

The distance between memory locations from which the data are fetched is different in each step, and ranges from p^{n-1} in the first step to p in the n-th step of the algorithm. Due to this, the time to perform a step is different from that for other steps. A discussion of the impact of that feature for the case of GF-transforms for ternary and quaternary functions is experimentally analyzed in [12, 38]. Similar conclusions can be derived for the RMF-transforms, since the difference between the GF- and RMF-transforms relevant for this consideration is just in the number of non-zero elements in the basic transform matrices. In the case of RMF-transforms, the number of non-zero elements is smaller.

The chief idea in constant geometry algorithms is to perform computations over data fetched from identical memory locations in all the steps. It follows that address arithmetic is simpler, since the addresses of memory locations to fetch data are computed in the first step and used later in other steps. This results in a speed up of computations, however, the drawback is that in-place computations are impossible and a memory twice as large is required to sore the input and output data of each step.

9.3.4 *Computing the RMF-Transform over Decision Diagrams*

In this section, we present the method for computing the RMF-spectrum over decision diagrams as the underlying data structure to represent the functions whose spectrum is computed. In this order, for the sake of completeness of presentation, we first introduced the concept of Multiple-place decision diagrams (MDDs) used to represent multiple-valued functions. MDDs are a generalization of Binary Decision Diagrams (BDDs) [6] that are defined in terms of the so-called Shannon expansion rules. Alternatively, BDDs can be viewed as graphical representations of the disjunctive normal forms, Sum-of-Product (SOP) expressions, viewed as particular examples of functional expressions for Boolean functions [33]. Therefore, we first introduce the corresponding concepts, the generalized Shannon expansion, and the multiple-valued counterpart of the disjunctive normal form.

9.3.5 *Multiple-Place Decision Diagrams*

We will illustrate the derivation of the generalized Shannon expression for multiple-valued functions by referring to this example.

Definition 9.3.1 (*Characteristic functions*) For a multiple-valued variable x_j taking values in the set $\{0, 1, \ldots, p-1\}$, $j = 0, \ldots, m-1$, the characteristic functions $J_i(x_j)$, $i = 0, 1, \ldots, p-1$ are defined as $J_i(x_j) = 1$ for $x_j = i$, and $J_i(x_j) = 0$ for $x_j \neq i$.

Table 9.3 Characteristic functions for $p = 3$ and $n = 2$

x_1x_2	$J_0(x_1)$	$J_1(x_1)$	$J_2(x_1)$	$J_0(x_2)$	$J_1(x_2)$	$J_2(x_2)$
00	1	0	0	1	0	0
01	1	0	0	0	1	0
02	1	0	0	0	0	1
10	0	1	0	1	0	0
11	0	1	0	0	1	0
12	0	1	0	0	0	1
20	0	0	1	1	0	0
21	0	0	1	0	1	0
22	0	0	1	0	0	1

Example 9.3.5 For $p = 3$ and $n = 2$, the characteristic functions $J_i(x_j)$ are given in Table 9.3.

Definition 9.3.2 (*Generalized Shannon expansion*) The generalized Shannon expansion for ternary logic functions is defined as

$$f = J_0(x_i)f_0 + J_1(x_i)f_1 + J_2(x_i)f_2, \tag{9.4}$$

where f_i, $i = 0, 1, 2$ are the co-factors of f for $x_i \in \{0, 1, 2\}$.

The recursive application of the generalized Shannon expansion rule to all the variables in a given function f results in a functional expression which when graphically represented yields to the decision diagrams that are an analogue of the Binary decision diagrams (BDDs) [27] and represent a generalization of this concept to the representation of multiple-valued functions. These diagrams are called Multiple-place decision trees (MTDTs) and Multiple-place decision diagrams (MDD) [29].

Example 9.3.6 For $p = 3$ and $n = 2$, by expanding a given $f(x_1, x_2)$ with respect to x_1,

$$f(x_1, x_2) = J_0(x_1)f(x_1 = 0, x_2) + J_1(x_1)f(x_1 = 1, x_2) + J_2(x_2)f(x_1 = 2, x_2).$$

After application of the generalized Shannon expansion with respect to x_2, it follows

$$\begin{aligned} f(x_1, x_2) = {} & J_0(x_2)(J_0(x_1)f(x_1 = 0, x_2 = 0) + J_1(x_1)f(x_1 = 1, x_2 = 0) \\ & + J_2(x_1)f(x_1 = 2, x_2 = 0)) + J_1(x_2)(J_0(x_1)f(x_1 = 0, x_2 = 1) \\ & + J_1(x_1)f(x_0 = 1, x_1 = 1) + J_2(x_0)f(x_0 = 2, x_1 = 1)) \\ & + J_2(x_2)(J_0(x_1)f(x_1 = 0, x_2 = 2) + J_1(x_1)f(x_1 = 1, x_2 = 2) \\ & + J_2(x_1)f(x_1 = 2, x_2 = 2) \end{aligned}$$

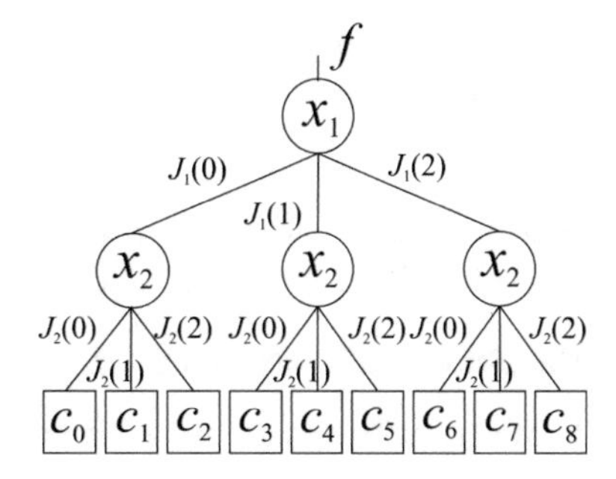

Fig. 9.4 MDT for f in Example 9.3.6

$$
\begin{aligned}
&= J_0(x_2)J_0(x_1)f(x_1 = 0, x_2 = 0) + J_0(x_2)J_1(x_1)f(x_1 = 1, x_2 = 0) \\
&\quad + J_0(x_2)J_2(x_1)f(x_1 = 2, x_2 = 0) + J_1(x_2)J_0(x_1)f(x_1 = 0, x_2 = 1) \\
&\quad + J_1(x_2)J_1(x_1)f(x_1 = 1, x_2 = 1) + J_1(x_2)J_2(x_1)f(x_1 = 2, x_2 = 1) \\
&\quad + J_2(x_2)J_0(x_1)f(x_1 = 0, x_2 = 2) + J_2(x_2)J_1(x_1)f(x_1 = 1, x_2 = 2) \\
&\quad + J_2(x_2)J_2(x_1)f(x_1 = 2, x_2 = 2).
\end{aligned}
$$

Figure 9.4 shows the Multiple-place decision tree (MDT) for ternary functions of two variables.

For a function of n variables, the decision tree has $(n + 1)$ levels, and each level consists of nodes to which the same variable is assigned. The first level has a single node called the root node. The level $(n + 1)$ consists of constant nodes showing the coefficients in the functional expressions whose graphical representation are the decision trees.

9.3.6 Reduction of Decision Trees

Decision diagrams are derived by the reduction of decision trees by eliminating the redundant information expressed in terms of the isomorphic subtrees, and the corresponding subdiagrams. Reduction is accomplished by sharing isomorphic subtrees and deleting any redundant information in the decision tree. It is assumed that two subtrees are isomorphic if

1. They are rooted in nodes at the same level,
2. The constant nodes of the subtrees represent identical subvectors $\mathbf{V}_i$ in the vector of values of constant nodes $\mathbf{V}$.

This definition includes different reduction rules used in different decision diagrams for either binary or multiple-valued functions, as well as bit-level and word-level decision diagrams [32].

The minimum possible isomorphic subtrees are equal constant nodes. In this case, the function represented has equal values at the points corresponding to these equal-valued constant nodes.

The maximum possible isomorphic subtrees are p equal subtrees rooted at the nodes pointed by the outgoing edges of the root node. In that case, the function f is independent of the variable assigned to the root node.

Definition 9.3.3 (*MDD reduction rules*)

1. If descendent nodes of a node are identical, then delete the node and connect the incoming edges of the deleted node to the corresponding successor. The label of this incoming edge is re-determined as the product of the label at the initial incoming edge with the sum of labels at the outgoing edges of the deleted node.
2. Share isomorphic subtrees, i.e., if there are isomorphic subtrees, keep a single subtree and redirect to it the incoming edges of all other isomorphic subtrees.

Definition 9.3.4 (*Cross points*) A cross point is a point where an edge longer than one crosses a level in the decision diagram.

Cross points are useful in expressing the impact of deleted nodes, which is important to take into account in computing over decision diagrams or performing the realizations of functions represented by decision diagrams. A cross point is illustrated by the decision diagram in Fig. 9.5 for the function f in Example 9.3.7.

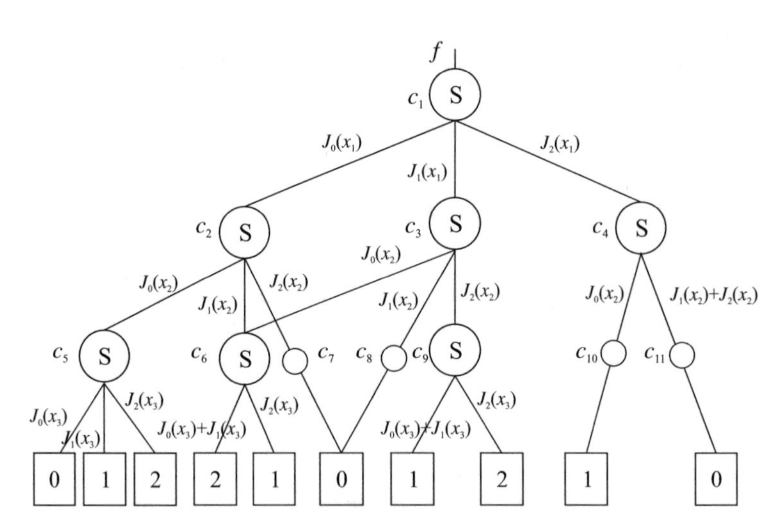

Fig. 9.5 MDD for the function f in Example 9.3.7

In a decision tree, edges connect nodes at successive levels, and we say that the length of such edges is 1. Due to the reduction, in a decision diagram, edges longer than one, i.e., connecting nodes at non-successive levels, can appear. For example, the length of an edge connecting a node at the $(i-1)$-th level with a node at the $(i+1)$-th level is two.

Nodes to which the same decision variable is assigned form a level in the decision tree or the diagram.

A path consists of nodes at different levels, with a single node at each level, from the root node to a constant node. Thus, each path connects the root node and a single node from each level, including the level of constant nodes, i.e., a path consists of edges connecting a single node per level, and the length of the path is the sum of the lengths of edges the path consists of.

9.3.7 Computing the RMF-Transform over MDDs

The same computations as in the above example can be performed assuming the decision diagram representations as the underlying data structure.

Example 9.3.7 Figure 9.5 shows the MDD for the function f specified by the function vector

$$\mathbf{F} = [0, 1, 2, 2, 2, 1, 0, 0, 0, 0, 0, 0, 1, 1, 2, 2, 2, 1, 1, 1, 1, 0, 0, 0, 0, 0, 0]^T .$$

In this figure, the symbol $a + b$ means that both edges labeled by a and b point to the same node.

The steps of the FFT-like algorithm described above can be performed over this MDD by processing each node in the diagram by the basic butterfly operation specified by $\mathbf{R}_3(1)$. The processing means that the inputs to the butterfly operation are subfunctions represented by the subdiagrams rooted at the nodes pointed by the outgoing edges of the processed node. For the clarity of presentation, we will express the impact of deleted nodes through cross points shown by small circles in Fig. 9.5 and viewed as crossings of a path in the diagram with an imaginary line connecting nodes at the same level in the diagram. In practical programming implementations, these computations are avoided and the procedure simplified by using properties of the performed transforms. In particular, the computations are reduced to transforming the related subfunctions and padding with zeros. It can be followed, depending on the transform, by the multiplication of the constant value of a terminal node or the subfunction pointed by the edge of the processed node by a constant value equal to the length of the path between these two nodes. The explanation for this implementation is the following. If a node is deleted from the MDD, this means that the outgoing edges of this node point to the identical subfunctions. Therefore, in the considered case of ternary functions, nodes have three outgoing edges and the subfunction represented by the deleted node has three identical parts. Thus, this node represents a

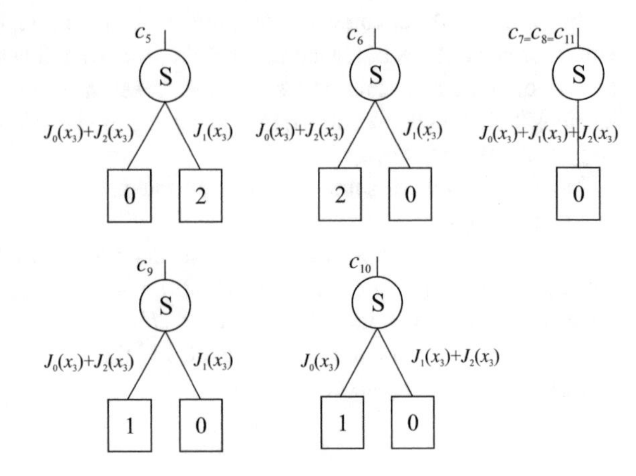

Fig. 9.6 MDDs for subfunctions c_i, $i = 5, 6, 7, 8, 9, 10, 11$ in Example 9.3.7

periodic subfunction or a constant. Then, due to the properties of the transforms, the spectrum of a constant function is the delta function, while the spectrum of a periodic function is the delta function of the length of a period Kronecker multiplied by the spectrum of the periodically repeated subfunction in the considered periodic functions.

If the nodes and cross points are labeled as in Fig. 9.5, the RMF-coefficients for the considered function f are computed as follows.

Step 1

$$\mathbf{c}_5 = \mathbf{R}_3(1)\begin{bmatrix}0\\1\\2\end{bmatrix} = \begin{bmatrix}0\\2\\0\end{bmatrix} \quad \mathbf{c}_6 = \mathbf{R}_3(1)\begin{bmatrix}2\\2\\1\end{bmatrix} = \begin{bmatrix}2\\0\\2\end{bmatrix}$$

$$\mathbf{c}_7 = \mathbf{R}_3(1)\begin{bmatrix}0\\0\\0\end{bmatrix} = \begin{bmatrix}0\\0\\0\end{bmatrix} \quad \mathbf{c}_8 = \mathbf{R}_3(1)\begin{bmatrix}0\\0\\0\end{bmatrix} = \begin{bmatrix}0\\0\\0\end{bmatrix}$$

$$\mathbf{c}_9 = \mathbf{R}_3(1)\begin{bmatrix}1\\1\\2\end{bmatrix} = \begin{bmatrix}1\\0\\1\end{bmatrix} \quad \mathbf{c}_{10} = \mathbf{R}_3(1)\begin{bmatrix}1\\1\\1\end{bmatrix} = \begin{bmatrix}1\\0\\0\end{bmatrix}$$

$$\mathbf{c}_{11} = \mathbf{R}_3(1)\begin{bmatrix}0\\0\\0\end{bmatrix} = \begin{bmatrix}0\\0\\0\end{bmatrix}.$$

Step 2

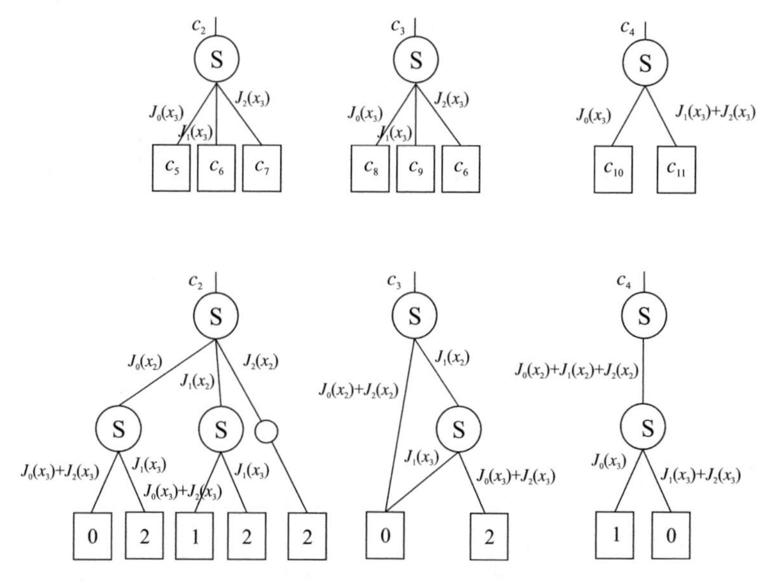

Fig. 9.7 MDDs for subfunctions c_i, $i = 2, 3, 4$ in Example 9.3.7

$$\mathbf{c}_2 = \mathbf{R}_3(1)\begin{bmatrix}\mathbf{c}_5\\ \mathbf{c}_6\\ \mathbf{c}_7\end{bmatrix} = \begin{bmatrix}\begin{bmatrix}0\\2\\0\end{bmatrix} \\ \begin{bmatrix}0\\2\\0\end{bmatrix} + 2\begin{bmatrix}2\\0\\2\end{bmatrix} \\ \begin{bmatrix}0\\2\\0\end{bmatrix} + \begin{bmatrix}2\\0\\2\end{bmatrix} + \begin{bmatrix}0\\0\\0\end{bmatrix}\end{bmatrix} = \begin{bmatrix}0\\2\\0\\1\\2\\1\\2\\2\\2\end{bmatrix}$$

$$\mathbf{c}_3 = \mathbf{R}_3(1)\begin{bmatrix}\mathbf{c}_8\\ \mathbf{c}_9\\ \mathbf{c}_6\end{bmatrix} = \begin{bmatrix}\begin{bmatrix}0\\0\\0\end{bmatrix} \\ \begin{bmatrix}0\\0\\0\end{bmatrix} + 2\begin{bmatrix}1\\0\\1\end{bmatrix} \\ \begin{bmatrix}0\\0\\0\end{bmatrix} + \begin{bmatrix}1\\0\\1\end{bmatrix} + \begin{bmatrix}2\\0\\2\end{bmatrix}\end{bmatrix} = \begin{bmatrix}0\\0\\0\\2\\0\\2\\0\\0\\0\end{bmatrix}$$

$$
\mathbf{c}_4 = \mathbf{R}_3(1)\begin{bmatrix}\mathbf{c}_{10}\\ \mathbf{c}_{11}\\ \mathbf{c}_{11}\end{bmatrix} = \begin{bmatrix}\begin{bmatrix}1\\0\\0\end{bmatrix}\\ \begin{bmatrix}1\\0\\0\end{bmatrix}+2\begin{bmatrix}0\\0\\0\end{bmatrix}\\ \begin{bmatrix}1\\0\\0\end{bmatrix}+\begin{bmatrix}0\\0\\0\end{bmatrix}+\begin{bmatrix}0\\0\\0\end{bmatrix}\end{bmatrix} = \begin{bmatrix}1\\0\\0\\1\\0\\0\\1\\0\\0\end{bmatrix}.
$$

Step 3

$$
\begin{aligned}
\mathbf{c}_1 &= \mathbf{R}_3(1)\begin{bmatrix}\mathbf{c}_2\\ \mathbf{c}_3\\ \mathbf{c}_4\end{bmatrix}\\
&= \begin{bmatrix}[0,2,0,1,2,1,2,2,2]^T\\ [0,2,0,1,2,1,2,2,2]^T+2\,[0,0,0,2,0,2,0,0,0]^T\\ [0,2,0,1,2,1,2,2,2]^T+[0,0,0,2,0,2,0,0,0]^T+[1,0,0,1,0,0,1,0,0]^T\end{bmatrix}^T\\
&= [0,2,0,1,2,1,2,2,2,0,2,0,2,2,2,2,2,2,1,2,0,1,2,0,0,2,2]^T.
\end{aligned}
$$

Each step of the computation can be represented by MDDs, which then can be combined into the MDD for the RMF-coefficients. From the spectral interpretation of decision diagrams, this MDD for the RMF-spectrum of f after conversion of the meaning of nodes and corresponding labels at the edges becomes the RMFDD for f

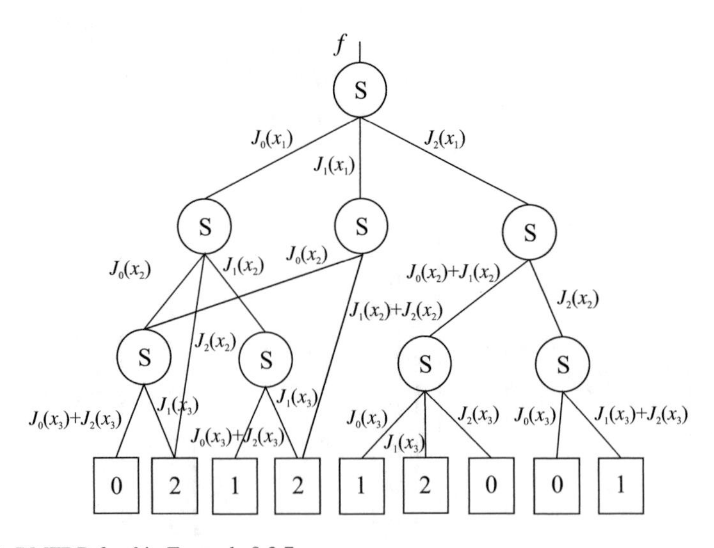

Fig. 9.8 RMFDD for f in Example 9.3.7

[33]. Figure 9.8 shows this RMFDD for the considered function f. It can be noticed that, counting from the left to the right, the third and fourth node for x_3, as well as the sixth and seventh node for the same variable, represent, respectively, subfunctions that are identical to each other multiplied by 2. Thus, the diagram can be simplified by allowing edges with multiplicative attributes in the same way as this is done in BDDs with negated edges. See for instance [27].

9.4 Algorithms Derived from Pascal Matrix Factorizations

In this section, we discuss various factorizations of the RMF-matrix based on its relationships with the Pascal matrix [10]. The Pascal matrix defined as the matrix of binomial coefficients is an infinite matrix. However, in finite domains, the term Pascal matrix $\mathbf{P}_i$ refers to an $(i \times i)$ matrix consisting of first i rows and i columns of the Pascal matrix.

We will distinguish the Pascal matrix and the relatively recently defined Pascal transform matrix [1], which is actually the Pascal matrix with a minus sign assigned to elements of every second column [1]. Note that in modulo p arithmetic, the negative sign of an integer can be interpreted as the multiplication by $p - 1$.

9.4.1 RMF-Matrix and Pascal Matrices

There is a strong relationship between the RMF-matrix and the Pascal matrix and Pascal transform matrix which can be expressed as follows.

Observation 9.4.1 *The $(p^n \times p^n)$ RMF-matrix is derived from the Pascal matrix of the same dimensions by multiplying every second column by $p - 1$ and reducing the elements of the resulting matrix modulo p. The RMF-matrix is derived from the Pascal transform matrix by replacing the sign minus with multiplication by $p - 1$ and reducing the elements of the resulting matrix modulo p.*

Notice that the second column of the $(p^n \times p^n)$ Pascal matrix is the sequence $\{0, 1, 2, \ldots, p^n - 1\}$, which, as remarked above, can be generated by the Gibbs exponentiation of the constant sequence all of whose elements are equal to $p - 1$. When elements of this sequence are computed modulo p, we get the n-th variable in p-valued functions. In [31], it is pointed out that the n-th variable can be used to generate the RMF-matrix, provided the multiplication by $p - 1$ of elements of every second column of the resulting matrix. The same observation holds in general for any p. From Observation 9.4.1, directly follows that a $(p^n \times p^n)$ Pascal matrix $\mathbf{P}_n$ can be converted into an RMF-matrix $\mathbf{R}_p(n)$ of the same dimensions by the following

Procedure 9.4.1 *1. Multiply elements of every second column of the matrix $\mathbf{P}_n$ by $p - 1$, i.e., perform the columnwise Hadamard product with the vector* $[1, (p - 1), 1, (p - 1), \ldots, (p - 1), 1]$.

2. *Compute elements of the produced matrix modulo p. The resulting matrix is the* $(p^n \times p^n)$ *RMF-matrix.*

The similar procedure can be applied to the Pascal transform matrix. Therefore, various factorizations proposed for the Pascal matrix and the Pascal transform matrix can be simply modified for the factorizations of the RMF-matrix, as will be illustrated by the following examples. Some of these factorizations can be useful for efficient computing of the RMF-spectra since they allow parallel implementations offering the exploitation of both data and task parallelism.

Example 9.4.1 For $p = 3$ and $n = 2$, the Pascal matrix is

$$\mathbf{P}_{3^2} = \mathbf{P}_9 = \begin{bmatrix} 1 & 0 & 0 & 0 & 0 & 0 & 0 & 0 & 0 \\ 1 & 1 & 0 & 0 & 0 & 0 & 0 & 0 & 0 \\ 1 & 2 & 1 & 0 & 0 & 0 & 0 & 0 & 0 \\ 1 & 3 & 3 & 1 & 0 & 0 & 0 & 0 & 0 \\ 1 & 4 & 6 & 4 & 1 & 0 & 0 & 0 & 0 \\ 1 & 5 & 10 & 10 & 5 & 1 & 0 & 0 & 0 \\ 1 & 6 & 15 & 20 & 15 & 6 & 1 & 0 & 0 \\ 1 & 7 & 21 & 35 & 35 & 21 & 7 & 1 & 0 \\ 1 & 8 & 28 & 56 & 70 & 56 & 28 & 8 & 1 \end{bmatrix}.$$

The columnwise Hadamard product multiplication $\odot$ with the vector $\mathbf{v} = [1, 2, 1, 2, 1, 2, 1, 2, 1]$, results in

$$\mathbf{v} \odot \mathbf{P}_9 = \begin{bmatrix} 1 & 0 & 0 & 0 & 0 & 0 & 0 & 0 & 0 \\ 1 & 2 & 0 & 0 & 0 & 0 & 0 & 0 & 0 \\ 1 & 4 & 1 & 0 & 0 & 0 & 0 & 0 & 0 \\ 1 & 6 & 3 & 2 & 0 & 0 & 0 & 0 & 0 \\ 1 & 8 & 6 & 8 & 1 & 0 & 0 & 0 & 0 \\ 1 & 10 & 10 & 20 & 5 & 2 & 0 & 0 & 0 \\ 1 & 12 & 15 & 40 & 15 & 12 & 1 & 0 & 0 \\ 1 & 14 & 21 & 70 & 35 & 42 & 7 & 2 & 0 \\ 1 & 16 & 28 & 112 & 70 & 112 & 28 & 16 & 1 \end{bmatrix},$$

which by computing its elements modulo 3 produces the RMF-matrix $\mathbf{R}_3(2)$ in Example 9.2.1.

Example 9.4.2 For $p = 5$ and $n = 1$, the Pascal matrix $\mathbf{P}_5$ is

$$\mathbf{P}_5 = \begin{bmatrix} 1 & 0 & 0 & 0 & 0 \\ 1 & 1 & 0 & 0 & 0 \\ 1 & 2 & 1 & 0 & 0 \\ 1 & 3 & 3 & 1 & 0 \\ 1 & 4 & 6 & 4 & 1 \end{bmatrix}.$$

The columnwise Hadamard product with $\mathbf{v} = [1, 4, 1, 4, 1]$ produces

$$\mathbf{v} \odot \mathbf{P}_5 = \begin{bmatrix} 1 & 0 & 0 & 0 & 0 \\ 1 & 4 & 0 & 0 & 0 \\ 1 & 8 & 1 & 0 & 0 \\ 1 & 12 & 3 & 4 & 0 \\ 1 & 16 & 6 & 16 & 1 \end{bmatrix}.$$

Computing elements of this matrix modulo 5 produce the RMF-matrix for $p = 5$.
The Pascal matrix $\mathbf{P}_5$ can be factorized as [48]

$$\mathbf{P}_5 = \begin{bmatrix} 1 & 0 & 0 & 0 & 0 \\ 0 & 1 & 0 & 0 & 0 \\ -2 & 1 & 1 & 0 & 0 \\ -5 & -1 & 2 & 1 & 0 \\ -9 & -6 & 1 & 3 & 1 \end{bmatrix} \cdot \begin{bmatrix} 1 & 0 & 0 & 0 & 0 \\ 1 & 1 & 0 & 0 & 0 \\ 2 & 1 & 1 & 0 & 0 \\ 3 & 2 & 1 & 1 & 0 \\ 5 & 3 & 2 & 1 & 1 \end{bmatrix}.$$

When these two matrices are computed modulo 5, we get the matrices whose product is the RMF-matrix for $p = 5$,

$$\mathbf{R}_5(1) = \begin{bmatrix} 1 & 0 & 0 & 0 & 0 \\ 0 & 1 & 0 & 0 & 0 \\ 3 & 1 & 1 & 0 & 0 \\ 0 & 4 & 2 & 1 & 0 \\ 1 & 4 & 1 & 3 & 1 \end{bmatrix} \cdot \begin{bmatrix} 1 & 0 & 0 & 0 & 0 \\ 1 & 1 & 0 & 0 & 0 \\ 2 & 1 & 1 & 0 & 0 \\ 3 & 2 & 1 & 1 & 0 \\ 0 & 3 & 2 & 1 & 1 \end{bmatrix}.$$

9.4.2 Computing Algorithms

Certain factorizations of Pascal matrices can be a basis to derive computing algorithms for the RMF-spectra.

Example 9.4.3 The Pascal matrix $\mathbf{P}_5$ can be factorized in terms of four binary matrices as [21]

$$\mathbf{P}_5 = \mathbf{V}_1 \cdot \mathbf{V}_2 \cdot \mathbf{V}_3 \cdot \mathbf{V}_4,$$

where

$$\mathbf{V}_1 = \begin{bmatrix} 1 & 0 & 0 & 0 & 0 \\ 0 & 1 & 0 & 0 & 0 \\ 0 & 0 & 1 & 0 & 0 \\ 0 & 0 & 0 & 1 & 0 \\ 0 & 0 & 0 & 1 & 1 \end{bmatrix}, \mathbf{V}_2 = \begin{bmatrix} 1 & 0 & 0 & 0 & 0 \\ 0 & 1 & 0 & 0 & 0 \\ 0 & 0 & 1 & 0 & 0 \\ 0 & 0 & 1 & 1 & 0 \\ 0 & 0 & 0 & 1 & 1 \end{bmatrix}, \mathbf{V}_3 = \begin{bmatrix} 1 & 0 & 0 & 0 & 0 \\ 0 & 1 & 0 & 0 & 0 \\ 0 & 1 & 1 & 0 & 0 \\ 0 & 0 & 1 & 1 & 0 \\ 0 & 0 & 0 & 1 & 1 \end{bmatrix}, \mathbf{V}_4 = \begin{bmatrix} 1 & 0 & 0 & 0 & 0 \\ 1 & 1 & 0 & 0 & 0 \\ 0 & 1 & 1 & 0 & 0 \\ 0 & 0 & 1 & 1 & 0 \\ 0 & 0 & 0 & 1 & 1 \end{bmatrix}.$$

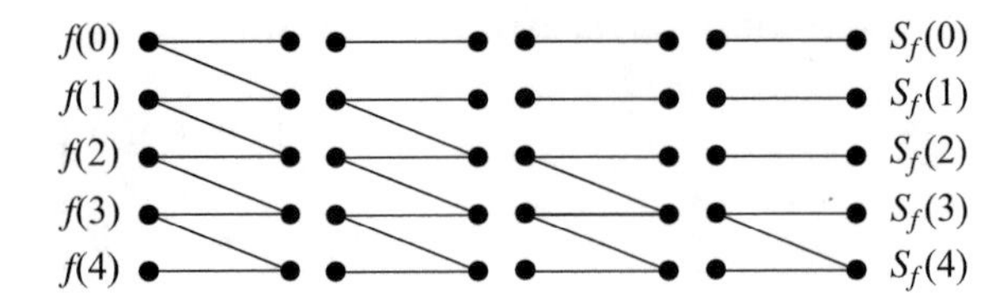

Fig. 9.9 Flow-graph of the algorithm in Example 9.4.3

Their product is

$$\mathbf{V}_{1234} = \begin{bmatrix} 1\,0\,0\,0\,0 \\ 1\,1\,0\,0\,0 \\ 1\,2\,1\,0\,0 \\ 1\,3\,3\,1\,0 \\ 1\,4\,6\,4\,1 \end{bmatrix}.$$

The columnwise Hadamrd product with $\mathbf{v} = [1, 4, 1, 4, 1]$, produces

$$\mathbf{V}_{1234-new} = \begin{bmatrix} 1 & 0\,0 & 0\,0 \\ 1 & 4\,0 & 0\,0 \\ 1 & 8\,1 & 0\,0 \\ 1 & 12\,3 & 4\,0 \\ 1 & 16\,6 & 16\,1 \end{bmatrix}.$$

This matrix, after computing its elements modulo 5, becomes the RMF-matrix $\mathbf{R}_5$. Since the columnwise Hadamard product with $\mathbf{v}$ can be transferred to the Hadamard product of $\mathbf{v}$ with the function vector $\mathbf{F}$ to be processed, this factorization leads to the computing algorithm for the RMF-transform. Figure 9.9 shows the flow-graph of this algorithm for the RMF-transform for $p = 5$ and $n = 1$.

The factorization in Example 9.4.3 can be extended to any p and n, as Example 9.4.4 illustrates for $p = 3$ and $n = 2$. A good feature of this factorization is that there is no multiplication and the addition is performed over neighboring elements of the function vector. A drawback of the algorithm is the number of steps, that is $p^n - 1$, which are performed serially. Another good feature is that computations can be performed in-place as in the case of Cooley-Tukey algorithms. Therefore, the algorithm is space efficient.

Example 9.4.4 The RMF-matrix $\mathbf{R}_3(2)$ can be factorized into 8 matrices $\mathbf{R}_i$, $i = 1, 2, \ldots, 8$, whose elements are 0 except elements on the main diagonal and elements on the subdiagonal in i rows counting from the bottom. In other words, these factor matrices are derived from the identity matrix by replacing with 1 the 0 value of elements of the left subdiagonal starting from the $p^n - i$-th row. For example, the factor matrix $\mathbf{R}_4$ is

$$\mathbf{R}_4 = \begin{bmatrix} 1&0&0&0&0&0&0&0&0 \\ 0&1&0&0&0&0&0&0&0 \\ 0&0&1&0&0&0&0&0&0 \\ 0&0&0&1&0&0&0&0&0 \\ 0&0&0&0&1&0&0&0&0 \\ 0&0&0&0&1&1&0&0&0 \\ 0&0&0&0&0&1&1&0&0 \\ 0&0&0&0&0&0&1&1&0 \\ 0&0&0&0&0&0&0&1&1 \end{bmatrix},$$

The other factor matrices are defined in the same way.

The matrix obtained by the product of these matrices $\mathbf{R}_i$ becomes the RMF-matrix $\mathbf{R}_3(2)$ after the multiplication of elements of every second column by 2.

As stated above, the recently introduced discrete Pascal transform is defined as a transform whose transform matrix is the Pascal matrix with the sign minus assigned to elements of every second column [1]. In [31], the RMF-transform matrix is generated as the matrix whose second column is the n-th p-valued variable. Other columns are obtained as the Gibbs exponentiation of the second column, with every second column multiplied by $p-1$. This multiplication corresponds to the sign minus in the case of the Pascal transform. Since by computing elements of the Pascal transform matrix modulo p we get the RMF-matrix, it follows, that the decompositions used to define the fast algorithms to compute the Pascal transform can be used to compute the RMF-transform.

Example 9.4.5 The Pascal transform matrix for $p = 4$ is

$$\mathbf{P}_4 = \begin{bmatrix} 1&0&0&0 \\ 1&-1&0&0 \\ 1&-2&1&0 \\ 1&-3&1&-1 \end{bmatrix}.$$

It can be factorized as [28]

$$\mathbf{P}_4 = \mathbf{Y}_1 \cdot \mathbf{Y}_2 \cdot \mathbf{Y}_3$$

where

$$\mathbf{Y}_1 = \begin{bmatrix} 1&0&0&0 \\ 0&1&0&0 \\ 0&0&1&0 \\ 0&0&1&-1 \end{bmatrix}, \mathbf{Y}_2 = \begin{bmatrix} 1&0&0&0 \\ 0&1&0&0 \\ 0&1&-1&0 \\ 0&0&1&-1 \end{bmatrix}, \mathbf{Y}_3 = \begin{bmatrix} 1&0&0&0 \\ 1&-1&0&0 \\ 0&1&-1&0 \\ 0&0&1&-1 \end{bmatrix}.$$

This is a factorization representing a modification of the factorization of the Pascal matrix corresponding to the modification of the Pascal matrix to define the Pascal transform matrix. Returning back to the modulo computations, we get a factorization for the RMF-matrix.

If the factor matrices $\mathbf{Y}_1$, $\mathbf{Y}_2$, and $\mathbf{Y}_3$ are computed modulo 4, matrices

$$\mathbf{W}_1 = \begin{bmatrix} 1\,0\,0\,0 \\ 0\,1\,0\,0 \\ 0\,0\,1\,0 \\ 0\,0\,1\,3 \end{bmatrix}, \mathbf{W}_2 = \begin{bmatrix} 1\,0\,0\,0 \\ 0\,1\,0\,0 \\ 0\,1\,3\,0 \\ 0\,0\,1\,3 \end{bmatrix}, \mathbf{W}_3 = \begin{bmatrix} 1\,0\,0\,0 \\ 1\,3\,0\,0 \\ 0\,1\,3\,0 \\ 0\,0\,1\,3 \end{bmatrix},$$

are obtained, whose product produces the RMF-matrix for $p = 4$.

Another factorization of the RMF-matrix can be derived from a particular way of defining the Pascal matrix as the matrix exponential of an appropriately selected matrix.

For a real or complex $(n \times n)$ matrix $\mathbf{X}$, the exponential of $\mathbf{X}$, $e^{\mathbf{X}}$, is the $(n \times n)$ matrix determined by the power series

$$e^{\mathbf{X}} = \sum_{r=0}^{\infty} \frac{1}{r!}\mathbf{X}^r = \mathbf{I} + \mathbf{X} + \frac{1}{2!}\mathbf{X}^2 + \frac{1}{3!}\mathbf{X}^3 + \cdots, \tag{9.5}$$

where $\mathbf{I}$ is the $(n \times n)$ identity matrix.

Consider the $(p^n \times p^n)$ matrix $\mathbf{X}$ whose main subdiagonal is the sequence $\{1, \ldots, p^n - 1\}$. This is a nilpotent matrix and, therefore, the power series of its exponential is finite having p^n terms. The resulting matrix is the $(p^n \times p^n)$ Pascal matrix [35].

The procedure of converting the Pascal matrix into a RMF-matrix discussed above can be applied to the representation of the Pascal matrix as the matrix exponential and it follows that the computation of the RMF-spectrum of a given function vector $\mathbf{F}$ can be performed as the sum of products of summands in $e^{\mathbf{X}}$ with the vector $\mathbf{F}$. This observation offers the possibility to perform the required computing over a parallel architecture as, for example, the Graphics Processing Unit (GPU) based architecture by exploiting task parallelism. A task is viewed as computing the product of a summand $\frac{1}{k!}\mathbf{X}^k$, $k = 1, 2, \ldots, n$ with the function vector $\mathbf{F}$. The addition of obtained vectors followed by the componentwise addition with $\mathbf{F}$ due to the appearance of $\mathbf{I}$ in (9.5) produces the RMF-spectrum. These tasks can be combined in various ways depending on the number of available processors. Due to the simplicity of summands $\frac{1}{k!}\mathbf{X}^k$, the computation can be fast. It should be noticed that referring to values of elements in summands, this way of computing resembles the computing of each RMF-spectral coefficient separately, which is also an acceptable way of computing spectral coefficients, especially when dealing with large functions, as discussed for the discrete Walsh transform in [7, 8, 11].

Example 9.4.6 For $p = 3$ and $n = 2$, the RMF-transform matrix can be obtained as

$$\mathbf{R}_3(2) = \mathbf{I}_3(2) + \mathbf{D}_3(2) + \frac{1}{2!}\mathbf{D}_3(2)^2 + \frac{1}{3!}\mathbf{D}_3(2)^3 + \frac{1}{4!}\mathbf{D}_3(2)^4$$
$$+\frac{1}{5!}\mathbf{D}_3(2)^5 + \frac{1}{6!}\mathbf{D}_3(2)^6 + \frac{1}{7!}\mathbf{D}_3(2)^7 + \frac{1}{8!}\mathbf{D}_3(2)^8,$$

where $\mathbf{I}_3(2)$ is the (9×9) identity matrix, and $\mathbf{D}_3(2)$ is a matrix of the same dimension all of whose elements are 0 except the elements on the main subdiagonal which take values from the ordered sequence $\{1, 2, 3, 4, 5, 6, 7, 8\}$. Thus,

$$\mathbf{D}_3(2) = \begin{bmatrix} 0\,0\,0\,0\,0\,0\,0\,0\,0 \\ 1\,0\,0\,0\,0\,0\,0\,0\,0 \\ 0\,2\,0\,0\,0\,0\,0\,0\,0 \\ 0\,0\,3\,0\,0\,0\,0\,0\,0 \\ 0\,0\,0\,4\,0\,0\,0\,0\,0 \\ 0\,0\,0\,0\,5\,0\,0\,0\,0 \\ 0\,0\,0\,0\,0\,6\,0\,0\,0 \\ 0\,0\,0\,0\,0\,0\,7\,0\,0 \\ 0\,0\,0\,0\,0\,0\,0\,8\,0 \end{bmatrix}.$$

After computing the matrices $\mathbf{D}_3(2)^i$ for $i = 2, 3, \ldots, 8$, we

1. Divide each matrix $\mathbf{D}_3(2)^i$ with $\frac{1}{i!}$, $i = 1, 2, 3, 4, 5, 6, 7, 8$
2. Perform a columnwise Hadamard multiplication of the produced matrices with $\mathbf{v} = [1, 2, 1, 2, 1, 2, 1, 2, 1]$,
3. Reduce elements of these matrices modulo 3.

The sum of these matrices produces the RMF-transform matrix and it follows that the RMF-spectrum of a given function f specified by its function vector $\mathbf{F}$ can be computed as the sum of products of these matrices with $\mathbf{F}$.

It should be notices that non-zero elements in the matrix $\mathbf{D}_3(2)^i$ are equal to the elements of the i-th subdiagonal in the RMF-matrix. In other words, the RMF-matrix is decomposed into the sum of p^n matrices with each matrix representing a subdagonal in it.

Since the involved matrices $\mathbf{D}_p(n)^i$ have a small number of non-zero elements, they can be combined to reduce the number of matrices to be multiplied with $\mathbf{F}$. The number of resulting matrices can be adjusted to the number or processors provided. In this way, we can produce a series of possible factorization offering a trade-off between the number of processors and speed of computing.

9.5 Concluding Remarks

Multiple-valued logic is an area where Professor Claudio Moraga has been making important contributions over more than 45 years. Spectral logic in general, and various application of spectral methods to the analysis of multiple-valued functions, synthesis of related circuits, and certain optimization problems, can especially be emphasized as topics of his research interest in this area. For that reason, in this chapter, we discussed the Reed-Muller-Fourier transform as a particular spectral transform tailored for the processing of multiple-valued functions sharing at the same time some important properties of related operators in a classical Fourier analysis and their generalizations.

As a continuation of recent discussions with Professor Moraga and a common friend, Professor Jaakko Astola, about relationships between the RMF-transform and Pascal matrices [37], we used these relationships to devise certain factorizations of the RMF-matrix which might potentially led to fast computing algorithms with efficient implementation on different many-core and multi-processor architectures.

9.6 Personal Note

After over 30 years of joint work with Claudio, enjoying all the time a gentle guidance almost invisible on the surface, but deep and very strong inside, and above all his friendship, I could write many pages on what we did, explored, traveled together, organized, and most importantly, learned. In spite of my desire to express all my gratitude to Claudio here, I am aware that I should spare the reader my reminiscences, and I will point out a single detail that I consider essentially important not just for me, but also as a message for future generations. Through it I have learn how an experienced researcher should support young researchers in their first steps into research world.

I have found and wanted to learn some publications by Claudio in 1976 while preparing my BSc thesis on Walsh and Haar discrete transforms. As it was a customary practice at that time, I sent him letters typed on a mechanical typewriter in my broken English asking for reprints. In this way we started our communication, and I got the impression that Claudio is a person ready to help and support youngsters in their attempts to learn. By continuously looking for his new publications, I realized that Claudio is regularly contributing to the international symposia on multiple-valued logic (ISMVL). In 1984, it happen that I was lucky and a paper of mine submitted to the ISMVL in Winnipeg, Canada, was accepted. A couple of persons close to me easily concluded that I am now in trouble since being so crazy to dare to submit something at a conference, without having the slightest idea how I can possibly manage to attend it if the paper eventually accepted. After some days of deep thinking and much wondering about, I addressed Claudio by sending a letter explaining shortly, but quite openly, the situation and with the paper and comments by reviewers enclosed. I directly and openly asked him that, if he possibly likes the idea discussed in the paper, could accept to be the co-author and present it. I easily assumed that he was going to attend the symposium knowing that he was a regular participant. The absence of any answer for some time, provoked many interesting, comical but friendly, comments on my (mis)understanding of the scientific and research world. I was greatly rewarded for all this ribbing when I received a letter by Claudio saying that he would accept to be the co-author and present the paper under the condition that I will accept his corrections in the manuscript. He attached a long list of improvements of various statements and reformulations, so many that the paper was basically completely rewritten. Just the main idea was preserved and emphasized in a very appropriated manner and expressed, much better that I could comprehended it myself. Further, enclosed was a research technical report that Claudio usually prepared and publishing at the Lehrstuhl 1 (Informatik) at the Dortmund University, Dortmund, Germany. The report contained the rewritten version of the paper as specified in the corrections list. It was easy to realize that this was a gentle push to direct me towards a proper attitude in the research work and I of course very gladly and gratefully followed it.

For all these years, I think this answer by Claudio was very fortunate for me. If it had been be some other, possibly less friendly answer, or equally important not so determined and decisive request on my part, accompanied by a complete solution of the matter, I am not sure how I might have reacted and responded. Without his strong support and the manner in which it was offered, it might easily have happened that I would stop submitting to international conferences and trying to find a way to participate in them.

That was the first lecture of Claudio gave me personally that determined my attitude towards research work and communicating with researchers. I am very grateful that I have continuously been learning from Claudio for many years and I wish to continue in the same way for a long time.

References

1. Aburdene, M.F., Goodman, T.J.: The discrete Pascal transform and its applications, *IEEE Signal Processing Letters*, Vol. 12 (7), 2005, 493–495.
2. Besslich, Ph. W.:Determination of the irredundant forms of a Boolean function using Walsh-Hadamard analysis and dyadic groups, *IEE J. Comput. Dig. Tech.*, Vol. 1, 1978, 143–151.
3. Besslich, Ph. W.:Efficient computer method for XOR logic design, *IEE Proc., Part E*, Vol. 129, 1982, 15–20.
4. Besslich, Ph.W.:Spectral processing of switching functions using signal flow transformations, in Karpovsky, M.G., (ed.):*Spectral Techniques and Fault Detection*, Academic Press, Orlando, Florida, 1985.
5. Besslich, Ph.W., Lu, T.: *Diskrete Orthogonaltransformationen*, Springer, Berlin, 1990.
6. Bryant, R.E.: Graph-based algorithms for Boolean functions manipulation, *IEEE Trans. Comput.*, Vol.C-35 (8), 1986, 667–691.
7. Clarke, E.M., McMillan, K.L., Zhao, X., Fujita, M.: Spectral transforms for extremely large Boolean functions, in Kebschull, U., Schubert, E., Rosenstiel, W., (eds.), *Proc. IFIP WG 10.5*

Workshop on Applications of the Reed-Muller Expansion in Circuit Design, 16–17.9.1993, Hamburg, Germany, 86–90.
8. Clarke, E.M., Zhao, X., Fujita, M., Matsunaga, Y., McGeer, R.: Fast Walsh transform computation with Binary Decision Diagram, *Proc. IFIP WG 10.5 Workshop on Applications of the Reed-Muller Expansion in Circuit Design*, September 16–17, 1993, Hamburg, Germany, 82–85.
9. Clarke, M., McMillan, K.L., Zhao, X., Fujita, M., Yang, J.: Spectral transforms for large Boolean functions with applications to technology mapping, in *Proc. 30th ACM/IEEE Design Automation Conference (DAC-93)*, IEEE Computer Society Press, 54–60.
10. Farina, A., Giompapa, S., Graziano, A., Liburdi, A., Ravanelli, M., Zirilli, F.: Tartaglia-Pascal's triangle – a historical perspective with applications, *Signal, Image and Video Processing*, Vol. 7 (1), 2013, 173–188.
11. Fujita, M., Chih-Yuan Yang, J., Clarke, E.M., Zhao, Z., McGeer, P.: Fast spectrum computation for logic functions using Binary decision diagrams, *Proc. IEEE Int. Symp. on Circuits and Systems ISCAS-94*, London, England, UK, 30 May-2 June 1994, 275–278.
12. Gajić, D., Stanković, R.S.: The impact of address arithmetic on the GPU implementation of fast algorithms for the Vilenkin-Chrestenson transform, *Proc. 43rd Int. Symp. on Multiple-Valued Logic*, Toyama, Japan, May 22–24, 2013, 296–301.
13. Gibbs, J.E.: Walsh spectrometry a form of spectral analysis well suited to binary digital computation, *NPL DES Repts.*, National Physical Lab., Teddington, Middlesex, England, 1967.
14. Gibbs, J.E.: Walsh functions and the Gibbs derivative, *NPL DES Memo.*, No. 10, 1973, ii + 13.
15. Gibbs, J.E.: Instant Fourier transform, *Electron. Lett.*, Vol. 13 (5), 122–123, 1977.
16. Good, I. J.: The interaction algorithm and practical Fourier analysis, *Journal of the Royal Statistical Society, Series B*, Vol. 20 (2), 1958, 361–372.
17. Good, I. J.: The relationship between two Fast Fourier transforms, *IEEE Trans. on Computers*, Vol. 20, 1971, 310–317.
18. Hurst, S.L.: *Logical Processing of Digital Signals*, Crane Russak and Edward Arnold, London and Basel, 1978.
19. Hurst, S.L., Miller, D.M., Muzio, J.C.: *Spectral Techniques for Digital Logic*, Academic Press, 1985.
20. Karpovsky, M.G., Stanković, R.S., Astola, J.T.: *Spectral Logic and Its Application in the Design of Digital Devices*, Wiley, 2008.
21. Lv, X.-G., Huang, T.-Z., Ren, Z.-G.: A new algorithm for linear systems of the Pascal type, *Journal of Computational and Applied Mathematics*, Vol. 225, 2009, 309–315.
22. Moraga, C., Stanković, R.S.: Properties of the Reed-Muller spectrum of symmetric functions, *Facta. Univ. Ser. Energ.*, Vol. 20 (2), 2007, 281–294.
23. Moraga, C., Stanković, M., Stanković, R.S.: Some properties of ternary functions with bent Reed-Muller spectra, *Proc. Workshop on Boolean Problems*, September 17–19, Freiberg, Germany, 2014,
24. Moraga, C., Stanković, M., Stanković, R.S.: Contribution to the study of ternary functions with a bent Reed-Muller spectrum, *45th Int. Symp. on Multiple-Valued Logic (ISMVL-2015)*, Waterloo, ON, Canada, May 18–20, 2015, 133–138.
25. Moraga, C., Stojković, S., Stanković, R.S.: Periodic behaviour of generalized Reed-Muller spectra", *Multiple Valued Logic and Soft Computing*, Vol. 15 (4), 2009, 267–281.
26. NVIDIA, OpenCL Programming Guide for the CUDA Architecture, 2011.
27. Sasao, T., Fujita, M., (eds.): *Representations of Discrete Functions*, Kluwer Academic Publishers, 1996.
28. Skodras, A.N.: Fast discrete Pascal transform, *Electronics Letters*, Vol. 42 (23), 2006, 1367–1368.
29. Srinivasan, A., Kam, T., Malik, Sh., Brayant, R.K.: Algorithms for discrete function manipulation, *Proc. Inf. Conf. on CAD*, 1990, 92–95.
30. Stanković, R.S.: A note on the relation between Reed-Muller expansions and Walsh transform, *IEEE Transactions on Electromagnetic Compatibility*, Vol. EMC-24 (1), 1982, 68–70.

31. Stanković, R.S.: Some remarks on Fourier transforms and differential operators for digital functions, *Proc. 22nd Int. Symp. on Multiple-Valued Logic*, May 27–29, 1992, Sendai, Japan, 365–370.
32. Stanković, R. S.: Unified view of decision diagrams for representation of discrete functions, *Multi. Val. Logic*, Vol. 8 (2), 2002, 237–283.
33. Stanković, R.S., Astola, J.T.: *Spectral Interpretation of Decision Diagrams*, Springer, 2003.
34. Stanković, R.S. Astola, J.T., Moraga, C.: *Representation of Multiple-Valued Logic Functions*, Claypool & Morgan Publishers, 2012.
35. Stanković, R.S, Astola, J.T., Moraga, C.: Pascal matrices, Reed-Muller expressions and Reed-Muller error correcting codes, *Zbornik Radova*, Matematički institut, Belgrade, Serbia, Vol. 18 (26), 2015, 145–172.
36. Stanković, R.S., Moraga, C., Astola, J.T.: Reed-Muller expressions in the previous decade, *Multiple-Valued Logic and Soft Computing*, Vol. 10 (1), 2004, 5–28.
37. Stanković, R.S, Astola, J.T., Moraga, C.: Pascal matrices, Reed-Muller expressions and Reed-Muller error correcting codes, *Zbornik Radova*, Matematički institut, Vol. 18 (26), 2015, 145–172.
38. Stankovi c, R. S., Astola, J. T., Moraga, C., Gajić, D. B.: Constant geometry algorithms for Galois field expressions and their implementation on GPUs, *Proc. 44th IEEE Int. Symp. on Multiple-Valued Logic*, Bremen, Germany, May 19–21, 2014, 79–84.
39. Stanković, R.S., Butzer, P.L., Schipp, F., Wade, W.R., Su, W., Endow, Y., Fridli, S., Golubov, B.I., Pichler, F., Onneweer, K.C.W.: *Dyadic Walsh Analysis from 1924 Onwards, Walsh-Gibbs-Butzer Dyadic Differentiation in Science*, Vol. 1 *Foundations, A Monograph Based on Articles of the Founding Authors, Reproduced in Full*, Atlantis Studies in Mathematics for Engineering and Science, Vol. 12, Atlantis Press/Springer, 2015.
40. Stanković, R.S., Moraga, C., Astola, J.T.: *Fourier Analysis on Finite Non-Abelian Groups with Applications in Signal Processing and System Design*, Wiley/IEEE Press, 2005.
41. Stanković, R.S., Moraga, C.: Fast algorithms for detecting some properties of multiple-valued functions, *Proc. 14th Int. Symp. on Multiple-valued Logic*, Winnipeg, Canada, May 1984, 29–31.
42. Stanković, R.S., Moraga, C.: Reed-Muller-Fourier expansions over Galois fields of prime cardinality, *U. Kebschull, E. Schubert, W. Rosenstiel, Eds., Proc. IFIP W.10.5 Workshop on Application of Reed-Muller expansion in Circuit Design*, 16.-17.9.1993, Hamburg, Germany, 115–124.
43. Stanković, R.S., Moraga, C.:An algebraic transform for prime-valued functions, *Proc. 5th Int. Workshop on Spectral Techniques*, 15.-17.3.1994, Beijing, China, 205–209.
44. Stanković, R.S., Stanković, M., Moraga, C., Sasao, T.:The calculation of Reed-Muller-Fourier coefficients of multiple-valued functions through multiple-place decision diagrams, *Proc. 24th Int. Symp. on Multiple-valued Logic*, Boston, Massachusetts, USA, 22.-25.5.1994, 82–88.
45. Thomas, L.H.: Using a computer to solve problems in physics, *Application of Digital Computers*, Boston, Mass., Ginn, 1963.
46. Yanushkevich, S.N.: *Logic Differential Calculus in Multi-Valued Logic Design*, Techn. University of Szczecin Academic Publishers, Poland, 1998.
47. Yanushkevich, S.N., Miller, D.M., Shmerko, V.P., Stanković, R.S.: *Decision Diagram Techniques for Micro- and Nanoelectronic Design Handbook*, CRC Press, Taylor & Francis, 2006.
48. Zhang, Z., Wang, X.: A factorization of the symmetric Pascal matrix involving the Fibonacci matrix, *Discrete Applied Mathematics*, Vol. 155, 2007, 2371–2376.

Chapter 10
Multiple-Valued Logic and Complex-Valued Neural Networks

Igor Aizenberg

"What a nice thing is Man when he is indeed a Man"
Menander

To my Dear and Great Friend Claudio Moraga

10.1 Instead of Introduction

This is an unusual and non-traditional paper. On the one hand, it is a regular paper presenting some interesting research results. But on the other hand, it is devoted to Claudio Moraga, my Great Friend and Colleague. This is not only because this book is devoted to him and not only because of his 80th birthday. But this is also because the results presented here could not be obtained without him and his support, without his ability to look at the future and see a beauty in mathematical abstractions, which then can be miraculously transformed into applications.

I am very happy and proud of saying that Claudio is my Great Friend. I am also happy simply because I had a great privilege to closely collaborate with him. I am also happy because when it was some period of heavy challenges in my life, Claudio was the first who was willing to help.

In October 1989, the International Workshop on Gibbs Derivatives was held in Kupari (Croatia). My dad, Professor Naum Aizenberg was invited there to give a lecture. For the first time in his life he got an opportunity to attend an international conference. He as a great majority of his colleagues in the communist Soviet Union was banned from any international travel until 1989. After coming back home, my dad was absolutely excited. He told me: "I met there a great, smart and highly intelligent

I. Aizenberg (✉)
Department of Computer Science, Manhattan College, Riverdale, NY, USA
e-mail: igor.aizenberg@manhattan.edu

R. Seising and H. Allende-Cid (eds.), *Claudio Moraga: A Passion for Multi-Valued Logic and Soft Computing*, Studies in Fuzziness and Soft Computing 349, DOI 10.1007/978-3-319-48317-7_10

colleague. We spend a couple of days in scientific discussions and we became friends, like we knew each other for many years". This colleague was Professor Claudio Moraga. And it is important to say that discussions he had with my dad during those 2 days in 1989 became very important for me because I am not sure that the results, which are a scientific part of this paper, would ever be obtained without those discussions.

So what they discussed? In late 60s—early 70s Naum Aizenberg and his co-authors developed a new approach to multiple-valued logic. It was suggested to encode values of k-valued logic not using traditional alphabet $K = \{0, 1, \ldots, k-1\}$ of non-negative integers, but using an alphabet created from the kth roots of unity $E_k = \left\{1, \varepsilon - k, \varepsilon_k^2, \ldots, \varepsilon_k^{k-1}\right\}$ ($\varepsilon_k = e^{i2\pi/k}$ is the primitive kth root of unity, i is the imaginary unit, and k is some positive integer). A one-to-one correspondence between K and E_k can easily be established by $j \to \varepsilon_k^j, j \in K, \varepsilon_k^j \in E_k$. This approach has been for the first time presented in [1] and then described in detail in [2]. Then a comprehensive presentation of this approach was done in [3]. It is important to say that both papers [1, 2], and then the book [3] were published in Russian. That time 99 % of Soviet researches were completely banned by the official regulations from publishing anything in English and from submitting any paper to international journals. The paper [2] was translated into English. However its English translation was practically not available that time because just a few libraries across the Globe were subscribed to English translations of Soviet scientific journals. But this was not an obstacle for Claudio because he can read scientific papers in Russian! After reading papers [1, 2], he suggested very interesting ideas in the development of multiple-valued logic over the field of complex numbers, involving spectral techniques there [4, 5].

So it is not necessary to say how excited my dad was when he met Claudio in 1989. As we can easily recall now, late 1980s was a time of new boost in artificial neural networks. When this area was boosted for the first time in late 1950s—early 1960s, it was closely related to Boolean threshold logic. In fact, that time an artificial neuron was basically the same as a threshold element or an element of Boolean threshold logic. So it could implement linearly separable Boolean functions or linearly separable mappings $[-1, 1]^n \to \{-1, 1\}$.

The works [1–3] were inspired by threshold logic and aspiration of its extension to the multiple-valued case in such a way that classical Boolean threshold logic should be a particular case of multiple-valued threshold logic. During that meeting in October 1989 Claudio just mentioned that it probably should be great to utilize ideas of multiple-valued logic over the field of complex numbers in a new artificial neuron, which should be significantly more functional than traditional neurons with the threshold and sigmoid activation functions. Then it would be possible to design a network from such neurons, which can also be more functional than traditional popular networks (first of all a multilayer feedforward neural network). Unfortunately, my dad did not have a chance to meet with Claudio again. But their meeting in 1989 had a continuation, which is very important and valuable for me.

Since 1991 I concentrated my research on the development of multi-valued neuron and its learning. In 1992, the multi-valued neuron (MVN) and its learning algorithm were suggested in [6]. It was also suggested in the same paper to use MVN as a basic neuron in cellular neural networks (CNN), which in turn can be used as an associative memory for storing gray-scale images. In 1996 I moved from my home country Ukraine. I got a grant and went to work in Belgium. And in November 1997 we met with Claudio for the first time. We kept contacts by e-mail and Claudio invited me to Dortmund, to give a seminar and to spend a couple of days in research discussions. I was very impressed by his hospitality, by his kindness, by a unique depth of his knowledge in different areas. Three days, which we spent in very intensive discussions, then resulted in our collaboration, which started those rainy November days in 1997 and has been developing since that time. We met again in Dortmund in 1999 when I attended "Fuzzy Days" there. Our collaboration on medical image processing using MVN-based cellular neural network resulted in the journal paper [7]. In 2001, again I attended "Fuzzy Days" in Dortmund and again we met and had very productive discussions on everything. It is impossible to overestimate how important these meetings and our discussions were for me. But our collaboration reached its peak when with Claudio's support I got a number of consecutive grants, which made it possible for me to work in Dortmund with Claudio in 2003–2005. I was not there permanently, but used to come and stay for several months. This was a time when I got my best research result. It was succeeded to develop a multilayer neural network with multi-valued neurons (MLMVN) and its derivative-free backpropagation learning algorithm [8]. It was also succeeded a bit earlier to generalize and modify MVN, employing continuous inputs and outputs [9]. This was perhaps the most productive time in my research career. I greatly appreciate Claudio's support. I will never forget our discussions and how we worked on our papers, Claudio's wise advices and very useful remarks.

We then continued our collaboration after I got a faculty position in the United States and moved there in 2006. It was still very productive and resulted in more publications.

Let me devote a scientific part of this paper to brief presentation of multiple-valued threshold logic over the field of complex numbers, MVN and MLMVN, which resulted from this model of multiple-valued logic. Thus Sect. 10.2 is devoted to the theoretical foundations of multiple-valued logic over the field of complex numbers and MVN. Section 10.3 is devoted to MLMVN and its derivative-free learning. Section 10.4 is devoted to personal things (like most of Sect. 10.1).

10.2 Multiple-Valued Threshold Logic over the Field of Complex Numbers and Multi-valued Neuron

10.2.1 Multiple-Valued Threshold Logic over the Field of Complex Numbers

In this section we will use a conceptual narrative of multiple-valued logic over the field of complex numbers as it is presented in [10].

As it was and is well known, the McCulloch-Pitts neuron [11], which is the first historically known artificial neuron, implements input/output mappings described by threshold Boolean functions. Later, when continuous inputs were introduced, a neuron still produced a binary output. A continuous output was introduced along with a sigmoid activation in 1980s. But what about multiple-valued input/output mappings? Can we consider a concept of multiple-valued threshold logic similarly to the one of Boolean threshold logic?

From the time when a concept of multiple-valued logic was suggested by Jan Łukasiewicz in 1920 [12], the values of multiple-valued logic are traditionally encoded by non-negative integers. While in Boolean logic there are two truth values ("False" and "True" or 0 and 1), in multiple-value logic there are k truth values. Thus, while in Boolean logic, truth values are elements of the set $K_2 = \{0, 1\}$, in k-valued logic, it was suggested to encode the truth values by elements of the set $K = \{0, 1, \ldots, k-1\}$. Thus, in classical multiple-valued (k-valued) logic, a function of k-valued logic is $f(x_1, \ldots, x_n) : K^n \to K$. At the same time, in many applications (particularly in neural networks), it is very convenient to consider a normalized Boolean alphabet, that is $E_2 = \{1, -1\}$ instead of $K_2 = \{0, 1\}$. For example, this is important in neural learning rules, where a value of an input is an essential multiplicative term participating in the weight adjustment.

Is it possible to create a normalized k-valued alphabet from $K = \{0, 1, \ldots, k-1\}$ similarly to the creation of normalized $E_2 = \{1, -1\}$ from $K_2 = \{0, 1\}$, which is not normalized? Moreover, is it possible to do this in such a way that for $k = 2$ $E_2 = \{1, -1\}$ should be a particular case of such a normalized multiple-valued alphabet?

A very beautiful answer to these questions was suggested by Naum Aizenberg in early 1970s. It was presented in [1, 2] and then summarized in [3].

Let M be an arbitrary additive group and its cardinality is not lower than k. Let $A_k = \{a_o, \ldots, a_{k-1}\}, A_k \subseteq M$ be a structural alphabet.

Definition 10.2.1 [2]. A function $f(x_1, \ldots; X_n) \mid f : A_k^n \to A_k$ of n variables (where A_k^n is the nth Cartesian power of A_k) is a function of k-valued logic over the group M.

It is very easy to check, that a classical definition of a function of k-valued logic follows from Definition 10.2.1. Indeed, the set $K = \{0, 1, \ldots, k-1\}$ is an additive group with respect to mod k addition. If K is taken as a group M and if the structural alphabet $A_k = K$, then any function $f(x_1, \ldots; X_n) : K^n \to K$ is a function of k-valued logic over the group K according to Definition 10.2.1.

Let us now take the additive group of the field of complex of complex numbers $\mathbb{C}$ as a group M. As it is well known from algebra, there are exactly k kth complex-valued roots of unit. The root $\varepsilon_k = e^{i2\pi/k}$ (i is an imaginary unit) is a primitive kth root of unity. The rest $k-1$ roots can be obtained from ε_k by taking its 0th, 2nd, 3rd, …, $k-1$st powers. Thus, we obtain the set $E_k = \left\{1, \varepsilon_k^0, \varepsilon_k, \varepsilon_k^2, \ldots, \varepsilon_k^{k-1}\right\}$, of all the kth roots of unity, all located on the unit circle (see Fig. 10.1).

Since the set E_k contains exactly k elements, we may use this set as a structural alphabet in terms of Definition 10.2.1. Thus, any function $f(x_1, \ldots, x_n) : E_k^n \to E_k$

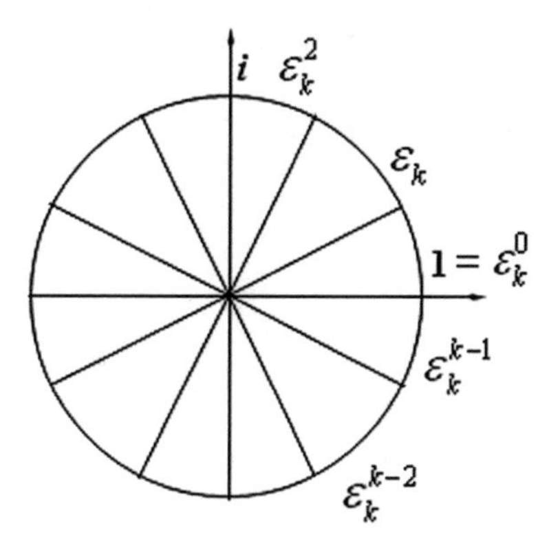

Fig. 10.1 kth roots of unity $1 = \varepsilon_k^0, \varepsilon_k, \varepsilon_k^2, \ldots, \varepsilon_k^{k-1}$ are located on the unit circle. They form a structural alphabet of multiple-valued logic over the field of complex numbers

is a function of k-valued logic over the additive group of the field of complex numbers or simply a function of k-valued logic over the field of complex numbers $\mathbb{C}$.

If $k = 2$, then $E_2 = \{1, -1\}$. Indeed, -1 is the primitive 2nd root of unit, and $1 = -1^0$ is the second of two 2nd roots of unit. As well as values of two-valued logic are normalized in E_2, values of k-valued logic are also normalized for any k in E_k. Evidently, there is a one-to-one correspondence between sets K and E_k, and any function $K^n \to K$ can be represented as a function $E_k^n \to E_k$ and vice versa.

The use of the alphabet E_2 in the threshold neuron is very important for definition of its activation function and for derivation of its learning algorithms. We will show now how important is that approach, which we have just presented, for multiple-valued threshold logic and multi-valued neurons.

Let us consider a function $f(x_1, \ldots, x_n) : T \to E_k; T \subseteq E_k^n$ of k-valued logic over the field of complex numbers. Evidently, $f(x_1, \ldots, x_n)$ is a fully defined function (if $T = E_k^n$) or a partially defined (if $T \subseteq E_k^n$).

Definition 10.2.2 [1–3] A function $f(x_1, \ldots, x_n) : T \to E_k; T \subseteq E_k^n$ of k-valued logic is a threshold function of k-valued logic (or multiple-valued (k-valued) threshold function) if there exist $n + 1$ complex numbers $w_o, w_1, \ldots, w_n$ such that for any $(x_1, \ldots, x_n) \in T$

$$f(x_1, \ldots, x_n) = P(w_o, w_1, \ldots, w_n) \tag{10.1}$$

where $P(z)$ is defined as follows:

$$P(z) = CSIGN(z) = \varepsilon_k^j, 2\pi j/k \le arg(z) \le 2\pi(j+1)/k \tag{10.2}$$

Function (10.2) is illustrated in Fig. 10.2 The complex plane is divided into k equal sectors by the lines passing through the origin and points on the unit circle corresponding to the kth roots of unity. Sectors are enumerated in the natural way: 0th, 1st, 2nd, ..., $k-1$st. The jth sector is limited by the boarders originating in the origin and crossing the unit circle at the points corresponding to the kth roots of

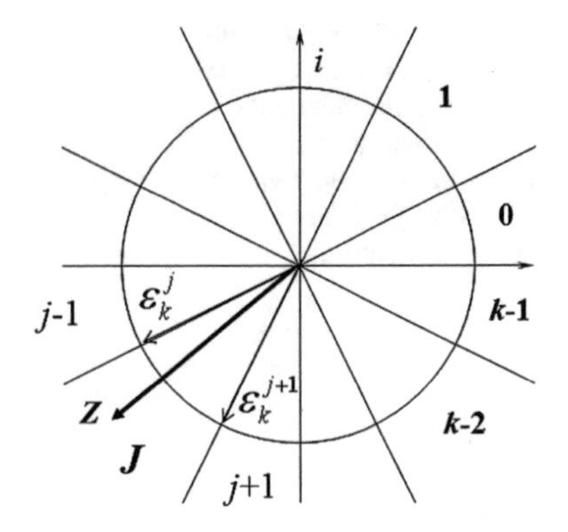

Fig. 10.2 Definition of multiple-valued activation function (10.2). $P(z) = e^{i2\frac{pij}{k}}$

unity ε_k^j and ε_k^{j+1}. If a complex number z is located in the jth sector, which means that $2\pi\frac{j}{k} \leq arg(z) \leq 2\pi\frac{(j+1)}{k}$, then $P(z) = e^{i2\frac{pij}{k}}$.

The vector $W = (w_0, \ldots, w_n)$ is called a *weighting vector* of the function f.

10.2.2 Multi-valued Neuron (MVN)

The discrete multi-valued neuron (MVN) was introduced in [6] as a generalization of an element of multiple-valued threshold logic considered in [2, 3]. This is a neuron with n inputs and one output (all belonging to the set $E_k = \{1, \varepsilon_k^0, \varepsilon_k, \varepsilon_k^2, \ldots, \varepsilon^k - 1_k, \}$) and activation function (10.2). The discrete MVN transforms n inputs $x_1, x_2, \ldots, x_n$ into its output according to (10.1). This means that an MVN input/output mapping is always a k-valued threshold function.

So like a neuron with the threshold activation function $sgn\,(z)$ implements an input/output mapping, which is a Boolean threshold function, MVN implements an input/output mapping, which is a k-valued threshold function. In [8, 9], it was suggested to consider MVN with continuous inputs and continuous output. Let O be the continuous set of the points located on the unit circle. Let $T \subseteq E_k$ (as we considered above) or $T \subseteq O$. Let MVN inputs belong to T, thus they can be kth roots of unity or arbitrary points located on the unit circle. In such a case, a discrete MVN input/output mapping is described by a function of n variables $f\,(x_1, \ldots, x_n)$, which is a function $f : T^n \rightarrow E_k$. Discrete MVN transforms n inputs $x_1, \ldots, x_n$ into its output according to (10.1).

To consider MVN with a continuous output, activation function (10.2) shall be modified. Function (10.2) divides the complex plane into k sectors whose angular size is $2\pi/k$. Evidently, $k \rightarrow \infty$ in the continuous case. But this means that the angular size of a sector approaches 0. In such a case, an MVN output can be determined as a projection of the weighted sum on the unit circle. Hence if $z = w_0 + w_1x_1 + w_2x_2$, then

$$P(z) = \frac{z}{|z|} = e^{Argz} \tag{10.3}$$

where $Arg z$ is a main value of the argument of the complex number z, is a continuous activation function of MVN. This activation function was suggested in [9]. It should also be mentioned that this idea of continuity was discussed earlier in [13] by George Georgiou, but that time it was not further developed.

The most efficient MVN learning algorithm is based on the error-correction learning rule. This rule is a generalization of the classical Rosenblatt's error-correction learning rule [14]. The MVN learning algorithm based on the error-correction learning rule was suggested in [15]. It is described most comprehensively in [15] where the convergence theorem is also proven.

According to the error-correction learning rule, the adjustment of the weights for both discrete and continuous MVN is completely determined by the neuron's error, which is the arithmetic difference $\delta = D - Y$ between the complex numbers D (a desired output) and Y (an actual output) located on the unit circle (see Fig. 10.3).

The error-correction learning rule is [10]

$$W_{r+1} = W_r + \frac{C_r}{(n+1)} \delta \bar{X} \tag{10.4}$$

and with a modification suggested in [8]:

$$W_{r+1} = W_r + \frac{C_r}{(n+1)\,|z_r|} \delta \bar{X} \tag{10.5}$$

where $\bar{X}$ is the input vector with the components complex-conjugated, n is the number of neuron inputs, r is the number of the learning step, W_r is the current weighting vector (to be corrected), W_{r+1} is the following weighting vector (after correction), C_r is the constant part of the learning rate (complex-valued in general), and $|z_r|$ is the absolute value of the weighted sum obtained on the rth learning step. A factor $\frac{1}{|z_r|}$ in (10.5) is a variable part of the learning rate. The error-correction rule (10.4)

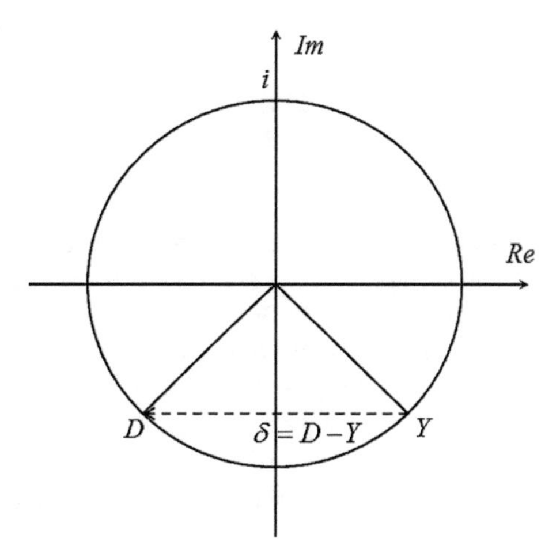

Fig. 10.3 Geometrical interpretation of the MVN learning rule

and its modification (10.5) ensure such a correction of the weights that the weighted sum moves exactly from the actual output Y to the desired output D or at least closer to D (see Fig. 10.3).

It should also be mentioned that it might be reasonable to calculate the error not as the difference between the desired ($D = \varepsilon_k^q$) and actual ($Y = \varepsilon_k^s$) outputs, but as the difference between the desired output D and the projection $\frac{z}{|z|}$ of the current weighted sum z on the unit circle. Evidently, in such a case $\delta = D - \frac{z}{|z|}$. This is useful for example, to learn highly nonlinear input/output mappings.

The discrete MVN with a periodic activation function was suggested in [16]. A periodic activation function makes it possible to increase the functionality of a single MVN. Let us consider the MVN input/output mapping described by some k-valued function $f(x_1, \ldots, x_n) : T \to E - K$, where $T \subseteq E_k$ or $T \subseteq O$. If this function $f(x_1, \ldots, x_n)$ is not a k-valued threshold function, it cannot be learned by a single MVN with the activation function (10.2).

Let us project the k-valued function $f(x_1, \ldots, x_n)$ into m-valued logic, where $m = kl$ and $l \geq 2$. To do this, let us define the following new discrete activation function for MVN:

$$P_l(z) = j \bmod k, \text{ } if \frac{2\pi j}{m} \leq argz \leq 2\pi \frac{(j+1)}{m} \tag{10.6}$$

$$j = 0, 1, \ldots, m-1; m = kl, l \geq 2. \tag{10.7}$$

This definition is illustrated in Fig. 10.4. The activation function (10.6) separates the complex plane into m equal sectors and $\forall d \in K$ there are exactly l sectors, in which the activation function (10.6) equals to d. This means that the activation function (10.6) establishes mappings from K into $M = \{0, 1, \ldots, k-1, k, k+1, \ldots m-1\}$, and from E_k into $E_m = \left\{1, \varepsilon_m, \varepsilon_m^2, \ldots, \varepsilon_m^{m-1}\right\}$, respectively.

Since $m = kl$, then each element from M and E_m has exactly l prototypes in K and E_k, respectively. In turn, this means that the neuron's output determined by (10.6) is equal to

$$\underbrace{\underbrace{0, 1, \ldots, k-1}_{0}, \underbrace{0, 1, \ldots, k-1}_{1}, \ldots, \underbrace{0, 1, \ldots, k-1}_{l-1}}_{lk=m} \tag{10.8}$$

depending on which one of the m sectors (whose ordinal numbers are determined by the elements of the set M) the weighted sum is located in. Hence, the MVN's activation function in this case becomes k-periodic and l-multiple.

It is important that in terms of multiple-valued logic, activation function (10.6) projects a k-valued function $f(x_1, \ldots, x_n)$ into m-valued logic. Evidently, $f(x_1, \ldots, x_n)$ is a partially defined function in m-valued logic because $K \subseteq M = \{0, 1, \ldots, m-1\}$, $E_k \subseteq E_m$ and $E_k^n \subseteq E_m^n$. Since $f(x_1, \ldots, x_n)$ is a k-valued function, it takes only k values out of m in m-valued logic. The projection established by

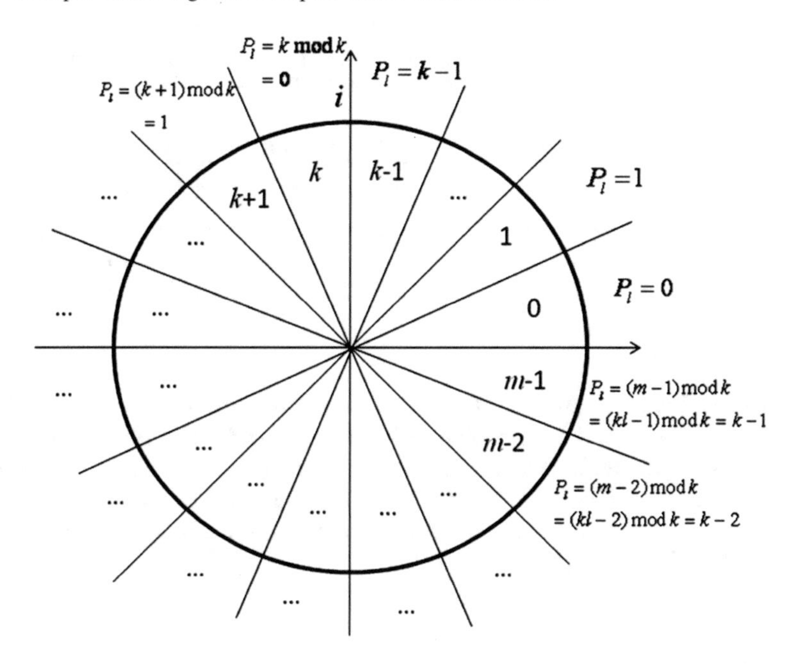

Fig. 10.4 Geometrical interpretation of the k-periodic and l-multiple discrete-valued MVN activation function (10.6)

(10.6) makes a great practical sense. On many occasions, a function $f(x_1, \ldots, x_n)$, being a non-threshold function in k-valued logic, can be at the same time a partially defined threshold function in m-valued logic and therefore it is possible to learn it using a single MVN with the activation function (10.6).

MVN with the activation function (10.6) is referred to as the multi-valued neuron with a periodic activation function (MVN-P) [16]. It is important to mention that if $l = 1$ in (10.6) then $m = k$ and the activation function (10.6) coincides with the activation function (10.2) accurate within the interpretation of the neuron's output (if the weighted sum is located in the jth sector then according to (10.2) the neuron's output is equal to $e^{ij2\pi/k} = \varepsilon^j \in E_k$, which is the jth of kth root of unity, while in (10.6) it is equal to $j \in K$), and the MVN-P becomes the regular MVN.

Learning of MVN-P is based on the same error correction learning rule (either (10.4) or (10.5)) as MVN learning, with a slight modification. Let $L \geq 2$ in (10.6) and $d \in \{0, 1, \ldots, k-1\}$ be the desired output. The activation function (10.6) determines the k-periodic and l-multiple sequence (10.8) with respect to sectors on the complex plane. Suppose that the current MVN-P's output is not correct and the current weighted sum is located in the sector $s \in M = \{0, 1, \ldots, m-1\}$, where $m = kl$. Since $l \geq 2$ in (10.6), there are l sectors on the complex plane, where function (10.6) takes a correct value (see also Fig. 10.4). Two of these l sectors are the closest ones to sector s (from right and left sides, respectively). From these two sectors, we choose sector q whose border is closer to the current weighted sum z in terms of the angu-

Table 10.1 MVN-P with the activation function (10.6) (where $k = 2, l = 2$) implements the $f(x_1, x_2) = x_1 XOR x_2$ function with the weighting vector $(0, i, 1)$ in the original 2-dimensional space

x_1	x_2	$z = w0 + w_1x_1 + w_2x_2$	$arg(z)$	$P_l(z)$	$f(x_1, x_2) = x_1 XOR x_2$
1	1	$i+1$	$\frac{\pi}{4}$	1	1
1	-1	$i-1$	$\frac{3\pi}{4}$	-1	-1
-1	1	$-i+1$	$\frac{7\pi}{4}$	-1	-1
-1	-1	$-i-1$	$\frac{5\pi}{4}$	1	1

lar distance. Then we take ε^q as the desired output and apply learning rule (10.4) or (10.5). MVN-P makes it possible to learn non-linearly separable input/output mappings using a single neuron. The simplest example is its ability to learn the XOR problem when $k = 2, l = 2$ in (10.6). This is illustrated in Table 10.1. A single MVN-P can easily learn such non-linearly separable problems as Parity n [10, 17] and mod k sum of n variables [10, 18].

10.3 Multilayer Neural Network with Multi-value Neurons (MLMVN)

MLMVN was first suggested in [9, 19] and then comprehensively presented and discussed in [8]. I am really happy that working on MLMVN, its concept and its learning algorithm, I very closely collaborated with Claudio. I sincerely believe that MLMVN is the best idea, which I ever developed. But I would say that I am not sure it would be possible to develop it without Claudio, his support and comprehensive discussions with him. I am proud of our joint publications [8, 9, 19] resulted from this great work. And I am especially proud of the fact that our paper [8] is my and Claudio's most cited journal paper!

MVN was nice, but the question on how it should be possible to design an MVN-based feedforward neural network was open for a long time. Since the functionality of a single MVN is higher than the one of a sigmoidal neuron, it could be expected that an MVN-based feedforward neural network should also be more functional than a multilayer feedforward neural network with sigmoidal neurons (MLF) also known as a multilayer perceptron (MLP) [20]. The bottleneck (as it was seen) was how to train an MVN-based feedforward neural network. The MLF learning [20] is based on solving the optimization problem of minimization of the error function. To solve this optimization problem, the differentiability of an activation function is of crucial importance. But MVN activation functions (10.2) and (10.3) are not differentiable as functions of a complex variable. So, what we should do?

A wonderful answer was found! To train MLMVN, it is possible to generalize the error-correction learning rule for a single MVN. In this way, MLMVN learning becomes derivative-free as MVN learning for a single MVN! MLMVN has exactly the same topology as MLF-MLP, but its learning algorithm is based on the same error-correction learning rule as the one for a single MVN, just modified in order to backpropagate the error. Thus the error backpropagation is the MLMVN learning algorithm differs from the one in the classical MLF-MLP learning algorithm. Let us describe here the MLMVN learning algorithm, which was for the first time comprehensively presented in [8].

MLMVN is a multilayer neural network with a standard feedforward topology where neurons are integrated into layers, and output of each neuron from the current layer is connected to the corresponding inputs of neurons from the following layer.

Let MLMVN contain one input layer, $m-1$ hidden layers and one output layer. Let us use the following notations. Let D_{jm} be a desired output of the jth neuron from the mth (output) layer; Y_{jm} be an actual output of the jth neuron from the mth (output) layer; w_i^{jr} be the weight corresponding to the ith input of the jrth neuron (jth neuron of the rth layer); Y_{jr} and z_{jr} be the actual output and the weighted sum of the jth neuron from the rth layer ($r = 1, \ldots, m$), respectively; $\tilde{Y}_{jr}$ be the updated (after the weights are corrected) output of the same neuron; N_r be the number of the neurons in the rth layer, and $x_1, \ldots, x_n$ be the network inputs. Evidently, the neurons from the $r+1$st layer have exactly $r+1$st inputs ($r = 1, \ldots, m$).

The global error of the network taken from the jth neuron of the mth (output) layer is calculated as follows:

$$\delta_{jm}^* = D_{jm} - Y_{jm};\, j = 1, N_m \tag{10.9}$$

The backpropagation of the global errors δ_{jm}^* through the network is used (from the mth (output) layer to the $m-1$st one, from the $m-1$st one to the $m-2$nd one, ..., from the 2nd one to the 1st one) to express the error of each neuron δ_{jr}, $r = 1, \ldots, m$ by means of the global errors δ_{jm}^* of the entire network.

The error backpropagation in MLMVN is based on the error sharing principle [8, 10]. This principle states the following. (1) The error of a single MVN must be shared among all the weights of the neuron. (2) The network error and the errors of each particular neuron in MLMVN must be shared among those neurons from the network that contribute to this error.

This principle is utilized for MLMVN as follows [8].

The global errors of the entire network are determined by (10.9). The local neurons errors are represented in the following way. First, we need to backpropagate the global errors of the network to the output layer neurons. The errors of the mth (output) layer neurons are:

$$\delta_{jm} = \frac{1}{t_m}\delta_{jm}^*;\, j = 1, N_m \tag{10.10}$$

where j_m specifies the jth neuron of the mth layer; $t_m = N_{m-1} + 1$, i.e. the number of all neurons in the preceding layer (layer $m-1$ where the error (10.10) will be then backpropagated to) incremented by 1. Thus, to obtain a local error of any output neuron, a global error of the network calculated for this output neuron, shall be divided by the number of neurons connected to this output neuron plus 1 (this output neuron itself).

The errors of the hidden layers neurons are then calculated using the following backpropagation procedure

$$\delta_{ir} = \frac{1}{t_r} \sum_{j=1}^{N_{r+1}} \delta_{j,r+1} (w_i^{j,r+1})^{-1} \tag{10.11}$$

where ir specifies the ith neuron of the rth layer ($r = 1, \ldots, m-1$); $t_r = N_{r-1} + 1$, $r = 2, \ldots, m$ is the number of all neurons in the layer $r-1$ incremented by 1, (n is the number of network inputs). Thus Eqs. (10.10)–(10.11) determine the error backpropagation for MLMVN.

The weights for all neurons of the network are corrected using the error-correction learning rules (10.4) and (10.5) adapted to MLMVN. For the neurons from the mth (output) layer (jth neuron of the mth layer),

$$\tilde{w}_i^{jm} = w_i^{jm} + C_{jm} \delta_{jm} \bar{\tilde{Y}}_{i,m-1}, i = 1, \ldots, N_{m-1} \tag{10.12}$$

$$\tilde{w}_0^{jm} = w_0^{jm} + C_{jm} \delta_{jm}, \tag{10.13}$$

for the neurons from the 2nd hidden layer till $m-1$st hidden layer (jth neuron of the rth layer ($r = 2, \ldots, m-1$),

$$\tilde{w}_i^{jr} = w_i^{jr} + \frac{1}{|z_{jr}|} C_{jr} \delta_{jr} \bar{\tilde{Y}}_{i,r-1}, i = 1, \ldots, N_{r-1} \tag{10.14}$$

$$\tilde{w}_0^{js} = w_0^{js} + \frac{1}{|z_{jr}|} C_{jr} \delta_{jr}, \tag{10.15}$$

and for the neurons from the 1st hidden layer

$$\tilde{w}_i^{j1} = w_i^{j1} + \frac{1}{|z_{j1}|} C_{j1} \delta_{j1} \bar{x}_i, i = 1, \ldots, n \tag{10.16}$$

$$\tilde{w}_0^{j1} = w_0^{j1} + \frac{1}{|z_{j1}|} C_{j1} \delta_{j1}, \tag{10.17}$$

The learning rate $C_{jr}; j = 1, \ldots, N_m; r = 1, \ldots, m$ in (10.12)–(10.16) is complex-valued in general, but to our best knowledge, in all actual applications described so far it was not used at all, so this means that in all known MLMVN applications $C_{jr} = 1; j = 1, \ldots, N_m; r = 1, \ldots, m$.

The MLMVN learning algorithm consists of the sequential checking for all learning samples whether an actual output of the network coincides with a desired output. If for some sample there is no coincidence, then the errors (10.9) must be backpropagated according to (10.10)–(10.11), and the weights must be then adjusted according to (10.12)–(10.16). The learning process continues either until the zero-error is reached or the root mean square error (RMSE) criterion is satisfied. For any output neuron in MLMVN with discrete activation function (10.2), RMSE should be applied to the errors in terms of the numbers of sectors (see Fig. 10.2), thus not to the elements of the set $E_k = \{\varepsilon_k^0, \varepsilon_k, \ldots, \varepsilon_k^{k-1}\}$, but to the elements of the set $K = \{0, 1, K-1\}$. The local errors for the rth learning sample in these terms are calculated as

$$\gamma_r = (\alpha_{j_r} - \alpha_r) \bmod k; \alpha_{j_r}, \alpha_r \in \{0, 1, \ldots, k-1\} \tag{10.18}$$

($\varepsilon_k^{\alpha_{jr}}$ is the desired output and $\varepsilon^{\alpha_{r_k}}$ is the actual output).

For any output neuron in MLMVN with continuous activation function (10.3), RMSE should be applied to the arguments of the output neurons' weighted sums and calculated in terms of the angular error. Let Δ_r be the square error of the network for the rth learning sample. For MLMVN with a single output neuron it is

$$\Delta_r = \gamma_r^2, r = 1, \ldots, N, \tag{10.19}$$

(γ_r is the local error taken from (10.18) and for MLMVN with N_m output neurons (m is the output layer index) it is

$$\Delta_r = \frac{1}{N_m} \sum_{j=1}^{N_m} (\gamma_{jr})^2; r = 1, \ldots, N. \tag{10.20}$$

The MLMVN learning process in such a case continues until RMSE drops below some pre-determined acceptable minimal value λ (N is the number of learning samples):

$$RMSE = \sqrt{MSE} = \sqrt{\frac{1}{N} \sum_{r=1}^{N} \Delta_r} \leq \lambda. \tag{10.21}$$

This learning algorithm was for the first time described in detail and the backpropagation learning rule (10.10)–(10.11) was justified in [8]. The convergence theorem for this learning algorithm for MLMVN containing arbitrary amount of layers and neurons in layers was proven in [10].

MLMVN has many remarkable applications. It significantly outperforms MLF-MLP in terms of functionality (a smaller MLMVN may learn input/output mappings, which cannot be learned even using a bigger MLF-MLP), speed of learning, and generalization capability. In a number of applications, MLMVN shows results comparable or even better than support vector machines (SVM). MLMVN was successfully used for example, in the following applications: generation of a genetic code as a function of multiple-valued logic [21] (it is very pleasant that this application was again developed in collaboration with Claudio!), recognition of type of blur and its parameters for image restoration [22], recognition of blurred images [23], decoding of signals in an EEG-based brain-computer interface [24], and different kinds of long-term time series prediction [10, 25, 26]. In [27], the learning algorithm for MLMVN with discrete outputs was modified by introducing soft margins. This makes it possible to avoid classification errors caused by closeness to each other of some learning samples belonging to different classes. A batch learning algorithm for MLMVN with a single hidden layer was proposed in [28]. It was then generalized in [29] for MLMVN with multiple output neurons. This algorithm converges much faster than a regular learning algorithm.

Fig. 10.5 Finland, May 2011. A tour during ISMVL'2011. My Great Friend Claudio Moraga

Fig. 10.6 Barcelona, Spain, July 2010, IEEE WCCI-2010. Claudio with me and my wife Ella

Fig. 10.7 Uzhhorod, Ukraine, August 2008. Claudio with me after a workshop session. It was a workshop devoted to scientific heritage and commemoration of Prof. Naum Aizenberg

Fig. 10.8 Barcelona, Spain, July 2010, IEEE WCCI-2010. A conference dinner

Fig. 10.9 Uzhhorod, Ukraine, August 2008. Claudio with me at the Ethnographic Museum

Thus MLMVN is a really powerful and very promising tool. We may hope it will find many new applications. It is very pleasant for me to say that Claudio's merit in its creation cannot be overestimated and I am really proud of the fact that I worked on MLMVN together with Claudio, feeling every day his great support.

10.4 Best Wishes to My Great Friend Claudio Moraga!

Claudio is a great personality. It is difficult to overestimate his great intellect. It is difficult to overestimate his great research contributions. But the most important is that Claudio is a Great Friend and a Great Colleague! I was happy to work with him. I know for sure that I could not achieve many things without his support and his influence. Claudio is one of the brightest personalities I ever met. And I am sure that many of his colleagues and former students share the same feelings. Let me forward my warmest wishes to my Great Friend Claudio Moraga! Let me conclude this paper with some pictures taken in different parts of our planet where we were together with Claudio (Figs. 10.5, 10.6, 10.7, 10.8 and 10.9).

References

1. N. N. Aizenberg, Yu. L. Ivaskiv, and D. A. Pospelov: About one generalization of the threshold function, Doklady Akademii Nauk SSSR (The Reports of the Academy of Sciences of the USSR), vol. 196, No 6, 1971, pp. 1287–1290 (in Russian).
2. N. N. Aizenberg, Yu. L. Ivaskiv, D. A. Pospelov, and G.F. Hudiakov: Multivalued Threshold Functions. Synthesis of Multivalued Threshold Elements, *Cybernetics and Systems Analysis*, vol. 9, No 1, 1973, pp. 61–77.
3. N. N. Aizenberg and Yu. L. Ivaskiv: *Multiple-Valued Threshold Logic*, Naukova Dumka Publisher House, Kiev, 1977 (in Russian).
4. C. Moraga: Complex Spectral Logic, Proceedings of the 8th IEEE International symposium on Multiple-valued Logic, IEEE Computer Society Press, 1978, pp. 149–156.
5. C. Moraga: On some Applications of the Chrestenson Functions in Logic Design and Data Processing, *Mathematics and Computers in Simulation*, vol. 27, Issues 5–6, October 1985, pp. 431–439.
6. N. N. Aizenberg and I. N. Aizenberg: CNN Based on Multi-Valued Neuron as a Model of Associative Memory for Gray-Scale Images, Proceedings of the Second IEEE Int. Workshop on Cellular Neural Networks and their Applications, Technical University Munich, Germany October 1992, pp. 36–41.
7. I Aizenberg, N Aizenberg, J. Hiltner, C. Moraga and E Meyer Zu Bexten: Cellular neural networks and computational intelligence in medical image processing, *Image and Vision Computing*, vol. 19, Issue 4, March 2001, pp. 177–183.
8. I. Aizenberg and C. Moraga: Multilayer Feedforward Neural Network Based on Multi-Valued Neurons (MLMVN) and a Backpropagation Learning Algorithm, Soft Computing, vol. 11, No 2, January 2007, pp. 169–183.
9. I. Aizenberg, C. Moraga, and D. Paliy: A Feedforward Neural Network based on Multi-Valued Neurons, In Computational Intelligence, Theory and Applications. Advances in Soft Computing, XIV, B. Reusch, Ed., Springer, Berlin, Heidelberg, New York, 2005, pp. 599–612.

10. I. Aizenberg: *Complex-valued neural networks with multi-valued neurons*, Springer, Hidelberg, 2011.
11. W. S. McCulloch and W. Pits: A Logical Calculus of the Ideas Immanent in Nervous Activity, *Bull. Math. Biophys.*, vol. 5, 1943. pp. 115–133.
12. J. Łukasiewicz: O logice trójwartościowej (in Polish), *Ruch filozoficzny* No 5, 1920, pp.170–171. English translation: On three-valued logic, in L. Borkowski (ed.), *Selected works by Jan Łukasiewicz*, North–Holland, Amsterdam, 1970, pp. 87-88.
13. G. M. Georgiou: The Multivalued and Continuous Perceptrons, Proceedings of 1993 World Congress on Neural Networks, vol. 4 , pp. 679–683.
14. F. Rosenblatt On the Convergence of Reinforcement Procedures in Simple Perceptron. Report VG 1196-G-4. Cornell Aeronautical Laboratory, Buffalo, NY, 1960.
15. N. N. Aizenberg, I.N. Aizenberg, and G.A. Krivosheev: Multi-Valued Neurons: Learning, Networks, Application to Image Recognition and Extrapolation of Temporal Serie, Lecture Notes in Computer Science, vol. 930 (J.Mira, F.Sandoval—Eds.), Springer, 1995, pp. 389–395.
16. I. Aizenberg: A Periodic Activation Function and a Modified Learning Algorithm for a Multi-Valued Neuron, *IEEE Transactions on Neural Networks*, vol. 21, No 12, December 2010, pp. 1939–1949.
17. I. Aizenberg: Solving the XOR and Parity n Problems Using a Single Universal Binary Neuron, *Soft Computing*, vol. 12, No 3, February 2008, pp. 215–222.
18. I. Aizenberg, M. Caudill, J. Jackson, and S. Alexander: Learning Nonlinearly Separable mod *k* Addition Problem Using a Single Multi-Valued Neuron With a Periodic Activation Function, Proceedings of the 2010 IEEE World Congress on Computational Intelligence—2010 IEEE International Joint Conference on Neural Networks, Barcelona, Spain, July 18-23, 2010, pp. 2577–2584.
19. I. Aizenberg and C. Moraga: Multi-Layered Neural Network based on Multi-Valued Neurons (MLMVN) and a Backpropagation Learning Algorithm, Technical Report No CI 171/04 (ISSN 1433-3325) of the Collaborative Research Center for Computational Intelligence of the University of Dortmund (SFB 531), available online at http://sfbci.cs.uni-dortmund.de/Publications/Reference/Downloads/17104.pdf, 2004.
20. D. E. Rumelhart, G.E. Hilton, R. J. Williams: Learning Internal Representations by Error Propagation. In: D.E. Rumelhart and J.L. McClelland (eds.): *Parallel Distributed Processing: Explorations in the Microstructure of Cognition*, Vol. 1, Chapter 8. MIT Press, Cambridge, Massachusetts, 1986, pp. 318–362.
21. I. Aizenberg and C. Moraga: The Genetic Code as a Function of Multiple-Valued Logic Over the Field of Complex Numbers and its Learning using Multilayer Neural Network Based on Multi-Valued Neurons, *Journal of Multiple-Valued Logic and Soft Computing*, No 4-6, November 2007, pp. 605–618.
22. I. Aizenberg, D. Paliy, J. Zurada, and J. Astola: Blur Identification by Multilayer Neural Network based on Multi-Valued Neurons, *IEEE Transactions on Neural Networks*, vol. 19, No 5, May 2008, pp. 883–898.
23. Aizenberg I., Alexander S., and Jackson J.: Recognition of Blurred Images Using Multilayer Neural Network Based on Multi-Valued Neurons, Proceedings of the 41st IEEE International Symposium on Multiple-Valued Logic (ISMVL-2011), to appear in the Proceedings, May 23-35, 2011, IEEE Computer Society Press, pp. 282–287.
24. Manyakov N.V., Aizenberg I., Chumerin N., and Van Hulle M.: Phase-Coded Brain-Computer Interface Based on MLMVN, in: A. Hirose (ed): *Complex-Valued Neural Networks: Advances and Applications*, Wiley, 2013, pp. 185–208.
25. Fink O., Zio E., and Weidmann U.: Predicting time series of railway speed restrictions with time-dependent machine learning techniques, *Expert Systems with Applications*, vol. 40(15), No 11, 2013, pp. 6033–6040.
26. Aizenberg I., Sheremetov L., Villa-Vargas L., and Martinez-Muñoz J.: Multilayer Neural Network with Multi-Valued Neurons in Time Series Forecasting of Oil Production, *Neurcomputing*, published "Online first" , December 2015, published in a hard copy Vol. 175, Part B, 29 January 2016, pp. 980–989.

27. I. Aizenberg: MLMVN with Soft Margins Learning, *IEEE Transactions on Neural Networks and Learning Systems*, vol. 25, No 9, September 2014, pp. 1632–1644.
28. I. Aizenberg. A. Luchetta, and S. Manetti: A modified Learning Algorithms for the Multilayer Neural Network with Multi-Valued Neurons based on the Complex QR Decomposition, *Soft Computing*, vol. 16, April 2012, pp. 563–575.
29. I. Aizenberg and E. Aizenberg: Batch LLS-based Learning Algorithm for MLMVN with Soft Margins, Proceedings of the 2014 IEEE Symposium Series of Computational Intelligence (SSCI-2014), December, 2014, pp. 48–55.

Chapter 11
Sequential Bayesian Estimation of Recurrent Neural Networks

Branimir Todorović, Claudio Moraga and Miomir Stanković

11.1 Introduction

The central problem of the sequential Bayesian estimation is to determine the probability density function of the hidden state of a dynamical system. The hidden state is defined as a vector of variables which evolution through time completely describe the behaviour of the dynamical system. The hidden state x_k with initial distribution $p(x_0)$ evolves as an unobserved first order Markov process according to the conditional probability density $p(x_k/x_{k-1})$. The observations y_k are conditionally independent given the state d are generalized according to the probability density $p(y_k/x_k)$. The state space model can be written as a set of system equations:

$$x_k = f(x_{k-1}, u_k, d_k) \tag{11.1}$$

$$y_k = h(x_k, v_k) \tag{11.2}$$

where d_k represents the dynamic (process) noise that drives the dynamic system through the nonlinear state transition function f, and v_k is the observation noise corrupting the observation of the state through nonlinear observation function h.

B. Todorović (✉)
Faculty of Mathematics and Sciences, Computer Science Department,
University of Nis, Nis, Serbia
e-mail: branimirtodorovic@pmf.ni.ac.rs

C. Moraga
Technical University of Dortmund, Dortmund, Germany

M. Stanković
Faculty of Occupational Safety, Department of Mathematics,
University of Nis, Nis, Serbia

R. Seising and H. Allende-Cid (eds.), *Claudio Moraga: A Passion for Multi-Valued Logic and Soft Computing*, Studies in Fuzziness and Soft Computing 349, DOI 10.1007/978-3-319-48317-7_11

The state transition density $p(x_k/x_{k-1})$ is fully specified by f and the process noise pdf $p(d_k)$, whereas h and the observation noise pdf $p(v_k)$ fully specify the observation likelihood $p(y_k/x_k)$.

In a sequential estimation framework, the state filtering probability density function (pdf) $p(x_k/y_{0:k})$, where $y_{0:k} = \{y_0, y_1, \ldots, y_k\}$ denotes the set of all observations, represents the complete solution. The optimal state estimate with respect to any criterion can be calculated based on this pdf. Sequential (or recursive) Bayesian estimation algorithm for determination of the filtering pdf consists of two steps: *prediction* and *update*.

In the first step (*prediction*) the previous posterior $p(x_{k-1}/y_{0:k-1})$ is projected forward in time through nonlinear probabilistic transition equation (11.1):

$$p(x_k/y_{0:k-1}) = \int p(x_k/x_{k-1})\, p(x_{k-1}/y_{0:k-1})\, dx_k \tag{11.3}$$

The state transition density $p(x_k/x_{k-1})$ is completely specified by $f(\cdot)$ and the process noise distribution.

In the second step (*update*), the predictive density is updated by incorporating the latest noisy measurement y_k using the observation likelihood $p(y_k/x_k)$, which is completely specified by observation function $h(\cdot)$ and observation noise distribution, to generate the posterior:

$$p(x_k/y_{0:k}) = \frac{p(y_k/x_k)\, p(x_k/y_{0:k-1})}{\int p(y_k/x_k)\, p(x_k/y_{0:k-1})\, dx_k} \tag{11.4}$$

The recurrence relations (11.3) and (11.4) are only conceptual solutions and the posterior density cannot be determined analytically in general. The restrictive set of analytical solutions includes the well-known Kalman filter [Kalman], which represents an optimal solution of (11.3) and (11.4) if the posterior density $p(x_k/y_{0:k})$ and dynamic and observation noise are Gaussian and $f(\cdot)$ and $h(\cdot)$ are known linear functions.

At the beginning of our research in this area, the framework of Bayesian sequential estimation was not new in the field of neural networks training. Kalman filter's counterpart for nonlinear systems—Extended Kalman Filter (*EKF*), have been repeatedly considered as on-line training algorithm for feed forward and re-current neural networks [1, 2]. *EKF* often outperformed gradient based algorithms like Back Propagation, for feed forward and Real Time Recurrent Learning (*RTRL*) or Back Propagation Through Time (*BPTT*) for recurrent neural net-works training, mainly because it uses the second order information in estimation [1]. Additionally, when Extended Kalman Filter was applied for simultaneous estimation of synaptic weights and neuron outputs of recurrent neural networks, it has been shown that it generalizes the well-known heuristic "teacher forcing" which improves speed and the stability of RNN training [2].

The derivation of EKF as Recurrent Neural Network (RNN) training algorithm is based on the assumption that a good approximation of optimal sequential (or recursive) Bayesian estimator can be obtained by propagating only first and second moment of probability density function of RNN hidden variables (synaptic weights and neuron outputs) through a linearized dynamics of RNN (which is obtained by applying first order Taylor expansion), instead of propagating the complete probability density function through nonlinear (in general) RNN dynamics. This assumption is of course violated when the dynamics is significantly non-linear or noise on data is non-Gaussian: heavy tailed or even multimodal. Therefore, in our research we have addressed the following problems.

1. Can we derive better recurrent neural network training algorithms if we apply more accurate approximations of nonlinear dynamics or more clever ways of propagating probability density function of RNN hidden states through nonlinear transformation?
2. What if the noise on data is non-Gaussian?
3. Can we use statistics (first and second moments) sequentially calculated by approximate Bayesian estimators to derive somehow the algorithm for structure adaptation of recurrent neural networks?

In order to be able to start working on this problems the first obvious step was to represent the dynamics of recurrent neural network in the form of the state space model, similar to the system of probabilistic equations (11.1) and (11.2).

11.2 State Space Models of Recurrent Neural Networks

We will give here, as an example, the state space models of three representative architectures of globally recurrent neural networks: Elman, fully connected, and NARX recurrent neural network. Experimental evaluation of these architectures in long term chaotic time series prediction is given in [3, 4].

11.2.1 Elman Network State Space Model

In Elman RNNs adaptive feedbacks are provided between every pair of hidden units. The network is illustrated in Fig. 11.1(a), and the state space model of the Elman network is given by equations

$$\begin{bmatrix} x_k^H \\ w_k^O \\ w_k^H \end{bmatrix} = \begin{bmatrix} f\left(x_{k-1}^H, w_{k-1}^H, u_k\right) \\ w_{k-1}^O \\ w_{k-1}^H \end{bmatrix} + \begin{bmatrix} d_{x_k^H} \\ d_{w_k^O} \\ d_{w_k^H} \end{bmatrix} \tag{11.5}$$

$$y_k = x_k^O + v_k, \;\; x_k^O = h\left(x_k^H, w_k^O\right) \tag{11.6}$$

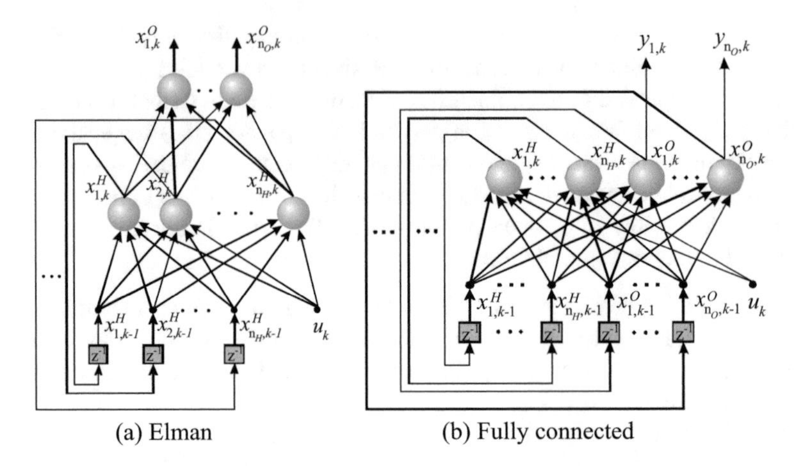

Fig. 11.1 Elman and fully connected RNN

where x_k^H represents the output of the hidden neurons in the k-th time step, x_k^O is the output of the neurons in the last layer, w_{k-1}^O is the vector of synaptic weights between the hidden and the output layer and w_{k-1}^H is the vector of recurrent adaptive connection weights. Note that in the original formulation of Elman, these weights were fixed. Random variables $d_{x_k^H}$, $d_{w_k^O}$, $d_{w_k^H}$ represent the process noises.

It is assumed that the output of the network $x_k^O = h\left(x_k^H, w_k^O\right)$ is corrupted by the observation noise ν_K.

11.2.2 Fully Connected Recurrent Network State Space Model

In fully connected RNNs adaptive feedbacks are provided between each pair of processing units (hidden and output). The state vector of a fully connected RNN consists of outputs (activities) of hidden x_k^H and output neurons x_k^O, and their synaptic weights w_k^H and x_k^O. The activation functions of the hidden and these output neurons are $f^H\left(x_k^O, x_k^H, w_{k-1}^H, u_k\right)$ and $f^O\left(x_k^O, x_k^H, w_{k-1}^O, u_k\right)$, respectively. The network structure is illustrated in Fig. 11.1(b).

The state space model of the network is given by:

$$\begin{bmatrix} x_k^O \\ x_k^H \\ w_k^O \\ w_k^H \end{bmatrix} = \begin{bmatrix} f^O\left(x_k^O, x_k^H, w_{k-1}^O, u_k\right) \\ f^H\left(x_k^O, x_k^H, w_{k-1}^H, u_k\right) \\ w_{k-1}^O \\ w_{k-1}^H \end{bmatrix} + \begin{bmatrix} d_{x_k^O} \\ d_{w_k^H} \\ d_{w_k^O} \\ d_{w_k^H} \end{bmatrix} \tag{11.7}$$

$$y_k = H \cdot \begin{bmatrix} x_k^O \\ x_k^H \\ w_k^O \\ w_k^H \end{bmatrix} + \nu_k, \qquad H = \left[I_{n_0 \times n_0} \; O_{n_0 \times (n_S - n_O)} \right] \tag{11.8}$$

The dynamic equation describes the evolution of neuron outputs and synaptic weights. In the observation equation, the matrix H selects the activities of output neurons as the only visible part of the state vector, where n_S is the number of hidden states which are estimated: $n_S = n_O + n_H + n_{W^O} + n_{W^H}$, n_O and n_H are the numbers of output and hidden neurons respectively, n_{W^O} is the number of adaptive weights of the output neurons, n_{W^H} is the number of adaptive weights of the hidden neurons.

11.2.3 NARX Recurrent Neural Network State Space Model

The non-linear AutoRegressive with eXogenous inputs (NARX) recurrent neural network has adaptive feedbacks between the output and the hidden units. These feedback connection and possible input connections are implemented as FIR filters. It has been shown [5] that NARX RNN often outperforms the classical recurrent neural networks, like Elman or fully connected RNN, in tasks that involve long term dependencies for which the desired output depends on inputs presented at times far in the past (Fig. 11.2).

The state vector of NARX RNN consists of outputs of the network in Δx time steps $x_k^O, x_{k-1}^O, \ldots, x_{k-\Delta_x+1}^O$ the output w_k^O, and hidden synaptic w_k^H weights. The dynamic equation of the NARX RNN state space model describes the evolution of network outputs and synaptic weights:

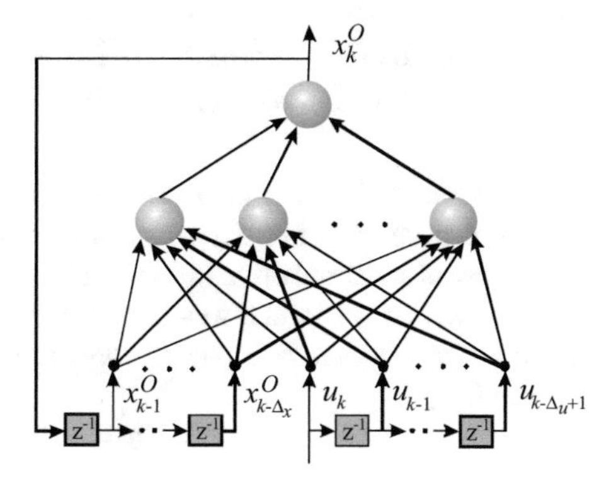

Fig. 11.2 NARX recurrent neural network

$$\begin{bmatrix} x_k^O \\ x_{k-1}^O \\ \vdots \\ x_{k-\Delta_x+1}^O \\ w_k^O \\ w_k^H \end{bmatrix} = \begin{bmatrix} f\left(x_{k-1}^O, \ldots x_{k-\Delta_x}^O, u_{k-1}, \ldots, u_{k-\Delta_u}, w_{k-1}\right) \\ x_{k-1}^O \\ \vdots \\ x_{k-\Delta_x+1}^O \\ w_{k-1}^O \\ w_{k-1}^H \end{bmatrix} + \begin{bmatrix} d_{x_k^O} \\ 0 \\ \vdots \\ 0 \\ d_{w_k^O} \\ d_{w_k^H} \end{bmatrix} \tag{11.9}$$

The observation equation selects the output neurons as observable:

$$y_k = H \cdot \begin{bmatrix} x_k^O \\ x_{k-1}^O \\ \vdots \\ x_{k-\Delta_x+1}^O \\ w_k^O \\ w_k^H \end{bmatrix} + v_k, \quad H = \left[I_{n_O \times n_O} \; O_{n_O \times (n_S - n_O)}\right] \tag{11.10}$$

where n_O represents the number of output neurons. n_S is the number of hidden states of the NARX RNN: $n_S = n_O + n_{W^O} + n_{W^H}$, n_{W^O} is the number of adaptive weights of output neurons, n_{W^H} is number of adaptive weights of hidden neurons.

All considered models have nonlinear hidden neurons and linear output neurons. In order to insure stability, in fully recurrent neural network, self-recurrent connections of linear output neurons are not allowed. Two types of nonlinear activation functions for hidden neurons have been often considered, the sigmoidal and the Gaussian radial basis function.

11.3 Gaussian Approximate Sequential Bayesian Estimation—Linear Minimum Mean Square Error (MMSE) Estimation

The well-known Kalman filter represent exact solution of the sequential Bayesian estimation problem if both dynamic and observation equations are linear, and initial state, process noise and observational noise are Gaussian random variables. However Kalman filter can be (and it was actually) derived in different way. Kalman assumed that the state estimator $\hat{x}_k$ can be represented as a linear function of the current observation y_k:

$$\hat{x}_k = A_k y_k + b_k \tag{11.11}$$

where matrix A_k and vector b_k are derived by minimizing mean square estimation error criterion:

$$R_k = \int\int \left(x_k + \hat{x}_k\right)^T \left(x_k - \hat{x}_k\right) \cdot p\left(x_k, y_k / y_{0:k-1}\right) dx_k dy_k \tag{11.12}$$

or, equivalently, by satisfying constraints that an estimator is unbiased:

$$\int\int\left(x_k - \hat{x}_k\left(y_k\right)\right)\cdot p\left(x_k, y_k/y_{0:k-1}\right)dx_k dy_k = 0 \tag{11.13}$$

and the estimation error is orthogonal to the current observation:

$$\int\int\left(x_k - \hat{x}_k\right)y_k^T\cdot p\left(x_k, y_k/y_{0:k-1}\right)dx_k dy_k = 0. \tag{11.14}$$

In its final form the estimator is given as:

$$\hat{x}_k = \hat{x}_{\bar{k}} + P_{x_k y_k}P_{y_k}^{-1}\left(y_k - \hat{y}_{\bar{k}}\right). \tag{11.15}$$

If assumptions on linearity and Gaussian distributions hold, the matrix Mean Square Error (MSE) corresponding to (11.15):

$$E\left[\left(x_k - \hat{x}_k\right)\left(x_k - \hat{x}_k\right)^T\right] = P_{x_k}^{-} - P_{x_k y_k}P_{y_k}^{-1}P_{x_k y_k}^T. \tag{11.16}$$

is the estimator covariance matrix P_{x_k}, otherwise it is an approximation. In previous equations

$$\hat{x}_{\bar{k}} = E\left[x_k/y_{0:k-1}\right] = \int x_k p\left(x_k/y_{0:k-1}\right)dx_k. \tag{11.17}$$

represents state prediction, and

$$P_{\bar{x_k}} = E\left[\left(x_k - \hat{x}_{\bar{k}}\right)\left(x_k - \hat{x}_{\bar{k}}\right)^T/y_{0:k-1}\right]. \tag{11.18}$$

is state prediction covariance matrix, while

$$\hat{y}_{\bar{k}} = E\left[y_k/y_{0:k-1}\right] = \int y_k p\left(y_k/y_{0:k-1}\right)dy_k \tag{11.19}$$

is observation prediction, and

$$P_{y_k} = E\left[\left(y_k - \hat{y}_{\bar{k}}\right)\left(y_k - \hat{y}_{\bar{k}}\right)^T/y_{0:k-1}\right] \tag{11.20}$$

is observation prediction covariance matrix, and finally

$$P_{x_k y_k} = E\left[\left(x_k - \tilde{x}_{\bar{k}}\right)\left(y_k - \hat{y}_{\bar{k}}\right)^T/y_{0:k-1}\right] \tag{11.21}$$

is cross covariance matrix between state and observation.

If the dynamic and the observation models are linear and process and observation noises are Gaussian, the linear MMSE estimator is optimal and exact solution of the recursive Bayesian estimation equations, it is the best MMSE estimator and is equal to the conditional mean $E\left[x_k/y_{0:k}\right]$, and it is also optimal Maximum Aposteriory (MAP) estimator, otherwise it is the best within the class of linear estimators.

The problem that remains to be solved is how to calculate (11.17), (11.18), (11.19), (11.20) and (11.21), which in general can be considered as propagating first and second order statistics of a random variable trough the nonlinear transformation.

11.3.1 Extended Kalman Filter

The extended Kalman filter uses the multidimensional Taylor series expansion to approximate the dynamic and observation equation of the RNN state space models. In our research we have considered only linear expansions.

Linearized dynamic equation is:

$$x_k = f\left(\hat{x}_{k-1}, u_k, \bar{d}_k\right) + F_k\left(x_{k-1} - \hat{x}_{k-1}\right) + G_k\left(d_k - \bar{d}_k\right), \tag{11.22}$$

where $F_k = \left.\frac{\partial f_k(x_{k-1}, u_k, d_k)}{\partial x_{k-1}}\right|_{\substack{x_{k-1} = \hat{x}_{k-1} \\ d_k = d_{\bar{k}}}}$, $\quad G_k = \left.\frac{\partial f_k(x_{k-1}, u_k, d_k)}{\partial x_k}\right|_{\substack{x_{k-1} = \hat{x}_{k-1} \\ d_k = d_{\bar{k}}}}$, $\hat{x}_{k-1} = E\left[x_{k-1}/y_{1:k-1}\right]$ represents the estimate of the state in time step $k-1$ and $\bar{d}_k = E\left[d_k\right]$ is process noise mean.

Prediction of the state $\hat{x}_{k-1} = E\left[x_{k-1}/y_{1:k-1}\right]$ and prediction covariance $P^-_{x_k} = E\left[\left(x_k - \hat{x}_{\bar{k}}\right)\left(x_k - \hat{x}_{\bar{k}}\right)^T/y_{0:k-1}\right]$ are obtained after applying (11.17) and (11.18) to linearized dynamic equation (11.22):

$$\hat{x}_{\bar{k}} = f\left(\hat{x}_{k-1}, u_k, \bar{d}:k\right) \tag{11.23}$$

$$P^-_{x_k} = F_k P_{x_{k-1}} F_k^T + G_k Q_k G_k^T \tag{11.24}$$

where $Q_k = E\left[\left(d_k - d_{\bar{k}}\right)\left(d_k - d_{\bar{k}}\right)^T\right]$ represents the process noise covariance.

After the linearization of the observation equation we obtain:

$$x_k = f_k\left(\hat{x}_{bark}, \bar{\nu}\right) + H_k\left(x_k - \hat{x}_{\bar{k}}\right) + L_k\left(\nu_k - \bar{\nu}\right), \tag{11.25}$$

where $H_k = \left.\frac{\partial h_k(x_k, \nu_k)}{\partial x_k}\right|_{\substack{x_k = \hat{x}_{\bar{k}} \\ \nu_k = \nu_{\bar{k}}}}$ and $L_k = \left.\partial h_k(x_{k-1}, \nu_k) / \partial \nu_k\right|_{\substack{x_k = \hat{x}_{\bar{k}} \\ \nu_k = \nu_{\bar{k}}}}$,
and $\nu_k = E[\nu_k]$ is the mean of the observation noise.

The prediction of the observation is given by:

$$\hat{y}_{\bar{k}} = h\left(\hat{x} - \bar{k}, \bar{\nu}_k\right) \tag{11.26}$$

and the prediction covariance is:

$$P_{y_k} = H_k P_{x_k}^- H_k^T + L_k R_k L_k^T \tag{11.27}$$

and cross covariance:

$$P_{x,y_k} = P_{x_k}^- H_k^T \tag{11.28}$$

where $R_k = E\left[\nu_k \nu_k^T\right]$ is the observation noise covariance.

11.3.2 Divided Difference Filter (DDF)

In [6] Nørgaard et al. proposed a new set of estimators based on a derivative free polynomial approximation of nonlinear dynamic and observation equation using Stirling's interpolation formula which uses central divided differences. We have considered only second order polynomial approximation, which is for arbitrary nonlinear function $f(x)$ given by:

$$f(x) \approx f(\bar{x}) + \tilde{D}_{\Delta x} f + \frac{1}{2!}\tilde{D}_{\Delta x}^2 f \tag{11.29}$$

where $\tilde{D}_{\Delta x} f$ and $\tilde{D}_{\Delta x}^2 f$ are the first and second order central divided difference operators acting on $f(x)$:

$$\tilde{D}_{\Delta x} f = (x - \bar{x}) \frac{f(\bar{x} + h) - f(h - h)}{2h} \tag{11.30}$$

$$\tilde{D}_{\Delta x}^2 f = (x - \bar{x}) \frac{f(\bar{x} + h) - f(h - h)}{h^2} \tag{11.31}$$

h is the central difference step size and $\bar{x}$, around which we expand $f(x)$, is the prior mean of random variable x.

Previous formulation can be extended to the multidimensional case by stochastic decoupling of random variable x:

$$z = R_x^{-T} x, \tag{11.32}$$

where R_x represents the upper triangular Cholesky factor of the covariance matrix $P_x = E\left[(x - \bar{x})(x - \bar{x})^T\right] = R_x^T R_x$. After decoupling we have:

$$f(x) = f\left(R_x^{-T} z\right) = \tilde{f}(z). \tag{11.33}$$

Individual components of random variable z are mutually uncorrelated, with unity variance $P_z = E\left[(z - \bar{z})(z - \bar{z})^T\right] = I$ and consequently we can apply the first and the second order central difference operators independently to the components of $\tilde{f}(z)$, in order to obtain the following multidimensional central difference operators:

$$\tilde{D}_{\Delta z}\tilde{f} = \frac{1}{h}\left(\sum_{i=1}^{n} \Delta_{z_i}\mu_i\delta_i\right)\tilde{f}(\bar{z}) \tag{11.34}$$

$$\tilde{D}^2_{\Delta z}\tilde{f} = 1/h^2\left(\sum_{i=1}^{n} \Delta^2_{z_i}\delta_i^2 + \sum_{i=1}^{n_x}\sum_{\substack{j=1 \\ j\neq i}}^{n_x} \Delta_{z_i}\Delta_{z_j}(\mu_i\delta_i)(\mu_i\delta_i)\right)\tilde{f}(\bar{z}) \tag{11.35}$$

where Δ_{z_i} represents the i-th component of $(z - \bar{z})$. Partial first and second order difference operators δ_i and δ_i^2, and the mean operator μ_i are defined as:

$$\delta_i\tilde{f}(z) = \tilde{f}\left(\bar{z} + \frac{h}{2}e_i\right) - \tilde{f}\left(\bar{z} - \frac{h}{2}e_i\right) \tag{11.36}$$

$$\delta_i^2\tilde{f}(z) = \tilde{f}(\bar{z} + he_i) + \tilde{f}(\bar{z} - he_i) - 2\tilde{f}(\bar{z}) \tag{11.37}$$

$$\mu_i\tilde{f}(z) = \frac{1}{2}\tilde{f}\left(\bar{z} + \frac{h}{2}e_i\right) + \tilde{f}\left(\bar{z} - \frac{h}{2}e_i\right) \tag{11.38}$$

where e_i is the i-th unit vector.

Approximation of the mean is given by:

$$\bar{y} \approx E\left[\tilde{f}(z) + \tilde{D}_{\Delta_x}\tilde{f} + \frac{1}{2!}\tilde{D}^2_{\Delta_x}\tilde{f}\right] \tag{11.39}$$

$$= \frac{h^2 - n_x}{h^2} f(\bar{x}) + \frac{1}{2h^2}\sum_{i=1}^{n_x}\left(f\left(\bar{x} + hR^T_{x,i}\right) + f\left(\bar{x} - hR^T_{x,i}\right)\right) \tag{11.40}$$

The approximation of the posterior covariance is:

$$P_y = \frac{1}{4h^2} \sum_{p=1}^{n_x} \left(f\left(\bar{x} + hR_{x,i}^T\right) - f\left(\bar{x} - hR_{x,i}^T\right)\right) \tag{11.41}$$

$$\cdot \left(f\left(\bar{x} + hR_{x,i}^T\right) - f\left(\bar{x} - hR_{x,i}^T\right)\right)^T \tag{11.42}$$

$$+ \frac{h^2 - 1}{4h^4} \sum_{p=1}^{n_x} \left(f\left(\bar{x} + hR_{x,i}^T\right) + f\left(\bar{x} - hR_{x,i}^T\right) - 2f\left(\bar{x}\right)\right) \tag{11.43}$$

$$\cdot \left(f\left(\bar{x} + hR_{x,i}^T\right) + f\left(\bar{x} - hR_{x,i}^T\right) - 2f\left(\bar{x}\right)\right)^T \tag{11.44}$$

where we select $h^2 > 1$, and consequently the covariance approximation well always be positive semi definite [6]. Nørgaard et al. have derived the alternative covariance estimate as well [6]:

$$P_y = \frac{h^2 - n_x}{h^2} \left(f\left(\bar{x}\right) - \bar{y}\right) \left(f\left(\bar{x}\right) - \bar{y}\right)^T \tag{11.45}$$

$$+ \frac{1}{2h^2} \sum_{p=1}^{n_x} \left(f\left(\bar{x} + hR_{x,i}^T\right) - \bar{y}\right) \left(f\left(\bar{x} + hR_{x,i}^T\right) - \bar{y}\right)^T \tag{11.46}$$

$$+ \frac{1}{2h^2} \sum_{i=1}^{n_x} \left(f\left(\bar{x} - hR_{x,i}^T\right) - \bar{y}\right) \left(f\left(\bar{x} - hR_{x,i}^T\right) - \bar{y}\right)^T \tag{11.47}$$

This estimate is less accurate than (11.44). Moreover, for $h^2 < n$ the last term becomes negative semi-definite with a possible implication that the covariance estimate (11.62) becomes non-positive definite. The reason why this estimate is considered here is to provide a comparison with the covariance estimate obtained by the Unscented Transformation described in the next subsection. The estimate of the cross-covariance matrix is:

$$P_{xy} = \sum_{p=1}^{n_x} \left(f\left(\bar{x} - hs_{x,p}\right) - \bar{y}\right) \left(f\left(\bar{x} - hs_{x,p}\right) - \bar{y}\right)^T \tag{11.48}$$

11.3.3 Unscented Kalman Filter (UKF)

The unscented transformation [7] is a method for calculating the statistics of a random variable which undergoes a nonlinear transformation. It is based on the intuition that is easier to approximate a probability distribution than arbitrary function.

Again we consider propagating the n_x-dimensional continuous random variable x with prior mean $\bar{x} = E[x]$ and covariance $P_x = E[(x - \bar{x})(x - \bar{x})]$ through nonlinear function $y = f(x)$. To calculate the first two moments of random variable y using unscented transformation we first select the set of $2n_x + 1$ samples $\mathcal{X}_i$ called sigma points with corresponding weights ω_i. The weights and sample locations are selected to accurately capture the prior mean and covariance of a random variable and to capture the posterior mean and covariance accurately up to the and including second order terms in the Taylor series expansion of the true quantities [7]. Sigma points and their weights which satisfy previous constraints are given by:

$$\mathcal{X}_0 = \bar{x}, \quad \omega_0 = \kappa / (n_x + \kappa), \quad i = 0 \tag{11.49}$$

$$\mathcal{X}_i = \bar{x} + \sqrt{n_x + \lambda} \cdot s_{x,i}, \quad \omega_i = 0.5 / (n_x + \kappa), \quad i = 1, \ldots, n_x \tag{11.50}$$

$$\mathcal{X}_{i+n_x} = \bar{x} - \sqrt{n_x + \lambda} \cdot s_{x,i}, \quad \omega_{i+n_x} = 0.5 / (n_x + \kappa), \quad i = 1, \ldots, n_x \tag{11.51}$$

where $\kappa \in \mathbb{R}$ is the scaling parameter, $s_{x,i}$ is the i-th row or column of the matrix square root of P_x. For weights associated with sigma points it holds $\sum_{i=0}^{2n_x} w_i = 1$.

Note that using this idea is possible to capture the higher order moments of posterior random variable, but at the cost of a larger set of sigma points [7].

After propagating sigma points through the nonlinear function $\mathcal{Y}_i = f(\mathcal{X}_i)$, approximations of the posterior mean, covariance and cross covariance are:

$$\bar{y} = \sum_{i=0}^{2n_x} \omega_i \mathcal{Y}_i \tag{11.52}$$

$$P_y = \sum_{i=0}^{2n_x} \omega_i (\mathcal{Y}_i - \bar{y}) (\mathcal{Y}_i - \bar{y})^T . \tag{11.53}$$

$$P_y = \sum_{i=0}^{2n_x} \omega_i (\mathcal{X}_i - \bar{x}) (\mathcal{Y}_i - \bar{y})^T . \tag{11.54}$$

The approximations are accurate to the second order of the Taylor series expansion of $f(x)$ (third order for Gaussian prior). Errors in the third and higher moments can be scaled by appropriate choice of scaling parameter κ. When prior random variable is Gaussian a useful heuristic is to select $\kappa = 3 - n_x$ [7].

It can be easily verified that for $h = \sqrt{n + \lambda}$, the estimates of the mean (11.40) and the covariance (11.47) obtained by applying Stirling's interpolation formula are equivalent to the estimates (11.52) and (11.53) obtained by unscented transformation [6].

In [6] Nørgaard showed that DDF has slightly smaller absolute error compared to UKF in the fourth order terms and also guarantees positive semi definiteness of the posterior covariance.

11.3.4 Square Root Implementation of Recursive Bayesian Estimators as RNN Training Algorithms [3, 4]

Straightforward implementation of Unscented Kalman Filter and Divided Difference Filter requires calculation of the prior state covariance matrix, which has $O\left(n^3/6\right)$ computational complexity. However it is the full covariance matrix of the estimate which is recursively updated. The square root implementations of UKF and DDF recursively update the Cholesky factors of the covariance matrices. Although the general complexity of the algorithms is still $O\left(n^3\right)$, they will have better numerical properties, comparable to the standard square root implementation of Kalman filter.

The square root implementations of EKF, UKF and DDF are based on three linear algebra algorithms: matrix orthogonal-triangular decomposition (triangularization), rank one update of a Cholesky factor and efficient solution of the over-determined least square problem.

The orthogonal-triangular decomposition of $m \times n$ matrix A ($[Q, R] = qr(A)$) produces $m \times n$ upper triangular matrix R, which is Choleskey factor of $A^T A$, and $m \times n$ unitary matrix Q such that $A = Q \times R$.

Rank one update of a Choleskey factor R, returns upper triangular Choleskey factor $\tilde{R}$ for which holds:

$$\tilde{R}^T \tilde{R} = R^T R \pm xx^T, \tag{11.55}$$

where x is column vector of appropriate length.

For overdetermined lest squares problem $AX = B$, if A is an upper triangular matrix, then X is simply computed by back substitution algorithm.

Using Cholesky decomposition of the prior covariance $P_{x_{k-1}} = R^T_{x_{k-1}} R_{x_{k-1}}$ we can represent the state and observation prediction covariance in EKF as:

$$P^-_{x_k} = F_k R^T_{x_{k-1}} R_{x_{k-1}} F^T_k + R^T_{d_k} R_{d-k} = \begin{bmatrix} F_k R_{x_{k-1}} & R^T_{d_k} \end{bmatrix} \cdot \begin{bmatrix} R_{x_{k-1}} F^T_k \\ R_{d_k} \end{bmatrix} = \left(R^-_{x_k}\right)^T R^-_{x_k} \tag{11.56}$$

$$P^-_{y_k} = H_k \left(R^-_{x_k}\right)^T R^-_{x_k} H^T_k + R^T_{\nu_k} R_{\nu_k} = \begin{bmatrix} H_k \left(R^-_{x_k}\right)^T & R^T_{\nu_k} \end{bmatrix} \cdot \begin{bmatrix} R^-_{x_k} H^T_k \\ R_{\nu_k} \end{bmatrix} = \left(R^-_{y_k}\right)^T R^-_{y_k}. \tag{11.57}$$

where R_{d_k} and R_{ν_k} represent the Choleskey factors of process and observation noise respectively.

We obtain the recursive update of the estimation covariance Choleskey factor using the numerically stable Joseph form of the covariance:

$$P_{x_k} = \left(I - K_k H_k \left(R^-_{x_k}\right)\right)^T R^-_{x_k} \left(I - K_k H_k\right)^T + K_k R^T_{\nu_k} R_{\nu_k} K^T_k \tag{11.58}$$

$$= \begin{bmatrix} \left(I - K_k H_k\right) \left(R^-_{x_k}\right)^T & K_k R^T_{\nu_k} \end{bmatrix} \cdot \begin{bmatrix} R^-_{x-k} \left(I - K_k H_k\right)^T \\ R_{\nu_k} K^T_k \end{bmatrix} = R^T_{x_k} R_{x_k} \tag{11.59}$$

11.4 Gaussian Sum Filters as RNN Training Algorithms

Gaussian filters, described in previous section, approximate propagation of state pdf through dynamic and observation equation by propagating first two moments. Such approximation of state pdf is not accurate enough if one wants to train RNN on noisy data, when noise pdf is multimodal or heavy tailed [8]. In order to deal with these problems, we have turned to the assumption that any probability density function can be approximated sufficiently accurately using finite Gaussian mixture [9, 10]:

$$p(x) \approx \sum_{i=1}^{n} P\{A_i\} p(x/A_i) = \sum_{i=1}^{n} w_i N(x; \bar{x}_i, P_i) \tag{11.60}$$

where A_i represents the event that the x is Gaussian distributed with mean $\bar{x}_i$ and covariance P_i, that is $A_i = \{x \sim N(\bar{x}_i, P_i)\}$. Events A_i are mutually exclusive $P\{A_i A_j\} = 0, \forall i \neq j$, and exhaustive $\sum_{i=1}^{N} P\{A_i\} = 1$, and $P\{A_i\} = w_i$. Based on previous assumptions we have derived equations of Gaussian Sum filter as recurrent neural network training algorithm [8, 11].

11.4.1 Gaussian Sum Filter Equations

To derive GS filter equations we will assume that the filtering and prediction densities as well as non-Gaussian noise densities can be represented as finite Gaussian mixtures.

$$p(x_{k-1}/y_{0:k-1}) = \sum_{j=1}^{n_{k-1}} P\{A_{k-1,j}/y_{0:k-1}\} p(x_{k-1}/y_{0:k-1}, A_{k-1,j}) \tag{11.61}$$

$$= \sum_{j=1}^{n_{k-1}} w_{k-1,j} N(x_{k-1}; \hat{x}_{k-1,j}, P_{x_{k-1,j}}) \tag{11.62}$$

$$p(d_k) = \sum_{j=1}^{n_{d_k}} P\{B_{k,j}\} p(d_k/B_{k,i}) = \sum_{j=1}^{n_{d_k}} wd_{k,i} N(d_k; \bar{d}_{k,i}, Q_{k,i}) \tag{11.63}$$

$$p(v_k) = \sum_{j=1}^{n_{v_k}} P\{C_{k,j}\} p(v_k/C_{k,i}) = \sum_{j=1}^{n_{v_k}} wv_{k,i} N(v_k; \bar{v}_{k,i}, R_{k,i}) \tag{11.64}$$

The predictive density is obtained as:

$$p\left(x_k/y_{0:k-1}\right) = \int p\left(x_k/x_{k-1}\right) p\left(x_{k-1}/y_{0:k-1}\right) dx_{k-1} \tag{11.65}$$

$$= \sum_{i=1}^{n_d} \sum_{j=1}^{n} P\left\{B_{k,i}\right\} P\left\{A_{k-1}/y_{0:k-1}\right\} \tag{11.66}$$

$$\cdot \int p\left(x_k/x_{k-1}, B_{k,i}\right) \cdot p\left(x_{k-1}/y_{0:k-1}, A_{k-1,i}\right) dx_{k-1} \tag{11.67}$$

If we introduce $D_{k,l}$ to denote a joint event $B_{k,i} \cap A_{k-1,j}$, we have:

$$p\left(x_k/y_{0:k-1}\right) = \sum_{l=1}^{n_{\bar{k}}} P\left\{D_{k,l}/y_{0:k-1}\right\} p\left(x_k/y_{0:k-1}, D_{k,l}\right) \tag{11.68}$$

where $l = (i-1) \cdot n + j$, $n_{\bar{k}} = n_{d_k} \cdot n_{k-1}$
and $P\left\{D_{k,l}/y_{0:k-1}\right\} = P\left\{B_{k,j}\right\} P\left\{A_{k-1,j}/y_{0:k-1}\right\}$ since $B_{k,i}$ are independent events.

Finally, based on assumptions (11.12) and (11.13) we obtain the predictive density as the finite Gaussian mixture:

$$p\left(x_k/y_{0:k-1}\right) = \sum_{l=1}^{n_{\bar{k}}} w_{\bar{k},l} N\left(x_k; \hat{x}_{\bar{k},l}, P_{\bar{x}_{\bar{k}},l}\right) \tag{11.69}$$

where $w_{\bar{k},l} \cdot w_{k\bar{-}1,j}$, and

$$\hat{x}_{\bar{k},l} = E\left[x_k/y_{0:k-1}, D_{k,l}\right], P_{\bar{x}_{\bar{k}},l} = E\left[\left(x_k - \hat{x}_{\bar{k},l}\right)\left(x_k - \hat{x}_{\bar{k},l}\right)^T / y_{0:k-1}, D_{k,l}\right] \tag{11.70}$$

The posterior state density is obtained as:

$$p\left(x_k/y_{0:k}\right) = \frac{p\left(y_k/x_k\right) p\left(x_k/y_{0:k-1}\right)}{\int p\left(y_k/x_k\right) p\left(x_k/y_{0:k-1}\right) dx_k} \tag{11.71}$$

$$= \frac{\sum_{l=1}^{n_k^*} P\left\{A_{k,l}/y_{0:k-1}\right\} p\left(y_k/y_{0:k-1}, A_{k,l}\right) p\left(x_k/y_{0:k}, A_{k,l}\right)}{\sum_{l=1}^{n_k^*} P\left\{A_{k,l}/y_{0:k-1}\right\} p\left(y_k/y_{0:k-1}, A_{k,l}\right)} \tag{11.72}$$

where $A_{k,l}$ denotes a joint event $C_{k,i} \cap D_{k,j}$ and $n_k^* = n_{d_k} \cdot n_{\bar{k}}$, $l = (i-1) \cdot n_{\bar{k}} + j$. It can be proved that:

$$p\left(A_{k,l}/y_{0:k}\right) = \frac{P\left\{A_{k,l}/y_{0:k-1}\right\} p\left(y_k/y_{0:k-1}, A_{k,l}\right)}{\sum_{l=1}^{n_k^*} P\left\{A_{k,l}/y_{0:k-1}\right\} p\left(y_k/y_{0:k-1}, A_{k,l}\right)} \tag{11.73}$$

therefore the posterior density is given by:

$$p\left(x_k/y_{0:k}\right) = \sum_{l=1}^{n_k^*} P\left\{A_{k,l}/y_{0:k}\right\} p\left(x_k/y_{0:k}, A_{k,l}\right) \tag{11.74}$$

The major drawback of the proposed algorithm is exponential growth of the number of components in a posterior density (11.74), since $n_k^* = n_{v_k} \cdot n_{\bar{k}} = n_{v_k} \cdot \left(n_{d_k} \cdot n_{k-1}\right)$. To solve this problem, after updating the posterior density $p\left(x_k/y_{0:k}\right)$ we apply a mixture reduction procedure to prevent exponential explosion of the number of mixture components.

11.4.2 Mixture Reduction

We apply the algorithm which clusters the components of the mixture and replace the cluster with a single Gaussian. The component of the mixture (11.74) with largest probability $w_{k,j}$ is selected as the principal component, and components that are close to it are clustered. Closeness of components is defined using Kull-back distance between two Gaussians:

$$D_i^2 = \frac{1}{2}\left(\hat{x}_c - \hat{x}_i\right)^T \left(P_c^{-1} + P_c^{-2}\right)\left(\hat{x}_c - \hat{x}_i\right) + \frac{1}{2} tr\left(P_1^{-1}P_2 + P_1 P_2^{-1} - 2I\right) \tag{11.75}$$

where w_c, $\hat{x}_c$ and P_c are the probability, mean and covariance of the principal component and w_i, $\hat{x}_i$ and P_i are the probability, mean and covariance of the ith component. A component for which $D_i^2 < T_{min}$ is selected as a class member. Threshold T_{min} defines the acceptable modification of the original distribution (11.75).

The cluster of components is approximated by a single Gaussian:

$$w_c = \sum_{i \in I_C} w_i \tag{11.76}$$

$$\hat{x}_c = \frac{\sum_{i \in I_C} w_i \hat{x}_i}{w_c} \tag{11.77}$$

$$P_c = \frac{\sum_{i \in I_C} w_i \left(P_i + \hat{x}_i \hat{x}_i^T\right) - \hat{x}_c \hat{x}_c^T}{w_c} \tag{11.78}$$

where I_C contains the indices of components close to the principle component.

The clustering procedure continues on the remaining components of the original mixture. If the number of components after clustering is below the user-defined maximal number N_{max} the mixture reduction is completed. Otherwise the minimum distance is incremented $T_{min} = T_{min} + \Delta T$ and the clustering procedure is repeated. ΔT is selected as a compromise between a number of iterations required and the possibility of clustering more components than necessary.

11.4.3 Implementation of Gaussian Sum Filters

The prediction density of Gaussian sum filter $p(x_k/y_{0:k-1})$ is obtained by solving the integral:

$$p(x_k/y_{0:k-1}, Dk, l) = \int p\left(x_k/x_{k-1,B_{k,i}}\right) \cdot p\left(x_{k-1}/y_{0:k-1}, A_{k-1,i}\right) dx_{k-1} \quad (11.79)$$

for $i = 1, \ldots, n_d$ and $j = 1, \ldots, n$, where n_d is the number of Gaussian components in dynamic noise pdf, and n is number of components in hidden state pdf. Solving integral (11.79) involves propagating Gaussian random variables $x_{k-1}/y_{0:k}, Ak - 1, i \sim N\left(\hat{x}_{k-1,i}, P_{x_{k-1}}\right)$ through nonlinear dynamic equation. Consequently, the posterior density of hidden state in Gaussian Sum filter is obtained by propagating random variables $x_k/y_{0:k}, A_{k,l} \sim N\left(\hat{x}_{\bar{k},l}, P_{x_k,l}\right)$ through observation equation. Since Extended Kalman Filter, Divided Difference Filter and Unscented Kalman Filter can be considered as tools for propagating Gaussian random variables through state space model equations, we have implemented Gaussian sum filter as a bank of parallel EKFs, or DDFs or UKFs.

11.5 On-line Adaptation of the Recurrent Radial Basis Function Network Structure

On line adaptation of the structure of Recurrent Radial Basis Function (RRBF) network is implemented by combining growing and pruning of hidden neurons and hidden and output connections [10, 12–17].

11.5.1 RRBF Network Growing

Growing of the network by adding new hidden neurons is performed when the adaptation of the connection weights is not sufficient to ensure tracking of the dynamics. The test for adding a new hidden neuron is obtained by applying Kalman filter consistency test which states that the innovations should be acceptable as zero mean and should have magnitude commensurate with the theoretical covariance as yielded by the filter [16, 17]. A new hidden neuron should be added if (a) the consistency test is not satisfied and (b) only *specialized* hidden neurons are activated by the current network input. A hidden neuron is referred to as *specialized* if its input and output parameters have accumulated certain level of knowledge, and new observations cannot significantly improve it. The moment of neuron specialization is determined based on the number of samples that have activated the neuron above certain threshold.

11.5.2 RRBF Network Pruning

During adaptation to a time-varying environment some connections or hidden neurons may become insignificant and should be pruned. A connection is insignificant if its parameter and the parameter change are both insignificant. The well-known pruning method OBS [9], ranks synaptic weights according to the saliency, defined as the change in the training error when the particular parameter is eliminated. The parameter with the smallest saliency is pruned. However, OBS was developed for the off-line trained networks with fixed training and test set. We have derived an analogous on-line pruning method for RRBF network [16, 17], by establishing the relation between the parameter saliency and the statistical significance of the parameter. Additional criterion is introduced in order to test the statistical significance of time-varying parameters. Since inverse of the Hessian of the cost function, needed for the significance test, is recursively updated by the Kalman filter, the pruning method does not significantly increase the overall computation complexity of the learning algorithm.

Parameter pruning implies elimination of the corresponding rows and columns in the Hessian of the cost function, which inverse is estimated as posterior covariance matrix. Using the partitioned matrix inversion lemma a new covariance matrix is obtained [16, 17]. The specialized or rarely activated hidden neurons are pruned if all of its output connections have statistically insignificant parameters.

11.6 Examples

In first two examples derived training algorithms are evaluated on tasks of long term prediction of chaotic time series. Long term prediction was obtained by iterating the neural network output, that is, by feeding the current output of the net-work back to the input through the recurrent connections. As a measure of the difference between original chaotic signal and long term (iterated) prediction of the recurrent neural network we have used the *Normalized Root Mean Squared Error* (NRMSE):

$$NRMSE = \sqrt{\frac{1}{\sigma^2 N} \sum_{k=1}^{N} \left(y_k - \hat{y}_k^-\right)^2} \tag{11.80}$$

In (11.80) σ stands for the standard deviation of clean time series, y_k is the true value of sample at time step k, and $\hat{yk}$ is prediction of a recurrent neural net-work.

11.6.1 Non-local Optimization

When Gaussian Sum (GS) filter is applied as the RNN training algorithm the pdf of RNN hidden variables (synaptic weights and neuron outputs), is approximated by the weighted sum of Gaussians. Each Gaussian component of such sum is rep-resented by the estimation of the mean and covariance of the RNN hidden variables and each component is updated in time using Gaussian filter: EKF, DDF or UKF. GS filters were implemented as banks of parallel Gaussian filters giving GS_EKF, GS_DDF and GS_UKF as training algorithms. The several different RNNs were trained in the same time (each component of the sum represents one RNN) and the estimates were combined (after being updated independently) in order to reduce the final number of components. Therefore, GS filters applied as training algorithms can be considered as non-local optimization techniques.

In this example we compare GS versus Gaussian filters as RNN training algorithms, applied in long term prediction of a Logistic Map chaotic behavior. The observations were obtained by iterating the difference equation:

$$x_k = 4 \cdot x_{k-1} \left(1 - x_{k-1}\right) \tag{11.81}$$

and scaling x_k into the range $[-1, 1]$.

NARX recurrent radial basis function network with 3 hidden neurons and 2 recurrent connections between output and hidden neuron was used as the basic architecture for training. In Table 11.1. we give the mean and variance of NRMSE as well as time needed to learn 2000 samples sequentially when Gaussian (EKF, DDF, UKF) and Gaussian sum filters (GS_EKF, GS_DDF, GS_UKF) were used as training algorithms. The results, were obtained for 30 independent runs with different initial values of adaptive parameters. The number of components in Gaussian mixture which represented the pdf of hidden variables (synaptic weights and delayed RNN output) was 10. Dynamic and observation noise were represented by Gaussian.

The mean of NRMSE for Gaussian sum filters is by two orders of magnitude smaller than for Gaussian filters and the rate of variances is even more impressive. The reason for such results is that GS filter starts with 10 different RNNs in parallel and

Table 11.1 NRMSE of NARX_RRBF (NARX recurrent neural network with Gaussian hidden neurons) long term prediction of Logistic Map chaotic time series

	mean(*NRMSE*)	*var*(*NRMSE*)
DDF	0.133	1.14e-2
UKF	0.136	1.07e-2
EKF	0.187	1.23e-2
GS_DDF	7.52e-3	1.26e-7
GS_UKF	7.86e-3	6.62e-7
GS_EKF	3.64e-2	7.3e-3

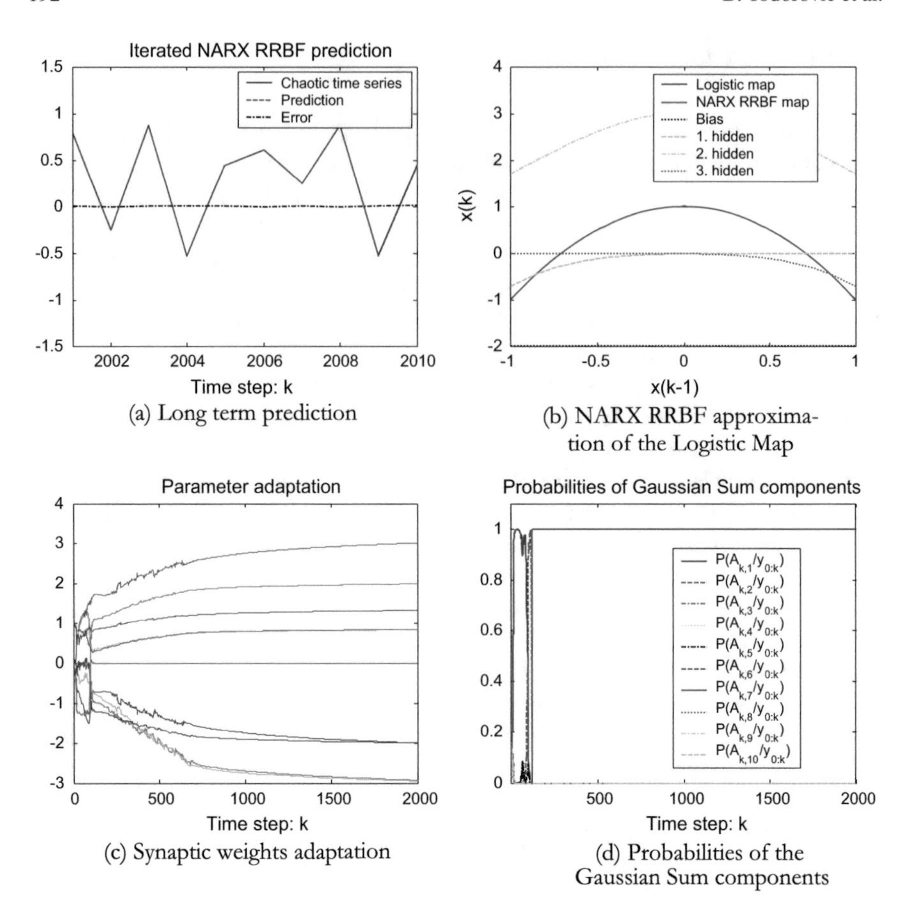

(a) Long term prediction

(b) NARX RRBF approximation of the Logistic Map

(c) Synaptic weights adaptation

(d) Probabilities of the Gaussian Sum components

Fig. 11.3 NARX RRBF (Recurrent Radial Basis Function) prediction of Logistic Map chaotic time series (Training algorithm GS_DDF)

updates each independently for every sample. In this case we do not have the growth of number of components and the condition that only components with significant probabilities should survive, proved to be very useful.

As it is shown in Fig. 11.3d after short period of training, algorithm selected only one recurrent neural network architecture as the best one (one with the highest probability) and training was continued with only one component in a mixture. That is the reason why training with Gaussian sum filters did not took significantly longer time than training with Gaussian filters [11].

11.6.2 Prediction of Noisy Chaotic Time Series (Non-gaussian Multimodal Noise)

In this example we have considered a long term-iterated prediction of Mackey Glass time series, corrupted with non-Gaussian multimodal observation noise. The chaotic Mackey-Glass differential delay equation:

$$\dot{x}(t) = \frac{0.2 \cdot x(t-\tau)}{1 + x(t-\tau)^{10}} - 0.1 \cdot x(t) \tag{11.82}$$

is integrated using a fourth-order Runge-Kutta method for $\tau = 30$ and sampled with period 6 to obtain the values of x at discrete time steps. The chaotic attractor of the clean time series is given in Fig. 11.4a.

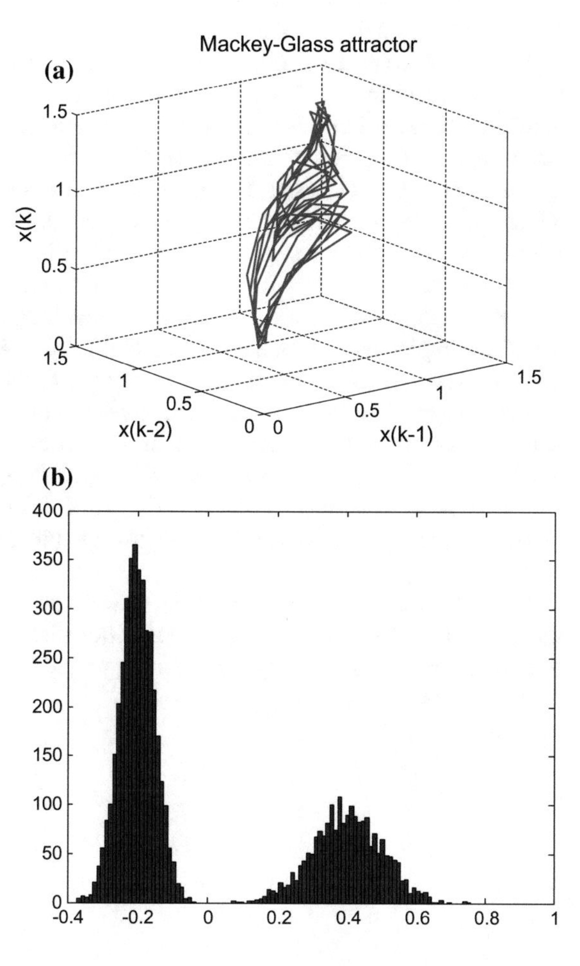

Fig. 11.4 **a** Chaotic Mackey-Glass attractor; **b** Histogram of multimodal observation noise (11.83)

Table 11.2 NRMSE of NARX_RMLP (NARX recurrent network with sigmoidal hidden neurons) iterated prediction of noisy Mackey-Glass chaotic time series *mean*(*NRMSE*) *var*(*NRMSE*)

	mean(*NRMSE*)	*var*(*NRMSE*)
GS_DDF	0.190	9.60e-3
GS_UKF	0.281	1.44e-2
GS_EKF	0.266	4.8e-2

Finally, the values of x were corrupted by multimodal observation noise to obtain the observations. The pdf of multimodal observation noise was Gaussian mixture given by:

$$p(v_k) = \alpha \cdot N\left(v_k; m_1, \sigma_1^2\right) + (1 - \alpha) \cdot N\left(v_k; m_2, \sigma_2^2\right) \tag{11.83}$$

where $\alpha = 2/3$, $m_1 = -0.2$, $\sigma_1 = 0.05$, $m_2 = 0.4$, $\sigma_2 = 0.1$. The histogram of (11.83) is given in Fig. 11.4b.

We have used GS_EKF, GS_UKF and GS_DDF to train NARX recurrent network with 6 recurrent inputs, 5 hidden sigmoidal neurons and one output (41 adaptive parameters). Gaussian filters failed to produce long term prediction be-cause the Gaussian approximation of multimodal observation nose was not good enough.

Performances of Gaussian sum filters for non-Gaussian noise are documented in Table 11.2. After sequential training on 2000 samples, recurrent neural network was iterated for next 100 time steps. The mean and variance of NRMSE is obtained from 30 independent runs with different initial values of adaptive parameters.

Figure 11.5 shows a typical result of iterated long term prediction for Gaussian sum filter implemented as a bank of parallel DDF's. All Gaussian sum filters used a five-component Gaussian mixture for state posterior, Gaussian process noise and two-component Gaussian mixture of observation noise. During training only noisy data were presented to the learning algorithm. In Fig. 11.5a attractor obtained from noisy data is shown. Attractor of the NARX RMLP (NARX recurrent network with sigmoidal hidden neurons) is shown in Fig. 11.5b. The evolution of the probabilities of the Gaussian mixture components is shown in Fig. 11.5c and long term prediction is shown in Fig. 11.5d.

Next two examples will illustrate the effectiveness of algorithm for on line structure adaptation. Algorithm uses statistics, sequentially estimated by Bayesian filters, to derive criteria for growing and pruning of hidden neurons and connections in recurrent neural networks.

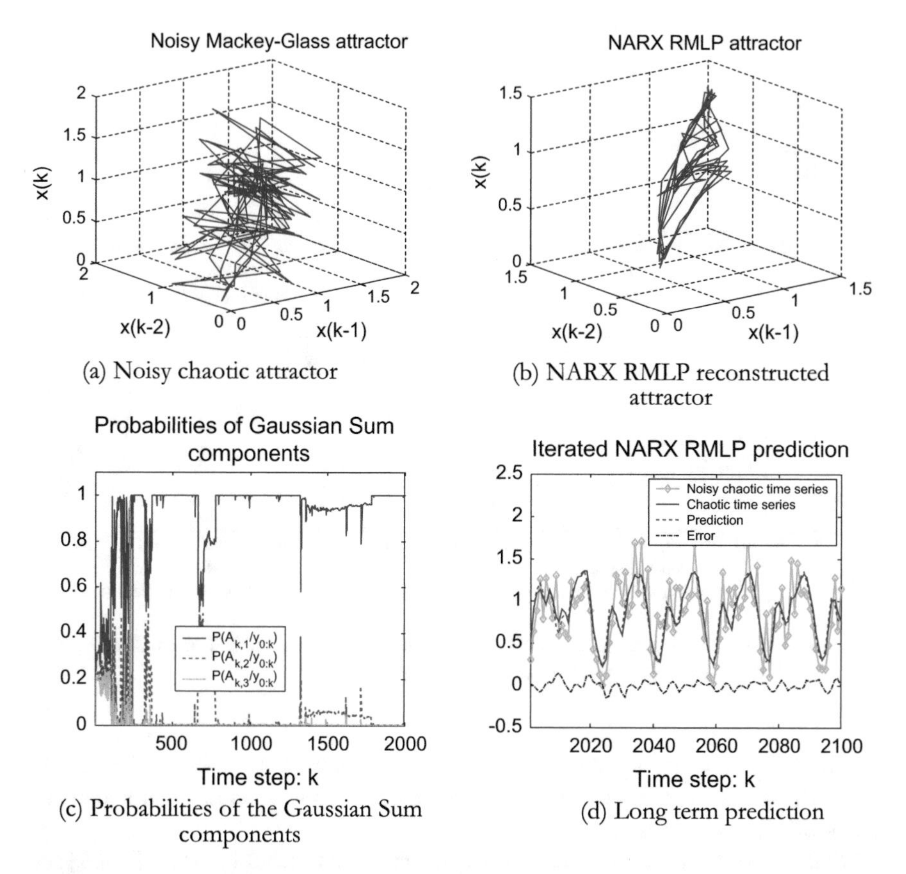

(a) Noisy chaotic attractor

(b) NARX RMLP reconstructed attractor

(c) Probabilities of the Gaussian Sum components

(d) Long term prediction

Fig. 11.5 NARX RMLP (NARX recurrent network with sigmoidal hidden neurons) prediction of noisy Mackey-Glass chaotic time series (Training algorithm GS_DDF)

11.6.3 On Line Estimation of State, Structure and Noise Variance

In this example, we have applied recurrent radial basis function network with adaptive structure for single step prediction of time series obtained from chaotic system described by Lorenz equations:

$$\dot{x} = -\beta x + yz \tag{11.84}$$

$$\dot{y} = \sigma(-y + z) \tag{11.85}$$

$$\dot{z} = -xy + \rho y - z \tag{11.86}$$

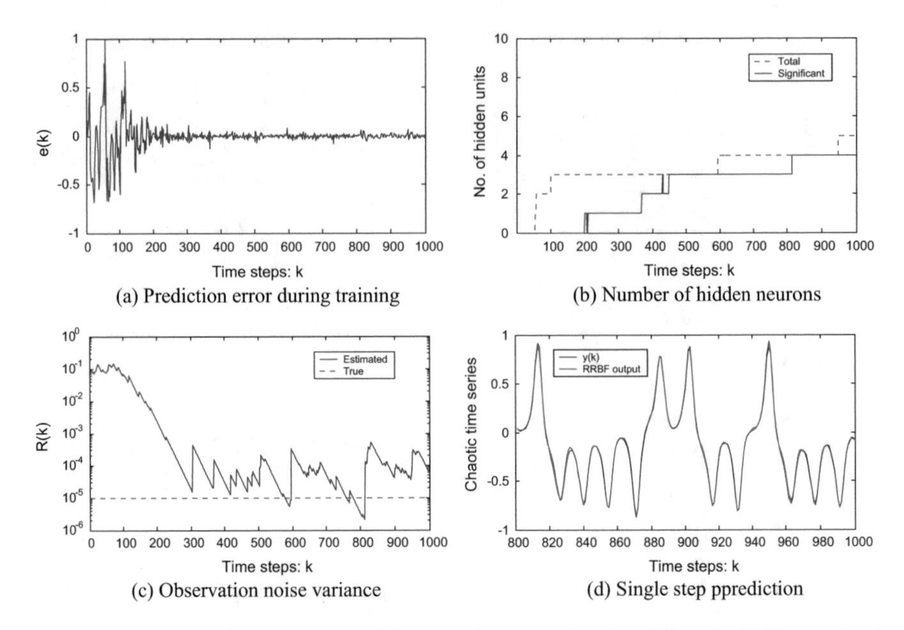

(a) Prediction error during training

(b) Number of hidden neurons

(c) Observation noise variance

(d) Single step pprediction

Fig. 11.6 On line learning of NARX RRBF state and structure (number of Gaussian hidden neurons) for single step prediction of Lorenz chaotic time series

where β, σ and ρ are adjustable parameters. The equations were iterated by the 4th order Runge Kutta method, where step size of 0.01 was used and the parameters are set at $\beta 8/3$, $\sigma = 10$ and $\rho = 28$.

State y (sampled at a period of 0.05 seconds) was scaled between $[-1, 1]$ and Gaussian noise $v_k \sim N\left(0, 1 \cdot e^{-5}\right)$ was added to obtain the observations. Time series of 1000 noisy observation was used for training. Each sample was presented only once to the RNN training algorithm derived from Extended Kalman Filter.

The results of on line adaptation of NARX RRBF parameters, states and structure for Lorenz time series single step prediction are illustrated in Fig. 11.6. The RRBF network output error (innovation $e\,(k)$) during training is shown in Fig. 11.6a. Growth pattern (Fig. 11.6b) shows the number of hidden neurons during training. On-line estimation of observation noise variance is shown in Fig. 11.6c, and comparison of original time series and RRBF network prediction is given in Fig. 11.6d.

All statistical tests for adding or pruning had the size $\alpha = 0.05$. Neurons were added with the following initial parameters: $\alpha_{n_H+1} = e\,(k), m_{n_H+1} = s(k)$ and $\sigma_{n_H+1} = \sigma_0$, where σ_0 is the initial width. Initial variances of parameters were chosen so that initial parameter values were statistically insignificant: $P_{pp}\,(0)^{-1/2}\,\hat{x}_p\,(0) < \gamma$. Hidden neuron was considered rarely activated if the ratio between the number of data samples that have activated a neuron above $\phi_{min} = 0.1$, and the total number of data samples from the moment when neuron was added, was less than 0.01.

11.6.4 Resolving Noise/Non-stationarity Dilemma

This example will illustrate the ability of the training algorithm to resolve dilemma whether novelty in a new training example comes from the nonstationarity of the problem, or because the level of noise on data has been increased. We consider the identification of the plant described by difference equation:

$$x_k = f\left(x_{k-1}, x_{k-2}, x_{k-3}, u_{k-1}, u_{k-2}\right) \tag{11.87}$$

where function f has the form:

$$f\left(x_1, x_2, x_3, x_4, x_5\right) = \frac{x_1 x_2 x_3 x_4 x_5 \left(x_3 - 1 - \alpha\right) + \left(1 - \beta\right) x_4}{1 + x_2^2 + x_3^2} \tag{11.88}$$

The parameters α and β ware equal to zero for the first 800 observations. For the next 400 samples parameters are $\alpha = \beta = 0.005k - 4$, and for the last 1300 observations $\alpha = \beta = 2$. The observations of the plant output are obtained for input:

$$u_k = sin\left(\frac{2\pi k}{250}\right) \text{ if } 2mN < k < \left(2m + 1\right) N, \tag{11.89}$$

$$u_k = 0.8sin\left(\frac{2\pi k}{250}\right) + 0.2\left(\frac{2\pi k}{25}\right) \text{ if } \left(2m + 1\right) N < k < \left(2m + 2\right) N, \tag{11.90}$$

where $m = 0, 1, 2$ and $N = 500$. For the first 1500 time steps, the variance of the measurement noise was chosen to give $SNR = 30dB$, and after that the variance was chosen to give the $SNR = 5dB$. The results of identification of a dynamic system (51), obtained by sequential adaptation of the RRBF network over 2500 data samples, are illustrated in Fig. 11.7.

The learning algorithm has detected the non-stationarity (for $800 < k < 1200$), by adding new hidden neurons, which soon became significant. The change of the measurement noise variance at $k = 1501$, also caused addition of new hidden neurons. However, they did not become significant and therefore they were susceptible to pruning (Fig. 11.7c). As a consequence, number of significant (permanent) neurons hasn't increased. At the same time the estimate of the measurement noise variance is updated to track the change (Fig. 11.7d)

11.7 Concluding Remarks

We considered on line training of recurrent neural networks as sequential (recursive) Bayesian estimation of synaptic weights, neuron outputs and structure. In their simplest form, approximate recursive Bayesian estimators can be considered as a second order on line optimization algorithms which utilize the recursive estimate

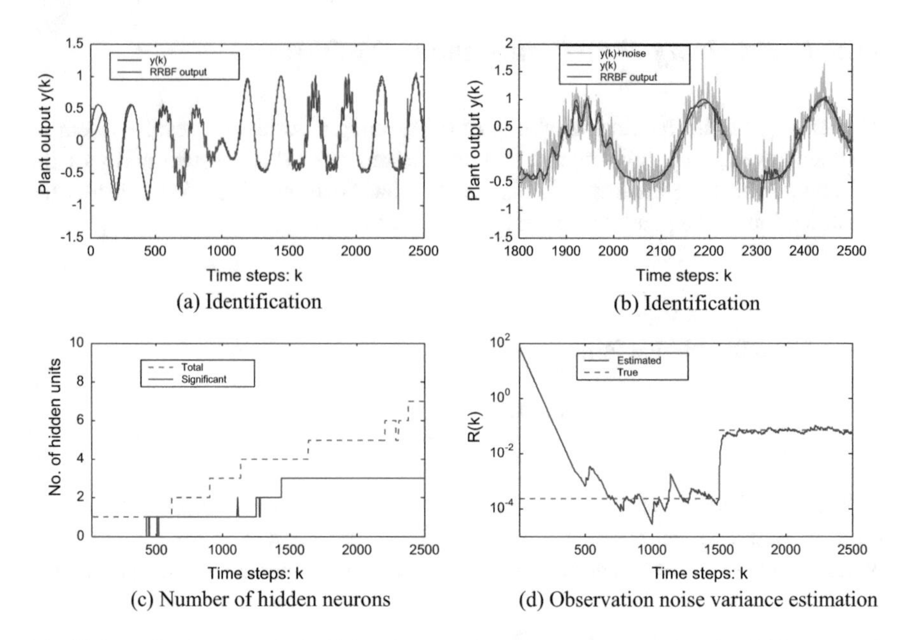

Fig. 11.7 On line learning of NARX RRBF state and structure (number of Gaussian hidden neurons) for non stationary dynamic system identification

of the inverse of the objective function hessian. Joint estimation of the parameters (synaptic weights) and states (neuron out-puts) generalizes the heuristic known as teacher forcing, by estimating the outputs of hidden neurons and by filtering out the noise from training data samples and using only the 'deterministic part' of the data for on line training.

Better approximation of state probability density function using Gaussian mixtures, enables deriving on line training algorithms for recurrent neural networks, which are capable to deal with non-Gaussian (multi modal or heavy tailed) noise on training data. Statistics, recursively updated during state estimation, can be efficiently used for deriving criteria for growing and pruning of synaptic connections and hidden neurons in recurrent neural networks. All derived algorithm have been empirically evaluated on problems of dynamic system modeling and short and long term time series prediction.

References

1. Williams, R. J. & Zipser, D.: Gradient-based learning algorithms for recurrent connectionist networks, TR NU_CCS_90-9, Boston: Northeastern University, CCS, 1990.
2. Williams, R. J.: Some observations on the use of the extended Kalman filter as a recurrent network learning algorithm, TR NU_CCS_92-1. Boston: Northeastern University, CCS, 1992.

3. Todorović B., Stanković M., Moraga C.: Derivative Free Training of Recurrent Neural Networks – A Comparison of Algorithms and Architectures, NCTA 2014 – Proceedings of the International Conference on Neural Computation Theory and Applications, part of IJCCI 2014, Rome, Italy, October, 2014, pp. 76–84.
4. Todorović B., Stanković M., Moraga C.: Recurrent Neural Networks Training Using Derivative Free Nonlinear Bayesian Filters, Computational Intelligence, in: Merelo, J. J.; Rosa, A.; Cadenas, J. M.; Dourado, A.; Madani, K.; Filipe, J. (eds.): *Computational Intelligence*, Proceedinga of the International Joint Conference, IJCCI 2014 Rome, Italy, October 22-24, 2014 Revised Selected Papers Berlin: Springer (Studies in Computational Intelligence, vol. 620), 2015, pp. pp 383–410.
5. Horne, B. G., Giles, C. L.: An experimental comparison of recurrent neural networks, *Advances in Neural Information Processing Systems*, vol. 7, 1995, pp. 697–704.
6. Nørgaard, M., Poulsen, N. K., and Ravn, O.: Advances in derivative free state estimation for nonlinear systems, Technical Report, IMM-REP-1998-15, Department of Mathematical Modelling, DTU, revised April 2000.
7. Julier, S. J., Uhlmann, J. K.: A new extension of the Kalman filter to nonlinear systems, Proceedings of AeroSense, The 11th international symposium on aero-space/defence sensing, simulation and controls, Orlando, FL, 1997.
8. Todorović, B., Stanković, M., Moraga C.: Nonlinear Bayesian Estimation of Recurrent Neural Networks, Proceedings of IEEE 4th International Conference on Intelligent Systems Design and Applications ISDA 2004, Budapest, Hungary, August 26–28, 2004, pp. 855–860.
9. Alspach, D. L. and Sorenson, H. W.: Nonlinear Bayesian Estimation using Gaussian Sum Approximation, *IEEE Transactions on Automatic Control*, vol. 17 (4), 1972, pp. 439–448.
10. Todorović,B., Stanković, M.: Sequential Growing and Pruning of Radial Basis Function network, Proceedings of IJCNN 2001, Washington DC, vol.3, 2001, pp., 1954–1959
11. Todorović B., Stanković M., Moraga C.: Gaussian sum filters for recurrent neural net-works training, NEUREL 2006: Eight Seminar on Neural Network Applications in Electrical Engineering, Proceedings, 2006, pp. 53–58.
12. Todorović, B., Stanković, M., Moraga C.: Modeling non-stationary dynamic systems using recurrent radial basis function networks, in Proc. of the 6th Seminar on Neural Network Applications in Electrical Engineering, NEUREL September 2002, Belgrade.
13. Todorović, B., Stanković, M., Moraga, C.: Extended Kalman Filter trained Recurrent Radial Basis Function Network in Nonlinear System Identification, in: Dorronsoro, José R. (ed.): *Artificial Neural Networks – ICANN 2002*, Proceedings of the International Conference, Madrid, Spain, August 2830, 2002, Berlin, Heidelberg: Springer (Lecture Notes in Computer Science Vol. 2415), pp 819–824.
14. Todorović, B., Moraga C., Stanković, M., Kovačević, B.: Neural Network training using Derivative Free Kalman filters, Proceedings of a Workshop on Computational Intelligence and Information Technologies, Nis, Serbia, October 13, 2003, pp. 39–46,
15. Todorović, B., Stanković, M., Moraga, C.: On-line Learning in Recurrent Neural Networks using Nonlinear Kalman Filters, in: Signal Processing and Information Technology, 2003. ISSPIT 2003. Proceedings of the 3rd IEEE International Symposium on Signal Processing and Information Technology (ISSPIT), Darmstadt, Germany, December 2003.
16. Todorović, B., Stanković, M., Moraga C.: On-line Adaptation of Radial Basis Function Networks using the Extended Kalman Filter, in: Sin k, P., Vaščák, J. and Hirota, K. (eds.): *Machine Intelligence: Quo Vadis? Advances in Fuzzy Systems-Applications and Theory*, vol. 21, World Scientific, pp 73–92, 2004.
17. Todorović, B., Stanković, M., Moraga C.: Extended Kalman Filter Based Adaptation of Time-varying Recurrent Radial Basis Function Networks Structure, in Machine In-telligence: Quo Vadis?, in: Sinčák, P., Vaščák, J. and Hirota, K. (eds.): *Machine Intelligence: Quo Vadis? Advances in Fuzzy Systems-Applications and Theory*, vol. 21, World Scientific, pp 115–124, 2004.

Chapter 12
Class-Memory Automata Revisited

Henrik Björklund and Thomas Schwentick

12.1 Introduction

Data words are an extension of traditional finite alphabet words, that have at each position, besides a symbol from some finite alphabet, a data value from some infinite domain. Data words and the corresponding model of data trees are being studied in database theory, e.g. as an abstraction of XML documents [3], and in verification, where data values might represent process ids [5, 9].

An early automata model for data words, nowadays usually called *register automata*, was introduced by Kaminski and Francez [11]. It has a polynomial-time membership problem [1], but its expressive power is somewhat limited. The non-emptiness problem and the combined complexity of the membership problem are **NP**-complete [16]. The inclusion problem is even undecidable [15].

In [2], *data automata* were introduced as a much more expressive model, e.g., capturing two-variable logic on data words, but still with decidable emptiness problem. In [1], *class-memory automata* were defined as an equally expressive model which, however, has a natural deterministic version. It was shown that class-memory automata are strictly more expressive than register automata.

Since then, several other automata models for data words have been proposed, e.g., in [4, 10, 13, 14, 17]. In [6], it was observed that there is a robust intermediate level of automata between register automata and class-memory automata, represented by class counting automata, non-reset history register automata, locally prefix-closed

H. Björklund
Umeå University, Umeå, Sweden

T. Schwentick (✉)
Theoretical Computer Science,
Technical University of Dortmund, Dortmund, Germany
e-mail: thomas.schwentick@tu-dortmund.de

R. Seising and H. Allende-Cid (eds.), *Claudio Moraga: A Passion for Multi-Valued Logic and Soft Computing*, Studies in Fuzziness and Soft Computing 349, DOI 10.1007/978-3-319-48317-7_12

data automata, and so-called weak class-memory automata. As expected, the complexity of decision problems for this intermediate level is intermediate as well: the emptiness problem for weak class-memory automata and containment and equivalence for deterministic class-memory automata are **EXPSPACE**-complete [6].

Whereas the inclusion of the intermediate level in the most expressive level is obvious since weak class-memory automata are a syntactic restriction of class-memory automata, the inclusion of the lower level in the intermediate level requires a proof. As mentioned above, it was shown in [1] that register automata are captured by class-memory automata and it is not hard to see that this proof actually only needs weak class-memory automata. Unfortunately, the presentation of this proof in [1] was a bit sketchy and it was brought to our attention by colleagues that some details were described in a misleading way. On the other hand, other work has relied on this proof, e.g., the proof of Proposition 22 in [17] uses the same proof technique and leaves the details to the reader.

The main purpose of this paper is to give a much more detailed and polished proof for a somewhat stronger result. We show in Sect. 12.3 that weak class-memory automata are strictly more expressive than the extension of register automata by the ability to guess data values.

In Sect. 12.4 we address another shortcoming of [1]. As pointed out in [7], the extension of class-memory automata by Presburger acceptance conditions proposed in [1] (seemingly) fails to capture the expressivity of class-memory automata. We therefore give a slightly modified definition of Presburger class-memory automata and show that it has the desired properties.

Acknowledgments. We are grateful to Luc Segoufin and Ahmet Kara for pointing us to the shortcomings in the presentation of the proof of Theorem 12.3.1 in [1]. We thank Mikołaj Bojańczyk, Ahmet Kara, Anca Muscholl and Luc Segoufin for many valuable discussions.

12.2 Preliminaries

Let Σ be a finite alphabet and Δ an infinite set. A **data word** is a finite sequence over $\Sigma \times \Delta$. A **data language** is a set of such words. For each data value d, the set of all positions with value d is called a **class** of w. Unless otherwise stated, data values can only be compared with respect to equality.

In the sequel, we will assume without loss of generality that all data languages and automata we investigate are defined over the same data set Δ, which contains all data values used in examples and proofs. In particular $\mathbb{N} \subseteq \Delta$. We will also talk about data languages over Σ, where Σ is a finite alphabet, implicitly assuming that the data set is Δ.

12.2.1 Register Automata

Register automata were introduced by Kaminski and Francez [11] and have later been studied in, e.g., [9, 15]. They were defined for sequences of data values only, but the generalization to data words is straightforward. Register automata are equipped with a constant number of registers in which they can store data values, which can later be compared with the data value of the current position.

The class of languages that RA recognize is closed under union, intersection, concatenation, and Kleene star, but not under complementation and reversal [11]. The nonemptiness problem and the combined complexity of the membership problem are **NP**-complete [16].

Partially in order to overcome the lacking closure under reversal, Kaminski and Zeitlin [12] defined *data guessing register automata*, an extension of register automata, that can nondeterministically guess new data values and assign them to registers.[1]

In the formal definition of this model, given next, the set S in a transition contains those registers into which new data values are loaded. These values need to be pairwise distinct and different from the current values stored in the registers. The definition of transition steps requires that in each step the data value d that is being read is already stored in some register i (and then $i \notin S$) or it is not yet there but newly stored in this step ($i \in S$). In this, our definition differs somewhat from the one in [12], but it is easily seen to be equivalent.

Definition 12.2.1 A **data-guessing register automaton (DGRA)** over finite alphabet Σ is a tuple $R = (Q, q_0, F, k, P)$, where Q is a finite set of states, q_0 is the initial state, F are the accepting states, k is the number of registers, and P is a finite set of transitions. A transition is a tuple (p, a, S, i, p'), where $p, p' \in Q$, $a \in \Sigma$, $S \subseteq \{1, \ldots, k\}$, and $i \in \{1, \ldots, k\}$.

A **configuration** of R is a pair (q, τ), where $q \in Q$ and $\tau : \{1, \ldots, k\} \to \Delta \cup \{\bot\}$ is a **register assignment**. The initial configuration is (q_0, τ_0), where $\tau_0(i) = \bot$ for all $i \in \{1, \ldots, k\}$.

We refer to the components of a transition $\delta = (p, a, S, i, p')$ by $\delta.from, \delta.symb$, $\delta.Set$, $\delta.reg$ and $\delta.to$, respectively. If $\delta.reg = i$, we say that δ *reads from register* i.

A sequence[2] $\rho \stackrel{\text{def}}{=} (p_0, \tau_0) \xrightarrow{\delta_1} \cdots \xrightarrow{\delta_n} (p_n, \tau_n)$ is a *run* of R for a data word $w = (a_1, d_1) \ldots (a_n, d_n)$, if the following holds.

(i) p_0 is the initial state of R,
(ii) $\tau_0(i) = \bot$, for every $i \in \{1, \ldots, k\}$, and
(iii) for every $t \in \{1, \ldots, n\}$, there are $i \in \{1, \ldots, k\}$ and $S \subseteq \{1, \ldots, k\}$ such that

(a) $\delta_t = (p_{t-1}, a_t, S, i, p_t)$,
(b) $\tau_t(i) = d_t$,

[1]These automata were called *look-ahead finite-memory automata* in [18] and *finite-memory automata with non-deterministic reassignment* in [12].

[2]Without the arrow notation, the sequence might be understood as $p_0, \tau_0, \delta_1, p_1, \tau_1, \ldots, \delta_n, p_n, \tau_n$.

(c) for every $j \in \{1, \ldots, k\} \setminus S$: $\tau_t(j) = \tau_{t-1}(j)$,
(d) for all $i, j \in S$ with $i \neq j$: $\tau_t(i) \neq \tau_t(j)$, and
(e) for every $i \in S$: $\tau_t(i) \neq \bot$ and $\tau_t(i)$ does not occur in $\tau_{t-1}(\{1, \ldots, k\})$.

The run is *accepting* if $p_n \in F$. A data word w belongs to the language $L(R)$ of the automaton if there is an accepting run of R for w.

It should be noted that other definitions of RAs allow a nonempty initial assignment. This makes it possible to consider a finite number of constants, an ability that isn't needed when we have a finite as well as an infinite set of data values.

The definition ensures that a data value can never occur in more than one register at the same time. In particular, this feature can be used to verify that the current data value is different from those in the registers.

Example 12.2.1 Consider the data language over the finite alphabet $\Sigma = \{a\}$ that contains all data words such that the first data value is unique, i.e., the value that appears at the first position doesn't appear anywhere else in the word. This language can be recognized by an RA (and thus also by a DGRA). The automaton just has to store the first data value it reads into a register and then verify that all subsequent values differ from it.

The *reversal* of this language, however, i.e., the language of all data words in which the *last* data value is unique, cannot be recognized by an RA. For a DGRA the reversal constitutes no problem. It just guesses the last value at the start and verifies that the guessed value appears nowhere in the word except at the last postition. □

12.2.2 Class-Memory Automata

We next recall the definition of class-memory automata.

Definition 12.2.2 ([1]) A **class-memory automaton** (CMA) C is a tuple $(Q, \Sigma, \delta, q_I, F_L, F_G)$, where Q is a finite set of states, Σ is a finite alphabet, q_I is the initial state,

- $\delta : (Q \times \Sigma \times (Q \cup \{\bot\})) \to \mathscr{P}(Q)$ is a **transition function**; and
- $F_G \subseteq F_L \subseteq Q$ are the sets of **globally and locally accepting states**, respectively.

The semantics of class-memory automata is defined through the notion of class-memory functions. Such a function simply assigns to every data value d the state of the automaton that was assumed after reading the last (previous) position with value d. More formally, a **class-memory function** is a function $f : \Delta \to Q \cup \{\bot\}$ such that $f(d) \neq \bot$ for only finitely many d. A **configuration** of C is a pair (q, f) where $q \in Q$ and f is a class-memory function. We call q the **global state** of C and $f(d)$ the **local state** of d. The initial configuration of A is (q_I, f_I), where $f_I(d) = \bot$ for all $d \in \Delta$. When reading a pair $(a, d) \in \Sigma \times \Delta$, the automaton can go from configuration (q, f) to (q', f') if (1) $q' \in \delta(q, a, f(d))$, (2) $f'(d) = q'$, and (3) for all $d' \neq d$, $f'(d') = f(d')$. The automaton accepts if, for the final configuration $(q, f), q \in F_G$ and $f(d) \in F_L \cup \{\perp\}$, for all $d \in \Delta$. A CMA is **deterministic** if each $\delta(p, a, q)$ is a singleton.

It should be noted that δ naturally induces a *transition relation* which is a subset of $(Q \times \Sigma \times (Q \cup \{\perp\})) \times Q$. We freely switch back and forth between these two points of view.

Example 12.2.2 As an example, consider processes sharing a printer with three kinds of events: a print job can be requested (r), start (s), and terminate (t). Let L_0 be the set of **valid traces**, i.e., the data words w with the following two properties.

(global)
: The string projection of w, that is, the string over Σ resulting from deleting all data values, matches the expression $(r^*sr^*tr^*)^*$.

(local)
: For each class of w, the string projection of the data word induced by this class matches $(rst)^*$.

We construct a CMA C that accepts the language L_0. The automaton C, depicted in Fig. 12.1, has four states

- p (the printer is printing for the current process),
- i (the current process is neither printing nor waiting for a print),
- wi (the current process is waiting for a print and the printer is idle)
- wb (the current process is waiting for a print but the printer is busy)

Edge label $(s, \{wi, wb\})$ indicates that the transition can be taken reading symbol s, in case the class-memory is wi or wb. Likewise for $(r, \{i, \perp\})$. The initial state is i and the accepting states are given by $F_G = F_L = \{i\}$.

To get a better understanding of the automaton let us have a look at the transitions leaving p. In state p the automaton just read some (s, d) reflecting the start of a print for the process number d. Thus, there are only two possible kinds of next data symbols: either (t, d) which ends the print of process d or (r, d') which moves process $d' \neq d$ into the waiting state. It should be noted that no (s, d') could be read next.

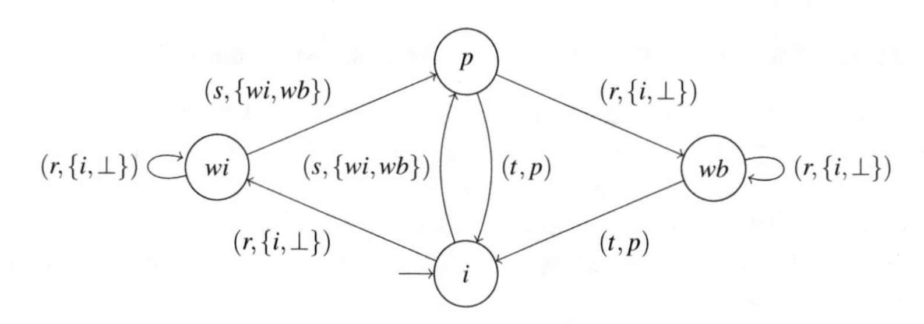

Fig. 12.1 A CMA for L_0. The labels are explained in Example 12.2.2

It is not hard to see that there is no DGRA for L_0. For the sake of a contradiction, assume that DGRA R accepts L_0. Let k be the number of registers of R. Let w be the data word $(r, 1) \cdots (r, k + 1)(s, 1)(t, 1) \cdots (s, k + 1)(t, k + 1)$. As $w \in L_0$, R has an accepting run ρ on w. After reading the first $k + 1$ positions of w, there is at least one data value $d \in \{1, \ldots, k + 1\}$ that does not occur in any register of R. Let $d' \in \mathbb{N}$ be a data value that does not occur in ρ at all. We can conclude that R also accepts the string $w' \notin L_0$ resulting from w by replacing $(s, d), (t, d)$ with $(s, d'), (t, d')$. Should d be guessed in ρ at some point between $(s, 1)$ and $(t, d - 1)$, then the new accepting run can guess d' instead. □

We will assume in the following that a DGRA R always guesses data values for all registers in its first step, i.e., for each transition (q_0, a, S, i, p') where q_0 is the initial state, $S = \{1, \ldots, k\}$. It is easy to see that a DGRA R' which does not meet this convention can be adapted accordingly: R can guess values for all registers and keep track of the set of registers, for which the original R' would not have guessed values. For each run of R' there is a corresponding run of R which guesses additional values that do not occur otherwise and therefore do not interfere. On the other hand, if an "incorrect" value is guessed initially, that interacts with a subsequently read position, then R' can just abandon that run.

Cotton-Barratt, Murawski and Ong suggested to study a restriction of class-memory automata which they called *weak* [6]. The restriction requires that $F_L = Q$, that is, that all local states are accepting and therefore the acceptance of a data word only relies on the (global) state at the end of a run. In a nutshell, this means that one can, in general, not require that all classes reach certain goals (states). However, classes can still be required to avoid certain behavior and therefore "safety conditions for classes" can be tested. It was shown in [6] that weak CMAs are exactly as expressive as Class Counting Automata [14], non-reset History Register Automata [17] and locally prefix-closed Data Automata [8].

12.3 Data-Guessing Register Automata and CMAs

In this section, we demonstrate that CMAs are strictly stronger than data-guessing register automata. Our goal is to establish the following theorem.

Theorem 12.3.1 *Weak CMAs are strictly more expressive than DGRAs.*

The strictness of the containment follows by an easy adaptation of Example 12.2.2. If we set $F_L = Q$ in C, then the automaton accepts the set L_1 of all prefixes of data words in L_0. However, the non-expressibility proof also works for L_1. The containment follows from Propositions 12.3.1 and 12.3.2 below. We will show that every DGRA R can be simulated by a CMA C. Before we describe the construction of C we need to define some notation.

A sequence $\pi \stackrel{\text{def}}{=} p_0 \stackrel{\delta_1}{\rightarrow} \cdots \stackrel{\delta_n}{\rightarrow} p_n$ is a *pre-run* of R for a data word $w = (a_1, d_1) \ldots (a_n, d_n)$, if it satisfies conditions (i) and (iiia) of the definition of a run from Definition 12.2.1. Each run induces a unique pre-run by dropping the register assignments τ_i. A position t whose transition δ_t reads register i is called an *i-position*.

An *interval* $I = (P, i, d)$ over a data word w consists of a contiguous set P of positions of w (the *position set of* I), a *register number* $i \in \{1, \dots, k\}$, and a *data value* $d \in \Delta$ (Fig. 12.2). We denote the three components of an interval I by $I.Pos$, $I.reg$, and $I.val$, respectively. We call I a j-interval, if $I.reg = j$, a d-interval if $I.val = d$ and a (d, j)-interval if it is both a d- and an j-interval. We refer to the minimal position $(\min(I.P))$ and the maximum position $(\max(I.P))$ of an interval I by $I.min$ and $I.max$, respectively.

We say that two intervals I and J *overlap* if $I.Pos \cap J.Pos \neq \emptyset$ and that J *meets* I if $J.max + 1 = I.min$.

An *interval structure* $\mathscr{I}$ of dimension k for a word w is a set of intervals over w such that, for every $i \in \{1, \dots, k\}$, the position sets of the i-intervals in $\mathscr{I}$ partition the set of positions of w.

A *witness structure* for a register automaton R with k registers and a data word w is a pair $(\pi, \mathscr{I})$ consisting of a pre-run π of R for w and an interval structure $\mathscr{I}$ of dimension k for w.

A witness structure $W = (\pi, \mathscr{I})$ for R and w is *valid* if it satisfies the following conditions.

(W1) The last state p_n of π is an accepting state of R.
(W2) For each i-interval I, we have $i \in \delta_t.Set$, where $t = I.min$. Further more, $i \notin \delta_s.Set$ for any $s \in \{I.min + 1, \dots, I.max\}$.
(W3) In every i-interval I, all transitions reading from register i read data value $I.val$. That is, for all $t \in \{I.min, \dots, I.max\}$, if $\delta_t.reg = i$ then $d_t = I.val$.
(W4) If I and J are intervals with the same data value then I and J do neither overlap nor meet.

Proposition 12.3.1 *For every DGRA R and every data word w it holds that $w \in L(R)$ if and only if there exists a valid witness structure W for R and w.*

Proof **(if)** From $W = (\pi, \mathscr{I})$ it is straightforward to construct a sequence $\rho \stackrel{\text{def}}{=} (p_0, \tau_0) \stackrel{\delta_1}{\rightarrow} \cdots \stackrel{\delta_n}{\rightarrow} (p_n, \tau_n)$: The p_t and δ_t can just be taken from π and the register assignments can be defined as follows: $\tau_0(i) \stackrel{\text{def}}{=} \bot$, for every register number i. For every $t \in \{1, \dots, n\}$ and register number i, $\tau_t(i)$ is just $I.val$ for the unique i-interval I from $\mathscr{I}$ with $t \in I.Pos$.

It remains to show that ρ is actually an accepting run of R on w. To verify this, we check that conditions (i)–(iii) from Definition 12.2.1 are satisfied. Condition (i) holds because π is a pre-run. Condition (ii) holds by construction of τ_0. We show that for each $t \in \{1, \dots, n\}$ conditions (iiia)–(iiie) hold (with $\delta_t.reg$ and $\delta_t.Set$ as i and S, respectively):

- Condition (iiia) holds since π is a pre-run.
- Condition (iiib) is guaranteed by (W3).
- Condition (iiic) follows from condition (W2): guess-transitions for a register i can only occur at first positions of i-intervals.

- Conditions (iiid) and (iiie) are guaranteed by (W4).

Finally, ρ is accepting due to (W1).

(only-if) It is straightforward to obtain a valid witness structure $W = (\pi, \mathscr{I})$ from an accepting run ρ for w. The pre-run π is just the pre-run induced by ρ. The interval structure $\mathscr{I}$ is constructed as follows.

There can be four kinds of i-intervals:

(A) from position 1 to position n, if $i \notin \delta_t.Set$, for every $t \in \{2, \ldots, n\}$;
(B) from position 1 to position $t-1$, if t is the minimal number in $\{2, \ldots, n\}$, for which $i \in \delta_t.Set$;
(C) from position t, for which $i \in \delta_t.Set$, to n, if t is the maximal number in $\{1, \ldots, n\}$, for which $i \in \delta_t.Set$;
(D) from a position t to a position $t' - 1 \geq t$, if $i \in \delta_t.Set$, $i \in \delta_{t'}.Set$ and $i \notin \delta_s.Set$, for every $s \in \{t+1, \ldots, t'-1\}$.

It is easy to see that $\mathscr{I}$ is an interval structure. Condition (W1) is satisfied because ρ is accepting. Condition (W2) holds by definition of $\mathscr{I}$. Condition (W3) is guaranteed by the semantics of DGRAs: inside an i-interval R always has to read the same data value from register i. Finally, (W4) is also guaranteed by the semantics of DGRAs: A data value that is guessed in step t is not allowed to be in any register after step $t-1$. □

In the following we describe how, given a DGRA R and a data word w, to construct a CMA C that tests whether there exists a valid witness structure for R and w.

Proposition 12.3.2 *From every DGRA R a weak CMA C can be computed such that, for every data word w, C accepts w if and only if there is a valid witness structure for R and w.*

Proof In principle, C will be constructed to guess a witness structure W for R and w and to verify that it is valid. However, C has to represent W in a somewhat implicit fashion, as will be detailed below.

We think of C as consisting of three layers. The first layer guesses a pre-run of R and tests its consistency. The second layer verifies that the interval structure induced by the pre-run and the data word w (in a way to be described soon), satisfies conditions (W2) and (W3). The third layer is responsible for the verification of condition (W4) with the help of two coloring schemes.

We will describe the three layers of C and argue their correctness, separately.

As already mentioned, the first layer of C guesses a pre-run of R for w. To this end, each state q of C has the following two components.

- $q.state$, a state of R.
- $q.trans \in P$, a transition (p, a, S, i, p') of R. For brevity,[3] we refer to the components of $q.trans$ by $q.from$, $q.symb$, $q.Set$, $q.reg$ and $q.to$, respectively.

[3]That is, we abbreviate, e.g., $q.trans.from$ by $q.from$.

It is straightforward that C can be defined such that it checks that its sequence of states indeed induces an accepting pre-run of R on the given data word w. Furthermore, the first layer succeeds if and only if (W1) holds.

The other two layers of C are built on top of this layer. In the following we will therefore assume that, for the input data word w, there is already a sequence $q_0, \ldots, q_n$ of "pre-states" of C with a, - and a *trans*-component. We will refer to the state after reading position t of w by q_t. In analogy to pre-runs of R we call a position t with $q_t.reg = i$ an *i-position*.

With a pre-run π of R, guessed by C, we associate an interval structure $\mathscr{I}_\pi$ as follows. First, for each register i, the position sets of i-intervals are determined just as in the (only-if)-proof of Proposition 12.3.1 (points (A)–(D)). We call an i-interval of $\mathscr{I}_\pi$ *idle* if it does not contain any i-positions.

For each non-idle i-interval I, we write $I.firstR$ for the first (smallest) i-position in I and $I.lastR$ for the last (largest) i-position in I.

For each non-idle interval I, its data value $I.val$ is defined as $d_{I.firstR}$. For idle intervals I, we let $I.val$ be some data value that does not occur in w and is not assigned to any other interval.[4]

It is easy to verify that $\mathscr{I}_\pi$ satifies condition (W2). Further more, if a witness structure for R and w exists at all, there also exists one for which the idle intervals have the same data values as in $\mathscr{I}_\pi$. The only property ever used of data values of an idle interval I is that they are different from all other values occurring in registers during I.

The second layer of C maintains, for every register $i \in \{1, \ldots, k\}$ a *mode*. More precisely, each state q has, for every register i, a component $q.i.mode$ which can be either *head*, *mid*, or *tail*. C makes sure that the mode-components of its state always respect the following two conditions.

(C1) For each register i and each non-idle i-interval I of $\mathscr{I}_\pi$, and each position t it holds

- $q_t.i.mode = head$ if $t \in \{I.min, \ldots, I.firstR - 1\}$,
- $q_t.i.mode = mid$ if $t \in \{I.firstR, \ldots, I.lastR - 1\}$, and
- $q_t.i.mode = tail$ if $t \in \{I.lastR, \ldots, I.max\}$.

For idle intervals I it holds $q_t.i.mode = tail$ for all $t \in I.Pos$.

(C2) For each register i and each i-position t,

- if $t = I.firstR$, for some i-interval I of $\mathscr{I}_\pi$, then $f_{t-1}(d_t) = \perp$ or there is a $j \in \{1, \ldots, k\}$ such that $f_{t-1}(d_t).reg = j$ and $f_{t-1}(d_t).j.mode = tail$,
- otherwise $f_{t-1}(d_t).reg = i$ and $f_{t-1}(d_t).i.mode = mid$.

We remark that towards condition (C1), C only needs to guess (and verify) for each register i, which i-positions are the last i-positions of their i-interval. Condition (C2) can be tested instantly as well. We recall that $f_{t-1}(d_t) = q_s$ where $s < t$ is the maximal position with $d_s = d_t$.

[4]For concreteness, we can assume that the set of data values is ordered and that the ℓ-th idle interval (in some canonical ordering) gets the ℓ-th data value that does not occur in w as its value.

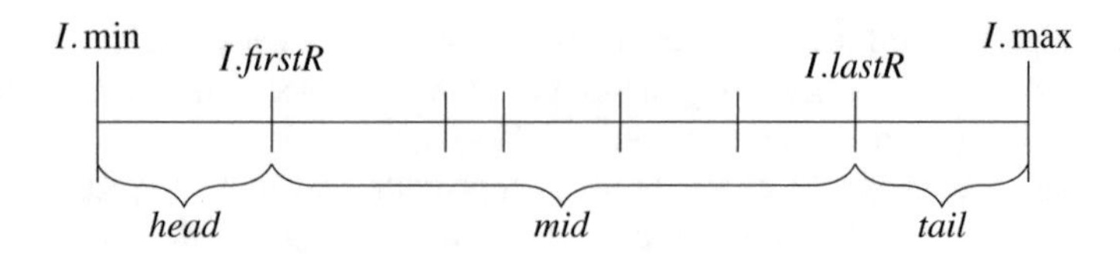

Fig. 12.2 Illustration of an interval

It is straightforward that if $\mathscr{I}_\pi$ is valid then C can guess the *mode*-components in a way that ensures (C1) and (C2).

Claim 1 *If (C1) and (C2) are satisfied then (W3) holds for* $\mathscr{I}_\pi$.

To prove this claim, we first show:

($*$) for each $t \in \{1, \ldots, n\}$ and each $i \in \{1, \ldots, k\}$,

(1) for every $d \in \Delta$, if $f_t(d).reg = i$ then $f_t(d).i.mode \in \{tail, mid\}$,
(2) there is at most one value $d \in \Delta$ such that $f_t(d).reg = i$ and $f_t(d).i.mode = mid$, and
(3) if such a d exists, then $q_t.i.mode = mid$.

Condition (1) holds by (C1) since for each i-position t of an i-interval I it clearly holds $t \geq I.firstR$. The proof of (2) and (3) is by induction on t with a straightforward induction basis $t = 1$.

For the induction step we consider a position $t > 1$. The functions f_{t-1} and f_t can only differ for d_t. We distinguish the three cases that (i) t is the first i-position of the i-interval I at t, (ii) t is some other i-position of I, or (iii) that t is not an i-position.

In case (i) we assume towards a contradiction that there is a value $d \neq d_t$ with $f_t(d).reg = i$ and $f_t(d).i.mode = mid$. As f_t and f_{t-1} agree besides d_t, it follows $f_{t-1}(d).reg = i$
and $f_{t-1}(d).i.mode = mid$. Then, by induction, $q_{t-1}.i.mode = mid$. However, from (C1) we can conclude that $q_{t-1}.i.mode = head$ if $I.\min < t$ or $q_{t-1}.i.mode = tail$ if $I.\min = t$, the desired contradiction. Therefore, we can conclude (2) and, since $f_t(d_t) = q_t$, also (3), in this case.

In case (ii) we conclude, by (C2), that $f_{t-1}(d_t).reg = i$ and $f_{t-1}(d_t).i.mode = mid$. The existence of a value $d \neq d_t$ with $f_t(d).reg = i$ and $f_t(d).i.mode = mid$ would yield $f_{t-1}(d).reg = i$ and $f_{t-1}(d).i.mode = mid$ and therefore contradict the induction hypothesis for $t-1$. Therefore, we can again conclude (2) and, since $f_t(d_t) = q_t$, also (3).

In case (iii), there would be two distinct data values d, different from d_t with $f_t(d).reg = i$ and $f_t(d).i.mode = mid$. However, both of them would also fulfill $f_{t-1}(d).reg = i$ and $f_{t-1}(d).i.mode = mid$ and therefore contradict the induction hypothesis for $t-1$. Thus (2) follows again and, by induction, also (3). This concludes the proof of ($*$).

Now we are ready to show that $\mathscr{I}_\pi$ satisfies condition (W3) if (C1) and (C2) hold. Towards a contradiction, let us assume that there is some i-interval I with two

i-positions $s < t$ such that $d_s \neq d_t$. Without loss of generality we can assume that there is no other i-position between s and t. By condition (C2), $f_{t-1}(d_t).reg = i$ and $f_{t-1}(d_t).i.mode = mid$. On the other hand, $q_s.reg = i$ and, by (C1), $q_s.i.mode = mid$. By our choice of s and t there are no i-positions $s' \in \{s+1, \ldots, t-1\}$. By (C2) there can be no position $s' \in \{s+1, \ldots, t-1\}$ with $d_{s'} = d_s$ either: the first such position would be a j-position for some $j \neq i$ and would require $f_{s'}(d_s).i.mode = tail$ thanks to (C2).

Therefore, $f_{t-1}(d_s) = q_s$ and thus there are two data values d (namely: d_s and d_t) with $f_{t-1}(d).reg = i$ and $f_{t-1}(d).i.mode = mid$, contradicting $(*)$. Thus, (W3) indeed holds and we have shown Claim 1.

It remains to describe the third layer of C. As mentioned before, it uses two coloring schemes to prevent that intervals with the same data value overlap or meet. To this end, C guesses, for each interval I of $\mathcal{I}_\pi$ an *overlap color*, which can be blackor yellowand a *meet color*, blueor white. These colors are guessed at the first position of I (that is: $I.\min$) and are stored in $q.i.oc$ and $q.i.mc$, respectively, for $i \stackrel{\text{def}}{=} I.reg$.

C then tests that the following conditions hold throughout $\mathcal{I}_\pi$.

(C3) For all intervals I, J, where J is the predeccessor interval of I, it holds: if M is the unique interval with $J.lastR \in M.Pos$ and $M.reg = I.reg$, then M and I have different overlap colors;

(C4) For all intervals I, J, where J is the predeccessor interval of I, it holds: if M is the unique interval with $I.min - 1 \in M.Pos$ and $M.reg = J.reg$, then M and J have different meet colors.

Here, J is the *predecessor interval of I* if $J.val = I.val$, $J.lastR < I.firstR$ and there is no position t with $d_t = I.val$ in $\{J.lastR + 1, \ldots, I.firstR - 1\}$.

Informally, (C3) states that the i-interval M containing the last reading position of the predeccessor interval J with the same data value as some i-interval I has a different overlap color than I, and hence must be different from I. Similarly, (C4) states that if the predeccessor interval J with the same data value as I has register j, the j-interval M containing the last position before I has a different meet color than J, and hence must be different from J.

Condition (C3) is tested by C whenever the first i-position t of an i-interval I is encountered. It holds if $f_{t-1}(d_t) = \bot$ or $f_{t-1}(d_t).i.oc \neq q_t.i.oc$.

Condition (C4) is also tested whenever the first i-position t of an i-interval I is met. However, at this point the j-interval M that contains $I.min - 1$ might already be closed. Therefore, C guesses, at the beginning of I (at position $s \stackrel{\text{def}}{=} I.min$) a register number $q_s.i.pr$ ($\stackrel{\text{def}}{=}$: j) and stores the meet color $q_{s-1}.j.mc$ of that register in $q_s.i.pmc$. Then it verifies at position t that $f_{t-1}(d_t) = \bot$ or (with $\ell \stackrel{\text{def}}{=} f_{t-1}(d_t).reg$) that $q_{t-1}.i.pr = \ell$ and $q_{t-1}.i.pmc \neq f_{t-1}(d_t).\ell.mc$.

It remains to show that the intervals of a valid witness structure can always be colored such that (C3) and (C4) hold and that, conversely, (C1–4) enforce (W4).

For the former, let us assume that $(\pi, \mathscr{I}_\pi)$ is valid. We call an interval I *oc-dependent on an interval* M, if I, M and the predeccessor J of I satisfy the precondition of (C3). Likewise, J is *mc*-dependent on M, if I, M and J satisfy the precondition of (C4).

It is easy to see that thanks to the validity of $(\pi, \mathscr{I}_\pi)$ an interval is never dependent on itself: indeed the assumption $I = M$ in the precondition of (C3) immediately yields that J and I overlap, contradicting (W4). Likewise, $J = M$ in the precondition of (C4) yields that I and J meet or overlap, again contradicting (W4).

Furthermore, if I is *oc*-dependent on M then $M.max < I.min$. Therefore, the directed graph of the "*oc*-depends on"-relation is acyclic and every interval can be *oc*-dependent on at most one other interval. Therefore, there is a straightforward way to assign overlap colors:

- color each interval black that is not *oc*-dependent on any other interval;
- then successively color each interval I that is *oc*-dependent on some colored interval M by the opposite overlap color of M.

Meet colors can be assigned analogously by observing that if J is *mc*-dependent on M then $M.min > J.max$.

Finally, we show that (C1–4) imply (W4). Towards a contradiction let us assume that $(\pi, \mathscr{I}_\pi)$ satisfies (C1–4) but there are intervals I, J and a data value d with $J.val = d = I.val$ such that I and J overlap or meet.

Let us first assume that I and J overlap, more specifically, that $J.firstR < I.firstR$ but $J.max \geq I.min$. Furthermore, we can assume that I and J are chosen such that $I.firstR$ is minimal. As all intervals with the same register number constitute a partition, we can concude that $I.reg \neq J.reg$, and we set $i \stackrel{\text{def}}{=} I.reg$ and $j \stackrel{\text{def}}{=} J.reg$.

There can be no position with data value d between two consecutive j-positions of J since the first such position t would have $f_{t-1}(d).reg = j$ and $f_{t-1}(d).j.mode = mid$, contradictioning (C2). In particular, there are no i-positions of I between any j-positions of J.

Since $J.firstR < I.firstR$, we can therefore conclude that $J.lastR < I.firstR$. Thanks to the minimal choice of I and J, there are no positions $t \in \{J.lastR + 1, \ldots, I.firstR - 1\}$ with $d_t = d$.

We distinguish two cases.

(1) I starts after the last j-position of J.
(2) The last j-position of J is in I.

We consider case (1) first. Since $J.firstR < I.firstR$ and $J.max \geq I.min$ it holds that $I.min - 1 \in J.Pos$. As $J.lastR < I.firstR$, the j-interval M in the precondition of (C4) is J, contradicting the fact that J and M have different meet colors—the desired contradiction. Similarly, case (2) yields a contradiction, as in this case the i-interval M in the precondition of (C3) is just I, again a contradiction as I and M have different overlap colors.

Finally, we assume that I and J meet, more specifically, that $J.max = I.min - 1$. However, similarly as before, the interval M in the precondition of (C4) is J, yielding a contradiction as M and J have different meet colors.

Therefore, (C1–4) guarantee (W4).

Altogether, we have shown that

1. C can verify conditions (C1–4),
2. if there is a valid witness structure for w then $(\pi, \mathscr{I}_\pi)$ and the colorings can be chosen such that (C1–4 hold, and
3. if (C1–4) hold then $(\pi, \mathscr{I}_\pi)$ is valid,

and therefore the proof of the proposition is complete by observing that C is indeed a weak CMA. □

12.4 A Robust Extension of Deterministic CMAs

As was shown in [1], the languages recognized by deterministic CMAs are closed under intersection, but not under union, complementation, concatenation, or Kleene star. In the same article, we therefore proposed an automaton model, called *Presburger CMA*. The class of languages recognized by deterministic Presburger CMA is closed under Boolean operations. As was pointed out by Decker et al. [7], Presburger CMA, as defined in [1] do not capture the same class of languages as CMA. Here, we remedy this situation by redefining Presburger CMA in such a way that

1. nondeterministic Presburger CMA have exactly the same expressive power as CMA, and
2. the languages accepted by *deterministic* Presburger CMA is closed under boolean operations.

Presburger CMA differ from CMA only with respect to the acceptance condition. We no longer use global and local accepting states, but rather require that the final configuration satisfies a limited Presburger formula Φ. A **limited Presburger formula** over a state set Q is a Boolean combination of atomic formulas of the following kinds, for states $q, q_1, \ldots, q_k$ from Q and constant numbers c, c'.

(1) q,
(2) $(q_1 + \cdots + q_k) = c$, and
(3) $(q_1 + \cdots + q_k \mod c') = c$.

Given a configuration (p, f) of a CMA with state set Q, we define $g(q)$ to be the number of data values d such that $f(d) = q$, i.e., $g(q) = |f^{-1}(q)|$. Configuration (p, f) satisfies

(1) q if $p = q$,
(2) $(q_1 + \cdots + q_k) = c$ if $g(q_1) + \cdots + g(q_k) = c$, and
(3) $(q_1 + \cdots + q_k \mod c') = c$ iff $(g(q_1) + \cdots + g(q_k) \mod c') = c$.

A run of a Presburger CMA is accepting if its final configuration satisfies Φ.

Proposition 12.4.1 *For each Presburger CMA there is an equivalent CMA and vice versa.*

Proof (*sketch*).It is easy to see that the acceptance condition of a CMA C can be mimicked by a Presburger CMA: the formula Φ can be chosen as a conjuction $\Phi_1 \wedge \Phi_2$, where $\Phi_1 = \bigvee_{q \in F_G} q$ and $\Phi_2 = \bigwedge_{q \notin F_L} (q = 0)$.

For the other implication, let C be a Presburger CMA with formula Φ. We assume, without loss of generality, that in Φ negation occurs only before atomic formulas. We construct a CMA C' that is equivalent to C. It simulates C while keeping track of which atoms of Φ are satisfied after each step of a run. The globally accepting states of C' are then simply those that represent a truth assignment to the atoms that satisfies Φ.

Atoms of the form q or $\neg q$ are easily dealt with – the automaton simply has to remember what state of C the simulation is currently in.

Atoms of the form $(q_1 + \cdots + q_k \mod c') = c$ or $(q_1 + \cdots + q_k \mod c') \neq c$ are dealt with by keeping a modulo c' counter that is incremented each time the simulation enters a state in $\{q_1, \ldots, q_k\}$ and decremented each time a data value "leaves" $\{q_1, \ldots, q_k\}$, i.e., when the simulated transition $(q, a, q') \to q''$ is such that $q' \in \{q_1, \ldots, q_k\}$.

For atoms of the forms $(q_1 + \cdots + q_k) = c$ and $(q_1 + \cdots + q_k) \neq c$ we apply a slightly more complicated construction. First, we construct a modified Presburger CMA $\hat{C}$ from C that has, for each state q of C, two states q and $\hat{q}$. For each transition $(p, a, p') \to q$ of C, there is an additional transition $(p, a, p') \to \hat{q}$. Furthermore, there are additional transitions $(\hat{p}, a, p') \to \hat{q}$ and $(\hat{p}, a, p') \to q$. That is, the copy $\hat{q}$ of a state q has the same behavior as q with one exception: it can only be assumed at the last position of a class, since there are no transitions of the form $(p, a, \hat{q}) \to q$.

To keep track of an atom $(q_1 + \cdots + q_k) = c$ it then suffices that C' counts in its state the number of classes in which a state from $\{\hat{q}_1, \ldots, \hat{q}_k\}$ is assumed, up to at most c, and that $q_1, \ldots, q_k$ become locally rejecting states.

An atom $(q_1 + \cdots + q_k) \neq c$ can be treated as the disjunction of the atoms $(q_1 + \cdots + q_k) = 0, \ldots, (q_1 + \cdots + q_k) = c - 1$ and $(q_1 + \cdots + q_k) > c$. The former can be tested as just described. To test $(q_1 + \cdots + q_k) > c$ it suffices that C' verifies that at least $c + 1$ times a state from $\{\hat{q}_1, \ldots, \hat{q}_k\}$ is assumed. □

Proposition 12.4.2 *The class of languages accepted by deterministic Presburger CMAs is closed under Boolean operations.*

Proof Let $A = (Q, \Sigma, \Delta, \delta, q_I, \Phi)$ and $B = (P, \Sigma, \Delta, \gamma, p_I, \Psi)$ be deterministic CMAs. To construct an automaton that accepts the complement of $L(A)$, we only need to negate Φ. For the intersection or the union of $L(A)$ and $L(B)$, we construct the product automaton $A \times B$ and use the conjunction (or the disjunction) of Φ and Ψ, where, e.g., in the atomic modulo formulas each q of A is replaced by the sum of those q' of $A \times B$ with q in their A-component. □

Acknowledgments We gratefully acknowledge the support of the Swedish Research Council grant 621-2011-6080 and the EU FP7 Media in Context (MICO) project.

References

1. H. Björklund and T. Schwentick. On notions of regularity for data languages. *Theor. Comput. Sci.*, 411(4-5):702–715, 2010.
2. M. Bojanczyk, C. David, A. Muscholl, T. Schwentick, and L. Segoufin. Two-variable logic on data words. *ACM Trans. Comput. Log.*, 12(4):27, 2011.
3. M. Bojanczyk, A. Muscholl, T. Schwentick, and L. Segoufin. Two-variable logic on data trees and XML reasoning. *J. ACM*, 56(3), 2009.
4. B. Bollig. An automaton over data words that captures EMSO logic. In *CONCUR 2011 - Concurrency Theory - 22nd International Conference, CONCUR 2011, Aachen, Germany, September 6-9, 2011. Proceedings*, pages 171–186, 2011.
5. B. Bollig, A. Cyriac, P. Gastin, and K. N. Kumar. Model checking languages of data words. In *Foundations of Software Science and Computational Structures - 15th International Conference, FOSSACS 2012, Held as Part of the European Joint Conferences on Theory and Practice of Software, ETAPS 2012, Tallinn, Estonia, March 24–April 1, 2012. Proceedings*, pages 391–405, 2012.
6. C. Cotton-Barratt, A. S. Murawski, and C. L. Ong. Weak and nested class memory automata. In *Language and Automata Theory and Applications - 9th International Conference, LATA 2015, Nice, France, March 2–6, 2015, Proceedings*, pages 188–199, 2015.
7. N. Decker, P. Habermehl, M. Leucker, and D. Thoma. Learning transparent data automata. In *Application and Theory of Petri Nets and Concurrency - 35th International Conference, PETRI NETS 2014, Tunis, Tunisia, June 23–27, 2014. Proceedings*, pages 130–149, 2014.
8. N. Decker, P. Habermehl, M. Leucker, and D. Thoma. Ordered navigation on multi-attributed data words. In *CONCUR 2014 - Concurrency Theory - 25th International Conference, CONCUR 2014, Rome, Italy, September 2–5, 2014. Proceedings*, pages 497–511, 2014.
9. S. Demri and R. Lazic. LTL with the freeze quantifier and register automata. *ACM Trans. Comput. Log.*, 10(3), 2009.
10. D. Figueira. Forward-XPath and extended register automata on data-trees. In *ICDT*, pages 231–241, 2010.
11. M. Kaminski and N. Francez. Finite-memory automata. *Theor. Comput. Sci.*, 134(2):329–363, 1994.
12. M. Kaminski and D. Zeitlin. Finite-memory automata with non-deterministic reassignment. *Int. J. Found. Comput. Sci.*, 21(5):741–760, 2010.
13. A. Manuel, A. Muscholl, and G. Puppis. Walking on data words. In *Computer Science - Theory and Applications - 8th International Computer Science Symposium in Russia, CSR 2013, Ekaterinburg, Russia, June 25–29, 2013. Proceedings*, pages 64–75, 2013.
14. A. Manuel and R. Ramanujam. Counting multiplicity over infinite alphabets. In *Reachability Problems, 3rd International Workshop, RP 2009, Palaiseau, France, September 23–25, 2009. Proceedings*, pages 141–153, 2009.
15. F. Neven, T. Schwentick, and V. Vianu. Finite state machines for strings over infinite alphabets. *ACM Trans. Comput. Log.*, 5(3):403–435, 2004.
16. H. Sakamoto and D. Ikeda. Intractability of decision problems for finite-memory automata. *Theor. Comput. Sci.*, 231(2):297–308, 2000.
17. N. Tzevelekos and R. Grigore. History-register automata. In *Foundations of Software Science and Computation Structures - 16th International Conference, FOSSACS 2013, Held as Part of the European Joint Conferences on Theory and Practice of Software, ETAPS 2013, Rome, Italy, March 16–24, 2013. Proceedings*, pages 17–33, 2013.
18. D. Zeitlin. Look-ahead finite-memory automata. Master's thesis, Department of Computer Science, Technion - Israel Institute of Technology, 2006.

Chapter 13
Ensemble Methods for Time Series Forecasting

Héctor Allende and Carlos Valle

13.1 Introduction

Time series forecasting have received a great deal attention in many practical data mining of engineering and science. Traditionally, the autoregressive integrated moving average (ARIMA) model has been one of the most widely used linear models in time series prediction. However, the ARIMA model cannot easily capture the nonlinear patterns. Artificial neural networks (ANN) and other novel neural network techniques have been successfully applied in solving nonlinear pattern estimation problems.

Extensive works in literature suggest that substantial enhancement in accuracies can be achieved by combining forecasts from different models. However, forecast combination is a difficult as well as a challenging task due to various reasons, and often simple linear methods are used for this purpose. In this work, we propose a nonlinear weighted ensemble mechanism for combining forecasts from time series models.

Ensemble techniques for time series forecasting have indispensable importance in many practical data mining applications. It is an ongoing dynamic area of research, and over the years various forecasting models have been developed in the literature [1, 2]. These methods consider the individual as well as the correlations in pairs of forecasts for creating the ensembles without reducing its efficiency, simplicity, robustness and flexibility. However, this is not at all an easy task and so far no single model alone can provide best forecasting results for all kinds of time series data [2].

H. Allende (✉) · C. Valle
Departamento de Informática,
Universidad Técnica Federico Santa María, Valparaíso, Chile
e-mail: hallende@inf.utfsm.cl

C. Valle
e-mail: cvalle@inf.utfsm.cl

R. Seising and H. Allende-Cid (eds.), *Claudio Moraga: A Passion for Multi-Valued Logic and Soft Computing*, Studies in Fuzziness and Soft Computing 349, DOI 10.1007/978-3-319-48317-7_13

Combining forecasts from conceptually different methods is a very effective way to improve the overall precisions in forecast. The earliest use in the practice started in 1969 with the important work of Bates on Granger [3]. Since then, numerous forecast combination methods have been developed in the literature [4]. The vital role of model combination in time series forecasting can be credited in the following facts: (i) by adequate ensemble techniques, the forecasting strengths of the participating models aggregate and their weaknesses diminish, thus enhancing the overall forecasting accuracy to a great extent, (ii) it is frequently a large uncertainty about the optimal forecasting model, and in such situations combination strategies are the most appropriate alternatives to use, and (iii) combining multiple forecasting can efficiently reduce errors arising from faulty assumptions bias, or mistakes in the data.

The ordinary average is the most simply used forecast ensemble technique. It is easy to understand implement and interpret. However, this method is frequently criticized because it does not utilize the relative performances of the contribution models and is quite sensitive to the outliers. For this reason, other forms of averaging, such as median, winsorized mean and trimmed mean, have been addressed in the literature [5]. Another method is the weighted linear combinations of individual forecasts in which the weights are determined from the past errors of the contributing models [6].

However, this method completely ignores the possible relationships between two or more participating models and hence is not adequate for combining nonstationary models, which have been addressed by researchers [7].

The focus of this chapter will be ensembles for combining time series forecasting. Design, implementation and application will be the main topics of this chapter. Specifically, what will be considered are the conditions under which ensemble based systems may be more beneficial than their single machine; algorithms for generating individual components of ensemble systems and various procedures through which they can be combined. Various ensemble-based algorithms will be analyzed: Bagging, Adaboost and negative correlation; as well as combination rules and decision templates.

Finally future research directions will be included like incremental learning, machine fusion and the others areas in which ensemble of machines have shown promise.

13.2 Time Series Analysis

13.2.1 Linear Models

The statistical approach for forecasting involves the construction of stochastic models to predict the value of an observation x_t using previous observations. This is often accomplished using linear stochastic difference equation models, with random input. By far, the most important class of such models is the linear autoregressive integrate moving average (ARIMA) model. Here we provide a very brief review of the linear ARIMA-models and optimal prediction for these models. A more comprehensive treatment may be found for example in [1]. The seasonal ARIMA $(p, d, q) \times (P, D, Q)^S$ model for such time series is represented by

$$\Phi_P(B^S)\phi_p(B)\nabla_S^D \nabla^d x_t = \Theta_Q(B^S)\theta_q(B)\varepsilon_t \quad (13.1)$$

where $\phi_P(B^S)$ is the nonseasonal autoregressive operator of order p, $\theta_q(B)$ is the nonseasonal moving average operator of order q, $\Phi_P(B^S)$, $\Theta_Q(B^S)$ are the seasonal autoregressive and moving average operator of order P and Q and the terms x_t and ε_t are the time series and a white noise respectively. Moreover it is assumed that $E[\varepsilon_t | x_{t-1}, x_{t-2}, \ldots] = 0$. This condition is satisfied for example when ε_t are zero mean, independent and identically distributed and independent of past $x_t s$. It is assumed throughout that ε_t has finite variance σ^2. The backshift operator B shifts the index of a time series observation backwards, e.g. $Bx_t = x_{t-1}$ and $B^k x_t = x_{t-k}$. The order of the operator is selected by Akaike's information criterion (AIC) or by Bayes information criterion (BIC) [8] and the parameters $\Phi_1, \ldots, \Phi_P, \phi_1, \ldots, \phi_p$, $\Theta_1, \ldots, \Theta_Q$ and $\theta_1, \ldots, \theta_q$ are selected from the time series data using optimization methods such as maximum likelihood [1] or using robust methods such as recursive generalized maximum likelihood [9]. The ARMA-models is limited by the requirement of stationarity and invertibility of the time series. In other words, the system generating the time series must be invariant and stable. In addition, the residuals must be independent and identically distributed [10].

The ARMA models require a stationary time series in order to be useful for forecasting. The condition for a series to be weak stationary is that for all t

$$E[x_t] = \mu; \quad V[x_t] = \sigma^2; \quad COV[x_t, x_{t-k}] = \gamma_k. \quad (13.2)$$

Diagnostic checking of the overall ARMA models is done by the residuals. Several tests have been proposed, among them the most popular one seems to be the so-called portmanteau test proposed by [11], and its robust version by [12]. These tests are based on a sum of squared correlations of the estimated residuals suitably scaled.

13.2.2 *Non-linear Models*

Theory and practice are mostly concerned with linear methods and models, such as ARIMA models and exponential smoothing methods. However, many time series exhibit features which cannot be explained in a linear framework. For example some economic series show different properties when the economy is going into, rather than coming out of, recession. As a result, there has been increasing interest in non-linear models.

Many types of non-linear models have been proposed in the literature. For example, see bilinear models [13], classification and regression trees [14], threshold autoregressive models [15] and Projection Pursuit Regression [16]. The rewards from using non-linear models can occasionally be substantial. However, on the other side, it is generally more difficult to compute forecasts more than one step ahead [17].

Another important class of non-linear models is that of non-linear ARMA models (NARMA) proposed by [18], which are generalizations of the linear ARMA models to the non-linear case. A NARMA model obeys the following equations:

$$x_t = h(x_{t-1}, x_{t-2}, \ldots, x_{t-p}, \varepsilon_{t-1}, \ldots, \varepsilon_{t-q}) + \varepsilon_t, \tag{13.3}$$

where h is an unknown smooth function, and as in Sect. 13.2.1 it is assumed that $E\left[\varepsilon_t | x_{t-1},\ x_{t-2}, \ldots\right] = 0$ and that variance of ε_t is σ^2. In this case the conditional mean predictor based on the infinite past observation is

$$\hat{x}_t = E[h(x_{t-1}, x_{t-2}, \ldots, x_{t-p}, \varepsilon_{t-1}, \ldots, \varepsilon_{t-q}) | x_{t-1}, x_{t-2}, \ldots]. \tag{13.4}$$

Suppose that the NARMA model is invertible in the sense that there exists a function ν such as

$$x_t = \nu(x_{t-1}, x_{t-2}, \ldots) + \varepsilon_t. \tag{13.5}$$

Then given the infinite past of observations $x_{t-1}, x_{t-2}, \ldots$, one can compute the ε_{t-1} in (13.3) exactly:

$$\varepsilon_{t-j} = \kappa(x_{t-j}, x_{t-j-1}, \ldots),\ j = 1, 2, \ldots, q. \tag{13.6}$$

In this case the mean estimate is

$$\hat{x}_t = h(x_{t-1}, x_{t-2}, \ldots, x_{t-p}, \varepsilon_{t-1}, \ldots, \varepsilon_{t-q}), \tag{13.7}$$

where the ε_{t-j} are specified in terms of present and past x_u's. The predictor of (13.7) has a mean square error σ^2.

Since we have only a finite observation record, we cannot compute (13.6) and (13.7). It seems reasonable to approximate the conditional mean predictor (13.7) by the recursive algorithm

$$\hat{x}_t = h(x_{t-1}, x_{t-2}, \ldots, x_{t-p}, \hat{\varepsilon}_{t-1}, \ldots, \hat{\varepsilon}_{t-q}), \tag{13.8}$$

$$\hat{\varepsilon}_{t-j} = x_{t-j} - \hat{x}_{t-j},\ j = 1, 2, \ldots, q, \tag{13.9}$$

with the following initial conditions

$$\hat{x}_0 = \hat{x}_{-1} = \cdots = \hat{x}_{-p+1} = \hat{\varepsilon}_0 = \cdots = \hat{\varepsilon}_{-q+1} = 0. \tag{13.10}$$

For the special case of non-linear autoregressive model (NAR), it is easy to check that (13.3) is given by

$$x_t = h(x_{t-1}, x_{t-2}, \ldots, x_{t-p}) + \varepsilon_t. \tag{13.11}$$

In this case, the minimum mean square error (MSE) optimal predictor of x_t given $x_{t-1}, x_{t-2}, \ldots, x_{t-p}$ is the conditional mean (for $t \geq p + 1$).

$$\hat{x}_t = E[x_t | x_{t-1}, \ldots, x_{t-p}] = h(x_{t-1}, \ldots, x_{t-p}). \tag{13.12}$$

This predictor has mean square error σ^2.

13.2.3 Concept Drift

A challenging scenario for time series is when the underlying law of probability changes. It is known as *concept drift* and it refers to the fact that data is obtained from a non-stationary environment [19]. In the classification setting, we can describe concept drift as any scenario where the posterior probability of a class c_k that a given instance belongs, $P(c_k|x) = P(x|c_k)P(c_k)/P(x)$, changes over time, i.e. $P_{t+1}(c_k|x) \neq P_t(c_k|x)$. $P(x)$ describes probability of an instance $x \in \mathcal{X}$. $P(c_k|x)$ is the likelihood of observing a data point x within a particular class. $P(c_k)$, defines the class prior probabilities, and relates class balance to the overall underlying law of probability. Note that an observation of change in $P(x)$ might be insufficient to characterize a concept drift, because of its independence of the class labels.

There are two ways to make a model adapt to a concept-drift. The first one consists in adapting a single model so that it can learn example by example and adapt to a changing pattern. Another approach is to use an ensemble of models. Zang et al. [20] conclude that both have the same capacity to handle stream data, including the presence of concept drift. Incremental single algorithms have a better performance in terms of efficiency, but the ensembles adapt better to the concept drift and are more stable.

13.3 Elements of the Ensemble Architecture

Ensemble methods are based on the idea of combining a set of individual predictors $\{f_1, f_2, \ldots, f_N\}$, building a decision function F by means of an aggregation operator that combines the individual forecasts instead of the popular keep-the-best (KTB) model [21, 22]. Ensemble methods can differ in three different stages: manipulating data, manipulating the base learner, and the function that the ensemble uses to generate the "consensus" output.

13.3.1 Data Manipulation

This is the most popular way to introduce diversity in a ensemble. There are different forms to manipulate the data: Implicit diversity can be introduced by using a random sample S^t to train each learner n according to a distribution D which can be uniform as in Bagging [23], or S^n can be created by using a weight distribution over the examples on the training set, as in Wagging [24]. Explicit diversity approaches, like Adaboost [25] and its variants, compute the distribution of instances for each learner according to a metric which depends on the previous ensemble.

Another way to encourage diversity is manipulating either the inputs or the outputs. For example, input smearing [26] enhances the diversity of Bagging by adding Gaussian noise to the inputs. The amount of noise, and therefore the amount of diversity, is controlled by a parameter. Experimental results show that this technique can outperform Bagging in terms of generalization error. Rodriguez et al. [27] encourage diversity by randomly splitting the feature set into K disjoint subsets. Next, Principal Component Analysis (PCA) is applied to each subset retaining all principal components. On the other hand, Breiman proposes output smearing [28] that adds noise to the targets in the same way as input smearing.

The next two sub-categories involve *sub-sampling*, which is a widely-known technique commonly used to cut down on the memory size and the computational cost of the training process. But also this technique contributes to implicitly encourage the diversity in the ensemble, hence, improving performance. For example, Zhang and Berardi [22] propose a neural network ensemble with both systematic and serial partitioning to build the sub-samples. The first partition method divides data into k subsamples, where k is the input lag that the network uses. Each data partition consists of a set of k-lag input vectors, selected from all k-lag input vectors throughout the original training sample, with non-overlapping examples to feed each ensemble member. According to the authors, this subsampling scheme diminishes the correlation among the ensemble predictors. The second data partitioning technique divides data into k disjoint subsamples based on the chronological time sequence. Thus, each neural network trains with one of the subsamples, obtaining an ensemble of k models. On the other hand, Subagging [29, 30] trains each predictor by using a subsample of size m from the original training set of M examples without replacement. This approach has performed similarly to Bagging for high-dimensional datasets. Crogging [31] uses cross-validation (CV) samples to train each forecasting model. The authors test different cross validation methods to encourage diversity among the neural networks. Crogging averages the learner outputs to obtain the aggregated prediction. Experiments show that this algorithm outperforms Bagging and KBT model.

Finally, unlike horizontal partitioning methods, vertical (Feature Set) partitioning techniques keep all the training examples in each subsample; however, each of them contains a subset of the original set of features. Thus, each classifier learns a different projection of the training set. This diversity can reduce the correlation among the predictors [32]. However, in classification setting this methods might lead to imbalanced class problems [33]. Several researchers have proposed algorithms which belong to this category, for example, Tumer and Oza [34] present an ensemble algorithm which trains each predictor with a subset of features selected depending on the correlations between individual features and class labels. The authors claim that this approach reduces dimensionality and improves generalization performance. Bryll et al. [32] propose attribute bagging (AB) which selects each subset of features by randomly selecting the features without replacement. Then, the number of features is selected by computing the classification performance of different sized random subsets of attributes.

13.3.2 Manipulating the Base Learner

Introducing differences among the learners is a well-known strategy to add diversity, and there are different forms to apply it. First, depending on the optimization method, diversity can be encouraged in the ensemble by starting the search in the hypothesis space from different points. Since back-propagation can lead to local optima, assigning different initial weights to the neural networks that compound the ensemble [35] can implicitly generate diversity. It is generally accepted that this way of introducing diversity obtains poor performance [36]. The second way is to manipulate the inducer's parameters, for example, using neural networks as base learners, and several attempts to change the network architecture have been proposed, as in [37, 38]. Opitz and Shavlik vary the number of the nodes or use different neural networks topology [39]. The first approach is not effective in practice, but the second method effectively encourages the diversity and the performance. The third group of techniques encourages explicit diversity by adding a regularization term in the optimization function of each learner. Negative correlation algorithms [40, 41] are the most studied approaches in this category. These methods will be studied in more detail in Sect. 13.5.3. Finally the fourth class of methods induce implicit diversity by using different base learners. The gist of this idea is to combine the best of each base learner. Even more ambitiously, these techniques hope that the synergistic effects of this combination may improve the performance of the ensemble [33]. Wichard et al. [42] state that a heterogeneous ensemble can reduce the variance because the individual errors are strongly uncorrelated. For instance, Woods et al. [43] combine four types of base learners: decision trees, neural networks, K-nearest neighbors and quadratic Bayes. Given a new instance from the testing-set, the output is obtained with model selection, that is, the algorithm chooses the learner classifier with the highest local accuracy in the feature space. Wang et al. [44] combine decision trees and neural networks in medical classification. Using different base learners implies that the bias of the learner has to match with the characteristics of the application domain [45]. Thus, every ensemble in this category needs an exhaustive tune in stage, in order to pick the appropriate base learners [46].

13.3.3 Aggregation Function

There are a wide variety of methods for combining the individual outputs of the predictors, and the most common in the literature are the following:

1. Linear combination: In regression settings, the joint decision F is commonly obtained by a linear combination of the individual hypotheses [6].

$$F(\mathbf{x}) = \sum_{n=1}^{N} \beta_n f_t(\mathbf{x}). \tag{13.13}$$

   The weight of each model is usually computed with some quality criteria: (i) it can be assigned to be the inverse of the previous forecast error (MSE, MAPE, MAE, etc.) [47]. This is called *error based* approaches. (ii) the optimal weights are obtained by minimizing the total sum of squared error (SSE). This is called variance-based method [47].
2. Simple Average: A particular case of the linear combination is the simple average, all hypothesis weights are equal, this is, $\beta_n = 1/N, \forall n = 1, \ldots, N$ [5, 48].
3. Trimmed mean: This method is an adaptation of the simple average. To predict a single point, it removes the largest and smallest individual predictions before computing the average [5].
4. Windorized Mean: This is another average-based approach. To obtain the prediction of a single instance, it replaces the smallest and largest individual predictions with the predictions closest to them. Next, simple average is computed over the modified set of predictions [48].

13.4 Why Ensemble?

Dietterich [49] identified statistical, computational and representational reasons that support these premises. Let us suppose that we have a set of predictors $\{f_1, f_2, \ldots, f_N\}$ with good performance in the training set S^m.

1. **Statistical**. We can pick one of them taking the risk that this single classifier might not necessarily be the best choice, since we do not know the exact probability function underlying the data. We can create a set of classifiers with zero training error [50], and therefore, all predictors will become indistinguishable.
2. **Computational**. Some training methods, such as hill-climbing or gradient descent, may fall into local optima and therefore, the initial parameters of each machine can affect the choice of the final hypothesis, obtaining a solution that is far from the optimal model.
3. **Representational**. The hypothesis space of the classifiers used to approximate the underlying function f_0 to the data may not contain the optimal prediction. For example, an ensemble of linear functions can approximate any decision function

with pre-determined performance [51]. Some base learners, such as neural networks, with a finite training sample select the first hypothesis that fits the training data. Thus, the effective space of hypotheses $\mathscr{H}'$ by the learner is substantially smaller than the hypotheses space $\mathscr{H}$. In addition, one can note the simplicity of combining weak predictors, in contrast to the use of a single highly complex machine. For this reason, ensemble approaches have been successful in solving real problems, which has made them an increasingly studied discipline.

The remarkable point is that with an appropriate design, the expected performance of the combined predictor can be better than the average performance of individual predictors, even if each of them is weakly good [52].

13.5 The Most Popular Ensemble Approaches

As we previously mentioned, different ensemble approaches are distinguished by the manner in which they manipulate the training data, by the way in which they select the individual hypotheses and by the way in how those are aggregated to the final decision. Now we will review the most commonly used ensemble approaches, namely Bagging, AdaBoost and Negative Correlation.

13.5.1 Bagging

Bagging, introduced by Breiman [23] for classification and regression settings, trains each learner using a bootstrap sample from the training set S^m and combines each output by uniform averaging. This technique can be viewed as an algorithm with only an implicit search for diversity, because no information about the predictions of the other models is incorporated to generate each individual predictor. According to Breiman, by averaging a set of predictors trained with a different bootstrap sample, this method might reduce the influence of the outliers [53, 54]. From this point of view, resampling has an effect similar to that of robust M-estimators in statistics [55], where the influence of sample points is bounded using appropriate loss functions, for example, Huber loss or Tukey's bi-square loss. Hence, this approach might considerably reduce the variance of the bagged prediction, especially when the base learner is unstable [23].

However, traditional bootstrap breaks the sequence of data points which may lead to a bad generalization in time series problems. In order to deal with this important issue, Inoue and Kilian adapt the way that Bagging construct its training samples [56]. Instead, the authors use resampling blocks of instances from the original training set with replacement, in order to encourage diversity, and at the same time, to capture the dependence among the examples. This procedure is called *block bootstrap*, and the block size is chosen to capture the dependence in error term [57]. A simulation study reveals that bagging considerably reduces the out-of-sample mean squared error of the predictions. In [58] the training samples keep the temporal order of the points, by introducing a weight for each example that corresponds to the number of times that the point is selected in the bootstrap sample. Thus, if an example has not been selected in the sample, its weight is equal to zero. Finally, the output of the ensemble for a single instance is computed as a weighted average, where the weight of each predictor is the weighted root mean squared error (WRMSE).

13.5.2 AdaBoost

In 1997, Freund and Schapire introduced Adaboost [25], a sequentially coupled approach that fits an additive model through a forward stagewise strategy [8], by training a single predictor to minimize the residual of the previous ensemble in each round. This algorithm has a theoretical background based on the "PAC" learning model [59]. The gist of this model is to create a method that combines a group of weak learners to produce a strong learning algorithm. This approach provides good generalization ability but it is computationally expensive since the learners are coupled in a sequential manner. The main idea of AdaBoost is to maintain a sampling distribution over the training set. When the algorithm begins, the distribution is uniform, that is, all weights are set to $w_m^1 = \frac{1}{N}, m = 1, \ldots, M$. Here, w_m^n is the weight of the mth example at round n. At each round of the algorithm, the weights of the incorrectly classified examples are increased, in such a way that in the following step the weak learner is forced to focus on the misclassified examples of the training set. Thus, the algorithm concentrates on the "hard" examples, instead of on the correctly classified examples. In normal scenarios, this approach usually outperforms Bagging. However, under noisy data it is the robust performance of Bagging that is commonly superior, rather than Adaboost which considerably decreases its generalization capacity [60].

In time series, Shresta and Solomatine introduce AdaBoost.RT [61], where the absolute relative error (ARE) for the learner n is computed by considering the error over a threshold ϕ. Also, the updating rule reduces the weights of the instances with errors lower than ϕ, and subsequently, the weights are then normalized to make them a distribution. Finally, F is obtained as the weighted average of the individual learners, with weights equal to $-\log(\beta_n)$. It is worth noting that both algorithms have no proof of convergence [61, 62]. Canestrelli uses Drucker's AdaBoost.R2 [62] for tide levels forecasting, which quantifies the loss for each training sample using one of the three possible functions:

1. Absolute percentage error (APE) $w_m^n = \frac{|f_n(\mathbf{x}_m)-y_m|}{\max_{m=1,\ldots,M}|f_n(\mathbf{x}_m)-y_m|}$.
2. Quadratic $w_m^n = (f_n(\mathbf{x}_m) - y_m)^2 \max_{m=1,\ldots,M} f_n(\mathbf{x}_m) - y_m)^2$.
3. Exponential: $1 - \exp \frac{-|f_n(\mathbf{x}_m)-y_m|}{\max_(m=1,\ldots,M)|f_n(\mathbf{x}_m)-y_m|}$.

Then, the confidence of the predictor is computed as a function of the weighted average error according to the weight distribution. And instances with higher errors increase its weights in order to have more chances to appear in the next training sample. Finally, the output of the ensemble is computed as the weighted median. Assaad et al. adapt this method for recurrent neural networks as the base learner [63]. They make two changes: first, in order to keep the temporal dependence of the training data, they weight the example errors in order to update the RNN weights instead of resampling. Second, an additional parameter is introduced to increase (even more) the weight of the wrong modeled examples. On the other hand, de Souza et al. [64] propose the BCC algorithm, an adaptation of Adaboost.R2 technique that uses the Genetic Programming (GP) as the base learner. It uses the exponential loss function to compute the error of each example. And, in order to update the weights, it includes a multiplicative factor which depends on the correlation between the learner predictions and the true targets. Finally, Goh et al. [65] introduce Modified Adaboost, which combines Ellman Networks using a weighted linear function. Besides, it computes the error of each instance by considering two loss functions widely used for comparing drug dissolution profiles.

According to the gradient descent perspective of Adaboost [8, 66], its goal is to add the i-th learner in the ensemble which minimizes

$$(\beta_n, f_n) = \text{argmin}_{(\beta, f)} \sum_{m=1}^{M} \ell(y_m, F(\mathbf{x}_m)), \tag{13.14}$$

where $F(\mathbf{x}_m) = F_{n-1}(\mathbf{x}_m) + \beta_n f_n(\mathbf{x}_m)$. At each round, the algorithm selects the hypothesis f_n which minimizes Eq. 13.14 over the sample distribution w^n. Then it performs a line search along the direction. From this point of view, Buhlmann and Yu [67] introduce a gradient boosting method with the quadratic loss function which is adapted for time series forecasting for load forecasting [68], finance [69], economics [70] and production [71].

13.5.3 *Negative Correlation Methods*

A very important feature of an ensemble is the diversity of their members, because a group of learners composed exclusively by exact replicas of the same hypothesis is clearly useless. A way to quantify this property is the so called *Bias-Variance-Covariance* decomposition [72][1] for the quadratic loss of an ensemble F obtained as a convex combination of a set of N predictors and the optimal prediction $f_0(\mathbf{x})$

---

[1]If the hypotheses are considered fixed, the expectations are taken based on the distribution of the inputs $\mathbf{x}$. If they are considered free, expectations are also taken with respect to the distribution of the sample(s) used to estimate them.

$$E[f_0(\mathbf{x}) - F(\mathbf{x})^2] = \text{bias}(F(\mathbf{x}))^2 + \sum_{t=1}^{T} w_t^2 \text{var}(f_n(\mathbf{x})) + E\left[\sum_i \sum_{i \neq j} w_i w_j \text{cov}(f_i, f_j)\right], \tag{13.15}$$

where $\text{bias}(F(\mathbf{x})) = E[F(\mathbf{x}) - f_0(\mathbf{x})] = \sum_{t=1}^{T} w_t E[f_n(\mathbf{x}) - f_0(\mathbf{x})]$, $\text{var}(f_n(\mathbf{x})) = E\left[(f_n(\mathbf{x}) - E[f_n(\mathbf{x})])^2\right]$, and $\text{cov}(f_i, f_j) = E\left[(f_i(\mathbf{x}) - E[f_i(\mathbf{x})])(f_j(\mathbf{x}) - E[f_j(\mathbf{x})])\right]$.

This decomposition suggests that selecting a set of hypotheses with negative covariance $\text{cov}(f_i, f_j)$ without increasing the individual biases and variances would be a very convenient scenario, since this would minimize the expected quadratic loss [8, 41, 72, 73]. Several methods based on the negative correlation among the learners emerge from this result. In the seminal work of Rosen [74], for example, the hypotheses are generated sequentially using a base learner which, at each step t, is provided with a different loss function that directly penalizes correlations with previous errors,

$$\ell_R(f_n(\mathbf{x}), y) = (y - f_i(\mathbf{x}))^2 + \lambda \sum_{j=1}^{t-1} d(t, j)(f_n(\mathbf{x}) - y)(f_j(\mathbf{x}) - y), \tag{13.16}$$

where λ is a scaling function when $t = j - 1$ and otherwise 0. Note that the learner always minimizes the error within the original set of examples. However, at each step, the learner minimizes a different loss function ℓ_R and therefore, the target hypothesis changes from round to round depending on the previous learning step.

In the Negative Correlation algorithm (NC) [41], the N hypotheses $f_1, \ldots, f_N$ are trained simultaneously. The loss function to be minimized by each individual hypothesis f_n is given by equation

$$\ell_{NC}(f_n(\mathbf{x}), y) = (y - f_n(\mathbf{x}))^2 + \lambda \sum_{j \neq n}^{n-1} (f_n(\mathbf{x}) - F(\mathbf{x}))(f_j(\mathbf{x}) - F(\mathbf{x})). \tag{13.17}$$

Note that both ℓ_R and ℓ_{NC} promote the individual accuracy by penalizing the differences $y - f_n(\mathbf{x})$. However, the correlations among the individual predictions $f_n(\mathbf{x})$ are no longer measured with respect to the target prediction y but with respect to the composite prediction $F(\mathbf{x})$. This is taking into account that equation (13.17) can be expressed as

$$\ell_{NC}(f_n(\mathbf{x}), y) = (y - f_n(\mathbf{x}))^2 + \lambda(f_n(\mathbf{x}) - F(\mathbf{x})) \sum_{j \neq n}^{n-1} (f_j(\mathbf{x}) - F(\mathbf{x})). \tag{13.18}$$

In each training epoch, $F(\mathbf{x})$ is computed as the uniform average of the predictions made for the whole ensemble in the previous timestep.

Note that $\sum_{j \neq n} (f_j(\mathbf{x}) - F) = -(f_n - F)$, since $F(\mathbf{x}) = \frac{1}{N}\left[f_n + \sum_{j \neq n} f_j(\mathbf{x})\right]$. Thus, ℓ_{NC} can be re-written as

$$\ell_{NC}(f_n(\mathbf{x}), y) = (y - f_n(\mathbf{x}))^2 - \lambda(f_n(\mathbf{x}) - F(\mathbf{x}))^2. \tag{13.19}$$

Rodan and Tiño [75] propose an ensemble of Echo State Networks [76] with diverse reservoirs, where the negative correlation term defined in equation (13.18) is used to compute their collective read-out.

Another point of interest is that the Negative Correlation algorithm can also be motivated by the ambiguity decomposition [77]

$$(y - F(\mathbf{x}))^2 = \sum_{n=1}^{N} w_n(y - f_n(\mathbf{x}))^2 - \sum_{n=1}^{N} w_n(f_n(\mathbf{x}) - F(\mathbf{x}))^2. \tag{13.20}$$

It states that the quadratic loss of the ensemble is the weight average of the individual errors, minus the weighted average of the individual deviations with respect to the ensemble output. The above-mentioned decomposition suggests that the ensemble error can be reduced by increasing the second term which is called the *ambiguity* term [73]. However, uncontrolled deviations can lead to increasing the first term in (13.20), hence, increasing the ensemble error. Note that the loss given in equation (13.18) is essentially the above mentioned decomposition corresponding to learner t.

Acknowledgments We thank to Professor Claudio Moraga for their continued support in the design and consolidation of the doctoral program in computer engineering of the Universidad Técnica Federico Santa María.
This work was supported in part by Research Project DGIP-UTFSM (Chile) 116.24.2 and in part by Basal Project FB 0821.

References

1. George EP Box, Gwilym M Jenkins, and Gregory C Reinsel. *Time series analysis: forecasting and control*. Prentice Hall Englewood cliffs nj, third edition edition, 1994.
2. Jan G De Gooijer and Rob J Hyndman. 25 years of time series forecasting. *International journal of forecasting*, 22 (3):443–473, 2006.
3. John M Bates and Clive WJ Granger. The combination of forecasts. *Journal of the Operational Research Society*, 20(4):451–468, 1969.
4. G Peter Zhang. Time series forecasting using a hybrid arima and neural network model. *Neurocomputing*, 50:159–175, 2003.
5. Lilian M de Menezes, Derek W. Bunn, and James W Taylor. Review of guidelines for the use of combined forecasts. *European Journal of Operational Research*, 120(1):190 – 204, 2000.
6. Ratnadip Adhikari and R. K. Agrawal. Combining multiple time series models through a robust weighted mechanism. In *1st International Conference on Recent Advances in Information Technology, RAIT 2012, Dhanbad, India, March 15-17, 2012*, pages 455–460. IEEE, 2012.
7. Hui Zou and Yuhong Yang. Combining time series models for forecasting. *International Journal of Forecasting*, 20(1):69–84, 2004.

8. T. Hastie, R. Tibshirani, and J. Friedman. *The Elements of Statistical Learning*. Springer Series in Statistics. Springer New York Inc., New York, NY, USA, 2001.
9. Hector Allende and Siegfried Heiler. Recursive generalized m-estimates for autoregressive moving average models. *Journal of Time Series Analysis*, 13(1):1–18, 1992.
10. Bruce L Bowerman and Richard T O'Connell. Forecasting and time series: An applied approach. 3rd. 1993.
11. Greta M Ljung and George EP Box. On a measure of lack of fit in time series models. *Biometrika*, 65(2):297–303, 1978.
12. H Allende and J Galbiati. Robust test in time series model. *J. Interamerican Statist. Inst*, 1(48):35–79, 1996.
13. T Subba Rao. On the theory of bilinear time series models. *Journal of the Royal Statistical Society. Series B (Methodological)*, pages 244–255, 1981.
14. Leo Breiman, Jerome Friedman, Charles J Stone, and Richard A Olshen. *Classification and regression trees*. CRC press, 1984.
15. *Vector Autoregressive Models for Multivariate Time Series*, pages 385–429. Springer New York, New York, NY, 2006.
16. Jerome H Friedman. Multivariate adaptive regression splines. *The annals of statistics*, pages 1–67, 1991.
17. Jin-Lung Lin and Clive WJ Granger. Forecasting from non-linear models in practice. *Journal of Forecasting*, 13(1):1–9, 1994.
18. Jerome T Connor, R Douglas Martin, and Les E Atlas. Recurrent neural networks and robust time series prediction. *IEEE transactions on neural networks*, 5(2):240–254, 1994.
19. Indre Zliobaite. Learning under concept drift: an overview. *CoRR*, abs/1010.4784, 2010.
20. Wenyu Zang, Peng Zhang, Chuan Zhou, and Li Guo. Comparative study between incremental and ensemble learning on data streams: Case study. *Journal Of Big Data*, 1(1), 2014.
21. Graham Elliott, Clive Granger, and Allan Timmermann, editors. *Handbook of Economic Forecasting*, volume 1. Elsevier, 1 edition, 2006.
22. P. G. Zhang and L. V. Berardi. Time series forecasting with neural network ensembles: an application for exchange rate prediction. *Journal of the Operational Research Society*, 52(6):652–664, 2001.
23. L. Breiman. Bagging predictors. *Machine Learning*, 24(2):123–140, 1996.
24. E. Bauer and R. Kohavi. An empirical comparison of voting classification algorithms: Bagging, boosting, and variants. *Machine Learning*, 36:105–139, 1999.
25. Y. Freund and R. E. Schapire. A decision-theoretic generalization of on-line learning and an application to boosting. *J. Comput. Syst. Sci.*, 55(1):119–139, 1997.
26. Eibe Frank and Bernhard Pfahringer. *Improving on Bagging with Input Smearing*, pages 97–106. Springer Berlin Heidelberg, Berlin, Heidelberg, 2006.
27. Juan J. Rodriguez, Ludmila I. Kuncheva, and Carlos J. Alonso. Rotation forest: A new classifier ensemble method. *IEEE Trans. Pattern Anal. Mach. Intell.*, 28(10):1619–1630, October 2006.
28. Leo Breiman. Randomizing outputs to increase prediction accuracy. *Machine Learning*, 40(3):229–242, 2000.
29. P. Bühlmann and Bin Yu. Analyzing bagging. *Annals of Statistics*, 30:927–961, 2002.
30. J.H. Friedman and P. Hall. On bagging and nonlinear estimation. *Journal of Statistical Planning and Inference*, 137, 2000.
31. D. K. Barrow and S. F. Crone. Crogging (cross-validation aggregation) for forecasting x2014; a novel algorithm of neural network ensembles on time series subsamples. In *Neural Networks (IJCNN), The 2013 International Joint Conference on*, pages 1–8, Aug 2013.
32. R. K. Bryll, R. Gutierrez-Osuna, and F. K. H. Quek. Attribute bagging: improving accuracy of classifier ensembles by using random feature subsets. *Pattern Recognition*, 36(6):1291–1302, 2003.
33. L. Rokach. Taxonomy for characterizing ensemble methods in classification tasks: A review and annotated bibliography. *Computational Statistics & Data Analysis*, 53(12):4046–4072, 2009.

34. K. Tumer and N. C. Oza. Input decimated ensembles. *Pattern Analysis and Applications*, 6(1):65–77, 2003.
35. J. F. Kolen, J. B. Pollack, J. F. Kolen, and J. B. Pollack. Back propagation is sensitive to initial conditions. In *Complex Systems*, pages 860–867. Morgan Kaufmann, 1990.
36. G. Brown, J. L. Wyatt, R. Harris, and Xin Yao. Diversity creation methods: a survey and categorisation. *Information Fusion*, 6(1):5–20, 2005.
37. D. Partridge and W. B. Yates. Engineering multiversion neural-net systems. *Neural Computation*, 8:869–893, 1995.
38. W. Yates and D. Partridge. Use of methodological diversity to improve neural network generalization. *Neural Computing and Applications*, 4(2):114–128, 1996.
39. D. W. Opitz and J. W. Shavlik. Generating accurate and diverse members of a neural-network ensemble. In *Advances in Neural Information Processing Systems*, pages 535–541. MIT Press, 1996.
40. R. Ñanculef, C. Valle, H. Allende, and C. Moraga. Training regression ensembles by equential target correction and resampling. *Inf. Sci.*, 195:154–174, July 2012.
41. Yong Liu and Xin Yao. Ensemble learning via negative correlation. *Neural Networks*, 12:1399–1404, 1999.
42. Ogorzalek M. Wichard JD, Christian M. Building ensembles with heterogeneous models. In *7th Course on the International School on Neural Nets IIASS*, 2002.
43. K. S. Woods, W. P. Kegelmeyer, and K. W. Bowyer. Combination of multiple classifiers using local accuracy estimates. *IEEE Trans. Pattern Anal. Mach. Intell.*, 19(4):405–410, 1997.
44. Wenjia Wang, P. Jones, and D. Partridge. Diversity between neural networks and decision trees for building multiple classifier systems. In Josef Kittler and Fabio Roli, editors, *Multiple Classifier Systems*, volume 1857 of *Lecture Notes in Computer Science*, pages 240–249. Springer, 2000.
45. P. Brazdil, J. Gama, and B. Henery. Characterizing the applicability of classification algorithms using meta-level learning. In F. Bergadano and L. De Raedt, editors, *ECML*, volume 784 of *Lecture Notes in Computer Science*, pages 83–102. Springer, 1994.
46. Niall Rooney, David Patterson, Sarab Anand, and Alexey Tsymbal. *Dynamic Integration of Regression Models*, pages 164–173. Springer Berlin Heidelberg, Berlin, Heidelberg, 2004.
47. Christiane Lemke and Bogdan Gabrys. Meta-learning for time series forecasting and forecast combination. *Neurocomputing*, 73(10-12):2006–2016, 2010.
48. Victor Richmond R. Jose and Robert L. Winkler. Simple robust averages of forecasts: Some empirical results. *International Journal of Forecasting*, 24(1):163 – 169, 2008.
49. T. G. Dietterich. Ensemble methods in machine learning. In *Proceedings of the First International Workshop on Multiple Classifier Systems*, MCS '00, pages 1–15, London, UK, UK, 2000. Springer-Verlag.
50. D. M. Hawkins. The Problem of Overfitting. *Journal of Chemical Information and Computer Sciences*, 44(1):1–12, 2004.
51. L. I. Kuncheva. *Combining Pattern Classifiers: Methods and Algorithms*. John Wiley and Sons, Inc., 2004.
52. L.K. Hansen and P. Salamon. Neural network ensembles. *IEEE Transactions on Pattern Analysis and Machine Intelligence*, 12(10):993–1001, 1990.
53. Y. Grandvalet. Bagging down-weights leverage points. In *IJCNN (4)*, pages 505–510, 2000.
54. Y. Grandvalet. Bagging equalizes influence. *Machine Learning*, 55(3):251–270, 2004.
55. P. J. Huber. *Robust Statistics*. Wiley Series in Probability and Statistics. Wiley-Interscience, 1981.
56. Atsushi Inoue and Lutz Kilian. Bagging time series models. *CEPR Discussion Paper No. 4333*, 2004.
57. Peter Hall and Joel L. Horowitz. Bootstrap Critical Values for Tests Based on Generalized-Method-of-Moments Estimators. *Econometrica*, 64(4):891–916, 1996.
58. Nikola Simidjievski, Ljupčo Todorovski, and Sašo Džeroski. Predicting long-term population dynamics with bagging and boosting of process-based models. *Expert Syst. Appl.*, 42(22):8484–8496, December 2015.

59. L. G. Valiant. A theory of the learnable. *Commun. ACM*, 27(11):1134–1142, 1984.
60. P. Melville, N. Shah, L. Mihalkova, and R. J. Mooney. Experiments on ensembles with missing and noisy data. In *In: Proceedings of the Workshop on Multi Classifier Systems*, pages 293–302. Springer Verlag, 2004.
61. D. L. Shrestha and D. P. Solomatine. Experiments with adaboost.rt, an improved boosting scheme for regression. *Neural Comput.*, 18(7):1678–1710, July 2006.
62. H. Drucker. Improving regressors using boosting techniques. In D. H. Fisher, editor, *ICML*, pages 107–115. Morgan Kaufmann, 1997.
63. Mohammad Assaad, Romuald Boné, and Hubert Cardot. A new boosting algorithm for improved time-series forecasting with recurrent neural networks. *Inf. Fusion*, 9(1):41–55, January 2008.
64. Luzia Vidal de Souza, Aurora Pozo, Joel Mauricio Correa da Rosa, and Anselmo Chaves Neto. Applying correlation to enhance boosting technique using genetic programming as base learner. *Applied Intelligence*, 33(3):291–301, 2010.
65. Wei Yee Goh, Chee Peng Lim, and Kok Khiang Peh. Predicting drug dissolution profiles with an ensemble of boosted neural networks: a time series approach. *IEEE Transactions on Neural Networks*, 14(2):459–463, 2003.
66. Jerome H. Friedman. Greedy function approximation: A gradient boosting machine. *Annals of Statistics*, 29:1189–1232, 2000.
67. Buhlmann P. and Yu B. Boosting with the l2 loss: Regression and classification. *Journal of the American Statistical Association*, 98:324–339, 2003.
68. Souhaib Ben Taieb and Rob Hyndman. A gradient boosting approach to the kaggle load forecasting competition. *International Journal of Forecasting*, 30(2):382–394, 2014.
69. Francesco Audrino and Peter Bühlmann. Splines for financial volatility. University of st. gallen department of economics working paper series 2007, Department of Economics, University of St. Gallen, 2007.
70. Klaus Wohlrabe and Teresa Buchen. Assessing the macroeconomic forecasting performance of boosting: Evidence for the united states, the euro area and germany. *Journal of Forecasting*, 33(4):231–242, 2014.
71. Nikolay Robinzonov, Gerhard Tutz, and Torsten Hothorn. Boosting techniques for nonlinear time series models. *AStA Advances in Statistical Analysis*, 96(1):99–122, 2012.
72. N. Ueda and R. Nakano. Generalization error of ensemble estimators. In *Proceedings of IEEE International Conference on Neural Networks.*, pages 90–95, Washington, USA, June 1996.
73. G. Brown. *Diversity in Neural Network Ensembles*. PhD thesis, The University of Birmingham, 2004.
74. B. Rosen. Ensemble learning using decorrelated neural networks. *Connection Science*, 8:373–384, 1996.
75. Ali Rodan and Peter Tiňo. Negatively correlated echo state networks. In *ESANN 2011, 19th European Symposium on Artificial Neural Networks, Bruges, Belgium, April 27-29, 2011, Proceedings*, 2011.
76. Mustafa C. Ozturk, Dongming Xu, and José C. Príncipe. Analysis and design of echo state networks. *Neural Comput.*, 19(1):111–138, January 2007.
77. A. Krogh and J. Vedelsby. Neural network ensembles, cross validation, and active learning. In *Advances in Neural Information Processing Systems*, pages 231–238. MIT Press, 1995.

Chapter 14
A Hierarchical Distributed Linear Evolutionary System for the Synthesis of 4-bit Reversible Circuits

Fatima Zohra Hadjam and Claudio Moraga

14.1 Introduction

Reversible technologies and the synthesis of reversible circuits are promising areas of research considering the expected further technological advances towards quantum computing, particularly to fulfill the increasing demand of low-power consumption applications. Benett showed in [1] that in order to limit power dissipation, it is necessary that all the computations have to be performed in a reversible way. An elementary reversible circuit, a "gate", realizes a bijection (one-to-one mapping) from the set of input tuples to the outputs. A general reversible circuit is made with a cascade of reversible gates, without fan-out (multiple use of a same signal is not permitted) and without feedback (the circuit should be cycle free). Quantum computing circuits are reversible.

So far, various methods for reversible circuits design have been proposed. Common synthesis approaches and recent comprehensive reviews are presented in [2, 3]. The various constraints imposed by the reversibility, as mentioned above, and the extended search space make their design very difficult comparing with the classical irreversible logic circuits design. This suggests that an evolutionary approach might provide an effective design alternative. There are very few publications on the evolutionary design of reversible circuits. Most of them use Genetic Algorithms (GAs) focusing on finding alternative realization of gates at the quantum level [4, 5] or on optimizing aspects of already available circuits, like e.g. reordering the outputs [6]. Papers on reversible circuits design using Genetic Programming are even more rare [7], although as shown in [8] very good results are obtained. There are important contributions of Genetic Programming to quantum computing [9], however at the algorithms level, not at the circuit level.

F.Z. Hadjam (✉) · C. Moraga
Department of Computer Science,
University Djillali Liabes of Sidi Bel Abbes, Sidi Bel Abbès, Algeria
e-mail: fatima.hadjam@googlemail.com

R. Seising and H. Allende-Cid (eds.), *Claudio Moraga: A Passion for Multi-Valued Logic and Soft Computing*, Studies in Fuzziness and Soft Computing 349, DOI 10.1007/978-3-319-48317-7_14

14.1.1 Related Works

Even limited to 4-bits reversible functions, the synthesis of optimal reversible circuits becomes an arduous task owing to the extremely large problem space ($16! = 244$ functions). According to [10], the output of a search alone, counting only the space required to list elementary gates for every function, would require over 100 terabytes of storage. To tackle this specific problem, several techniques have been deployed. The best and most relevant are presented in [11–21]. Due to space constrains, details on these methods will not be included.

14.1.2 Motivation

All the methods, whose references were mentioned above, are all based on the "lookup" in a library, a database, or a hash-table of partial solutions. Several approaches have been proposed to alleviate the storage space and the search time. To avoid confusions, in this paper we distinguish between "optimum" and "optimal". A solution is called optimum if it has formally been proven that no better solution exists. A solution is called optimal if no better solution is known. Is it possible to implement optimal or suboptimal 4-bit reversible functions without relying on existing libraries? Answering this question will be the subject of the current paper. For this purpose we propose a linear genetic programming based method to design a large set of benchmarks found in the literature cited above. Reversible Improved Mutli-Expression Programming, RIMEP2, is such a method. RIMEP2 has been introduced in [8, 22]. Optimal competitive solutions were reported for a group of 30 selected benchmarks (n-lines reversible functions, with $n = 3 \ldots 15$), where a quantum cost reduction up to 96.13 % was reached with an average of 41.08 %. The meaning of "quantum cost" will be explained later in Sect. 14.2.2.

To prepare an evolutionary design of the 69 4-bit reversible functions as benchmarks, aggregated from the references mentioned above, we introduce a new aspect in RIMEP2: distributed computation. At a first sight, distribution is frequently considered as parallelizing some inner aspects of the evolutionary algorithm such as distributing the fitness calculation or using an island model concept. The idea presented here is completely different. A new hierarchical topology of multiple populations is used and a new communication policy is introduced. Such architecture helps the evolutionary algorithm to explore and exploit the search space in an efficient way (see Sect. 5.3). The obtained results (see Sect. 5.4) outperformed most (or matched a few) of the already optimized and published 4-bit reversible circuits, reaching a quantum cost reduction of up to 62.71 % with an average of 10.79 % for the benchmarks where the quantum cost is considered as a performance measure. A gate count reduction up to 60 % was achieved with an average of 16.82 % for the benchmark groups where the gate count was considered. The rest of the paper is structured as follows:

RIMEP2 is introduced in Sect. 14.2. Section 14.3 describes the concept of distribution for RIMEP2 (leading to DRIMEP2) and explains its mechanism. Section 14.4, presenting a comparative view of design results using methods of the early references and DRIMEP2, illustrates its effectiveness. Some conclusions will close the paper.

14.2 Introduction to RIMEP2

As mentioned earlier, RIMEP2 is a linear graph-coded genetic programming system able to represent any multiple inputs-outputs mathematical function in the same chromosome (individual). In our case, n-bit reversible functions. As in any evolutionary approach, we first explain the encoding of the chromosome, the fitness calculation and then introduce the different used genetic operators.

14.2.1 The Chromosome Encoding

A RIMEP2 chromosome is a list of expressions. Each sub-expression represents a k-bit gate, where $1 \leq k \leq n$. The whole chromosome represents a cascade of gates which constitutes (or covers) the implemented reversible function. To clarify this encoding, consider an example: a full subtractor. Its irreversible specification is shown in Table 14.1a. RIMEP2 has evolved a set of different solutions (reversible circuits having different implementations and costs) using $\{P, NP, Cnot\}$ as function set (called later "library"). P, NP and $Cnot$ indicate respectively Peres, Negated control Peres (the first control is negated) and control Not (see [17, 23, 24]). The graphic representation of the cited gates is given in Fig. 14.1c. Two solutions have been selected, which encodings are shown in Table 14.1b. The encodings given in the Table 14.1b represent cascades of sub-expressions (gates). The chromosomes are traversed from left to right. Each column corresponds to a sub-expression (gate).

Table 14.1 The specification of the full subtractor

Inputs			Outputs	
a	b	c	o_1	o_2
0	0	0	0	0
0	0	1	1	1
0	1	0	1	1
0	1	1	1	0
1	0	0	0	1
1	0	1	0	0
1	1	0	0	0
1	1	1	1	1

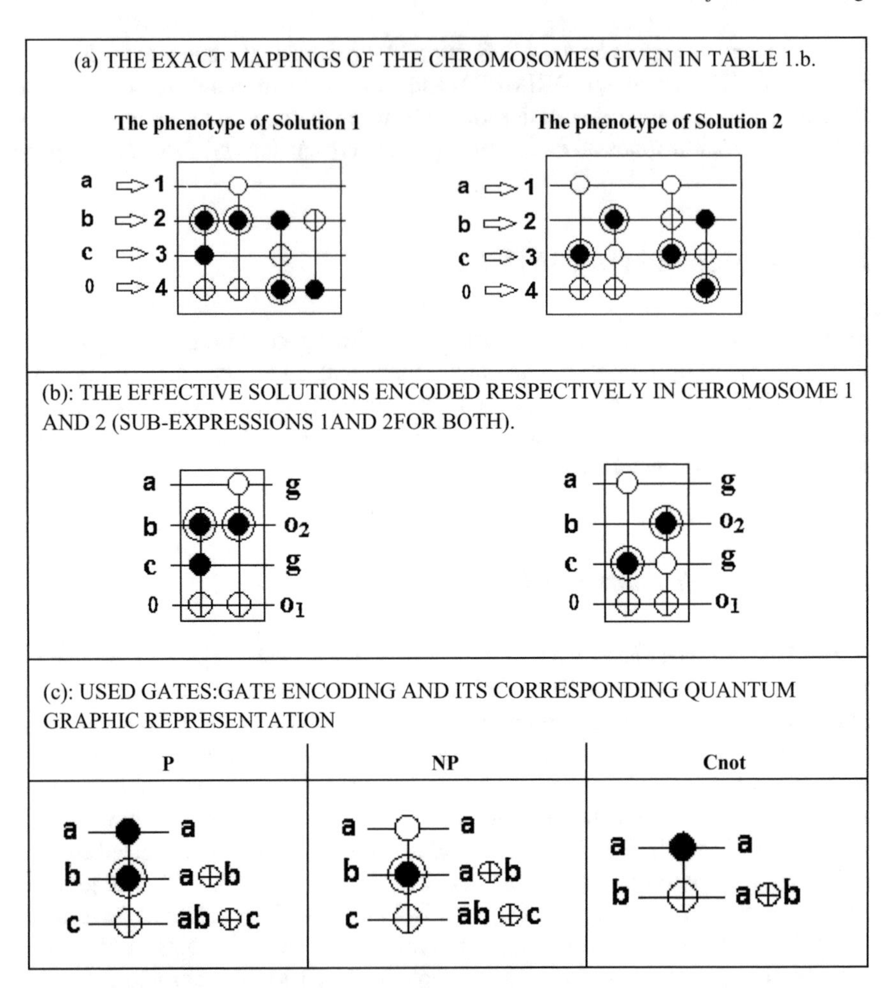

Fig. 14.1 Two evolved realizations of the full subtractor

Consider the solution 1, the first column: P(3, 2, 4). The top of the column indicates the code of the corresponding gate. The remaining scalars represent line numbers. Each line is related to a main input of the circuit being evolved. According to [25], to make such example reversible, 2 outputs are added to assure the one-to-one mapping. Recall that a reversible circuit has as many inputs as outputs, thus, an additional constant line "0" (ancillary bit) was added to the original inputs (lines). The drawings of Fig. 1.b represent the effective implementations of the subtractors encoded by the chromosomes (1 and 2 respectively) of the Table 14.2b. The drawings of Fig. 14.1a represent the complete mappings of the same chromosomes.

When traversing the chromosome, the algorithm stops once the full specification of the reversible function is met to avoid accumulating additional gates and therefore, increasing the size of the circuit (the number of gates). The order number of the cor-

Table 14.2 The encodings of the evolved solutions

The chromosome of solution 1			
P	NP	P	Cnot
3	1	2	4
2	2	4	2
4	4	3	–
The chromosome of solution 2			
NP	NP	NP	P
1	3	1	2
3	2	3	4
4	4	2	3

responding sub-expression will indicate the size of the solution. For both solutions, RIMEP2 stopped at the second sub-expression, thus, the size of the corresponding circuits are both equal to 2. The outputs of the circuit do not appear in the chromosome; they are calculated when evaluating the chromosome (see the point 4 of Sect. 2.2). During the fitness calculation, if a line is not among the inputs of a given sub-expression (gate), it continues unchanged. RIMEP2 guarantees that any evolved reversible circuit is syntactically correct, cycle and fan-out free simply by avoiding the double use of the same line for a given gate. Precautions should be taken when the mutation occurs. See Sect. 14.2.3.

14.2.2 Computing the Fitness

RIMEP2 is a fitness driven evolutionary algorithm: for a given individual i (evolved circuit i), the Fitness(individual i) is a vector $(f_1, f_2, f3)$, where f_1, f_2 and f_3 indicate respectively the number of errors, the quantum cost and the size of the circuit. The Fitness must be minimized in all its three components.

1. The evolved reversible circuit should first and mainly match the behavioural specification of the target circuit (the truth table in our case). The value of f_1 is given by the number of not matching values, i.e., by the number of errors and has to be minimized. A zero-f_1 (error) would indicate a perfect fit.
2. f_2 and f_3 aim to minimize, respectively, the quantum cost and the size of the evolved circuits (when different error and fan-out free expressions with the same quantum cost are met, then the one with the lowest number of gates is selected). For example: Given 2 individuals i and j with: Fitness $(i) = (f_{i1}, f_{i2}, f_{i3})$ and Fitness $(j) = (f_{j1}, f_{j2}, f_{j3})$,

   (a) If $f_{i1} < f_{j1}$ then individual i is considered to be better than individual j.
   (b) If $f_{i1} = f_{j1}$ and $f_{i2} < f_{j2}$, then individual i is considered to be better than individual j.

Table 14.3 Quantum cost of some reversible gates

Gate name	Abbreviation	Quantum cost
Not	NOT	1
Controlled NOT	Cnot	1
Toffoli	T	5
Peres	P	4
Negated Peres	NP	4

(c) If $f_{i1} = f_{j1}$ and $f_{i2} = f_{j2}$ but $f_{i3} < f_{j3}$, then i is considered to be better than individual j.

3. Each reversible gate can be built using several elementary quantum gates (the cost of each elementary quantum gate is assumed to be equal to 1). Accumulating these costs will constitute the quantum cost of the whole evolved circuit (see [23, 24]). More information is given in Sect. 2.2 of [8]. Quantum cost of the considered gates is given in Table 14.3.

14.2.3 RIMEP2 Genetic Operators

(a) The selection A tournament selection with a size of 2 was used in RIMEP2. The size of the tournament was chosen based on the fact that low population size values are used to evolve with RIMEP2.
(b) The crossover The 'uniform crossover' (multi-point crossover) was used to evolve the whole group of selected benchmarks. No post-processing is needed to guarantee that functional and syntactically correct circuits are produced. This constitutes a feature of RIMEP2. An example of the crossover operation is illus-

Parent 1

1	2	3	4
T	Cnot	T	T
1	2	2	4
2	4	3	3
4	-	4	2

Parent 2

1	2	3	4
Cnot	P	Cnot	P
4	2	2	1
2	1	1	3
-	3	-	2

Offspring 1

T	P	Cnot	T
1	2	2	4
2	1	1	3
4	3	-	2

Offspring 2

Cnot	Cnot	T	P
4	2	2	1
2	4	3	3
-	-	4	2

Fig. 14.2 The crossover operation

Individual 1 before mutation | **Individual 1 after mutation**

(a): MUTATION TYPE 1

1	2	3	4
T	P	**Cnot**	T
1	2	2	4
2	1	1	3
4	3	-	2

1	2	3	4
T	P	**P**	T
1	2	2	4
2	1	1	3
4	3	**4**	2

(b): MUTATION TYPE 2

T	P	Cnot	T
1	2	2	4
2	1	**1**	3
4	3	-	2

T	P	Cnot	T
1	2	2	4
2	1	**3**	3
4	3	-	2

(c): MUTATION TYPE 3

T	P	Cnot	T
3	2	2	4
2	1	1	3
4	3	-	2

T	P	Cnot	T
4	2	2	4
3	1	1	3
2	3	-	2

(d): MUTATION TYPE 4

T	P	**Cnot**	T
1	2	**2**	4
2	1	**1**	3
4	3	**-**	2

Cnot	P	**T**	T
2	2	**1**	4
1	1	**2**	3
-	3	**4**	2

Fig. 14.3 The four types of mutation

trated in Fig. 14.2. The numbers appearing on the top the Fig. 14.2 indicate the order of the sub-expressions in the chromosomes. The operation of the crossover is repeated PopSize/2 times, where PopSize denotes the size of the population. Two selected individuals (parents) are candidates to cross with a predefined probability (see Table 14.3).

(c) The mutation Four types of mutation are considered each, randomly selected and assigned a probability below a predefined threshold:

(1) Mutating the operator from the proposed library. A missing address (input) should be filled when the new operator needs more inputs than the previous one. RIMEP2 controls that no input (line) already in use is assigned, thus avoiding a fanout problem. See Fig. 14.3a.
(2) Mutating an address (an input of a gate): the selected line (input) to be mutated should not be replaced by another line that is already related to one of the inputs of the current gate, in order to keep the circuit fanout free. See Fig. 14.3b.
(3) Circular shifting of the address (inputs) in the same expression (gate). See Fig. 14.3c.
(4) Exchanging the order of two randomly selected expressions (gates). See Fig. 14.3d.

14.3 Distributed RIMEP2

This section describes the concept of distribution for RIMEP2 leading to DRIMEP2.

14.3.1 Motivation

RIMEP2 has proved, in many previous studies, its ability to find competitive solutions to well known benchmarks. Motivated by the emergence and the availability of many parallel computer architectures, a new MPI-based hierarchical distributed RIMEP2 was developed, focusing on keeping the search space diverse at all times. Distributed evolutionary algorithms (EA) have been extensively used in the literature. They have been developed, studied and applied starting over 20 years ago. Some selected references are given in [2, 26–30] where a survey of early literature and history of parallel EAs and their classifications are presented.

The Hierarchical topology (called also master-slave, see for example [31]), in itself, is not new when specifying the distributed architecture. The master stores the population, executes genetic operations, and distributes individuals to the slaves who have to evaluate their fitness and send the evaluated fitness back to the master. In the current study, both the master and the slave are evolutionary algorithms each with its own population wherein the communication is established in a one way from the slaves called workers to the master called main unit. More details are given in Sect. 14.3.2.

14.3.2 Hierarchically Distributed RIMEP2

As we mentioned before, a new topology with a new communication policy is presented in order to boost the current RIMEP2 to explore and exploit more regions of the whole search space in an efficient manner. The new idea may be explained as follows:

(1) One obstacle that one may meet, when using EAs, is "premature convergence". It happens when high ranked individuals quickly dominate the population making the genetic operators unable to produce better individuals (offspring) and to enhance solution quality. The algorithm is, then, trapped in a local optimum. An alternative to avoid such a situation is to preserve population diversity during the evolution. One should point out that high diversity can also prevent sufficient exploitation and therefore slows down the convergence by making the search random.

(2) Many approaches have been proposed and implemented to prevent the premature convergence (see [8, 32–34]).

(3) The new idea disclosed in this paper is the following: The proposed new system is multi-population based.

(a) We distinguish a main computation unit (MU) representing the main population with a standard size of a sequential RIMEP2.
(b) The rest of the populations are evolving on other units called workers (WU). Each worker has a small population size.
(c) The whole set of populations evolves in an isolated manner (independently of the others). Except for the encoding and the length of the chromosome that are common for all the units, the parameter setting may differ from one population to another such as using only crossover or only mutation or both, using different rates of crossover and mutations or may be using different types of selection. Different libraries can be assigned to different population with different random generator seeds to assure the maximum of diversity.
(d) The main job of the workers is to evolve independently and shortly in different regions of the search space. The jumping from one region to another one is assured by a repetitive population reinitialization (partial or quasi total). The population reinitialization will be called a '*jump*' for the rest of the paper.
(e) The communications occur in one way; from the workers to the main unit every two jumps. The workers have to evolve for a short time. Then, a predefined number of best individuals will migrate from these WUs to the MU. After that, a new jump will happen.
(f) The MU receives migrants from the workers. In this way the diversity is supported. The MU continues its evolution until the desired solution is encountered or the limit time is reached.
(g) The topology of the proposed system has a tree structure where the root constitutes the MU. The structure may be recursive which means that each worker can be a MU and posses workers. More explanations are given in Fig. 14.4. The word hierarchical in the naming of the new proposed system reflects the hierarchical aspect of the proposed topology. One should emphasize that given two successive levels i and $i+1$ in the presented topology, the workers of the level i have (Population size and maximum number of generations will be abreviated respectively as PopSize and MaxGenera):
    - PopSize(level $i+1$) = PopSize (level i)/factor $_1$
    - MaxGenera (level i+1) = MaxGenera(level i)/factor $_2$
    - Factor $_1$ and factor $_2$ are predefined values. Factor$_2$ indicates the number of periods when the migration to the MU should happen and the reinitialization of the WU population takes place.
(h) There are other influencing parameters:
    - How many individuals migrate? This parameter is hard to determine, however, it is correlated with the size of the population receiving these migrants and also how varied are these migrants. We want to introduce diversity in the receiving population without over disturbing its evolution. We have imposed a restriction that similar individuals will not be sent

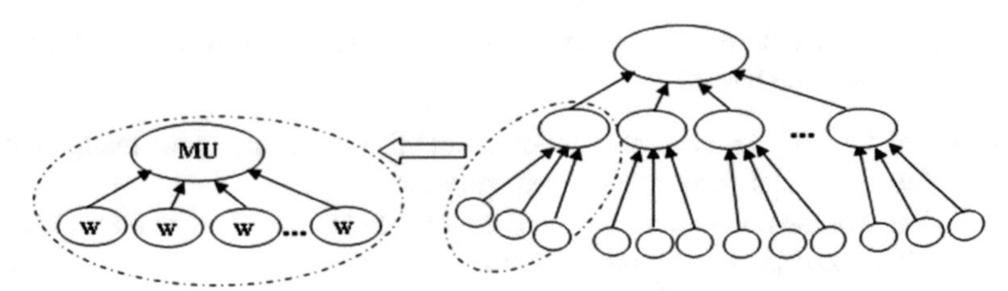

The root of the structure represents the MU. Each of the rest of the nodes represents either a worker for the upper level or a main unit for the lower level. MU refers to Main unit and WU refers to Worker unit.

Fig. 14.4 The proposed hierarchical topology to distribute RIMEP2

twice even from two different worker i.e. if two similar individuals attempt to migrate from two different workers, the main unit will receive only one of them. Two individuals are similar if they have equal fitness (similar phenotypes).

- Which individuals in the MU populations have to be replaced? The most popular way is to replace the worst individuals. According to our experiments, an individual is categorized as worse if it has the highest fitness (remember that our aim is to minimize the fitness; reduce the numbers of errors) but it does not mean that this individual cannot hold good genetic material capable to improve the best solution. First the duplicates (individuals who have similar phenotypes) are replaced and then if necessary, the worse individuals.
- As reported earlier, the populations of the workers are partially or quasi totally reinitialized every period indicated by $factor_2$. Let $factor_3$ denote the number of preserved individuals after the reinitialization. Notice that a high value of $factor_3$ will dominate the new generated population and zero value will lead to a blind search.
- These factors, as the list of the parameter setting of any optimization approach, are generally fixed using heuristics (experiment-based). Various related works, in the literature, have addressed this problematic from meta-evolutionary algorithms, statistic-based to static and dynamic parameter tuning (see e.g. [35]).

(i) The background behind choosing such a structure is:

- Our experiments revealed that even the workers were able to encounter an optimal solution. It is probably due to the possibility of finding the right place where to search.
- Jumping allows sweeping large regions of the search space while the micro-evolution develops the initialized new individuals. In the meantime the MU continues the main evolution.
- The results reveal very interesting aspects. More details about the experiments are given in the following section.

14.4 Experiments: Performance and Results

In this section, we will present the experimental aspect of the study. Two performance statistics are used to compare the DRIMEP2 results to their homologous in the literature cited below: Gate count or quantum cost. The chosen benchmarks are divided in 4 sub-groups according respectively to the published results in [10, 16] where the first statistic (gate count) is considered, and in [19, 36] where the second statistic (quantum cost) is used. 69 benchmarks were used to evaluate the performance of DRIMEP2.

14.4.1 DRIMEP2 Parameter Settings

The parameters of the DRIMEP2 system are separated in two sets. The first set considers the parameter of the sequential RIMEP2 such as the probabilities of crossover and mutation, the length of the chromosome, the size of the population and the maximum number of generations. These parameters have been fixed according to previous experiments published in [35]. The second set of parameters concerns the policy of the new proposed distributed RIMEP2 namely:

(a) The number of levels in the hierarchical structure of the new proposed parallel architecture of DRIMEP2, and the number of workers associated to every level (see Fig. 14.4).
(b) The total number of workers for the whole levels additionally to the main unit will constitute the total number of islands.
(c) $Factor_1$ and $factor_2$ explained earlier in the third sub-section of Sect. 5.3.
(d) The number of migrants from a worker to a main unit,
(e) The $factor_3$ which indicates how many individuals are preserved when the reintialization of a worker occurs.

Table 14.4 The parameter setting

Sequential RIMEP2 parameters		Distributed RIMEP2 parameters	
Parameter	Value	Parameter	Value
Crossover probability	0.7	Number of levels	2
Mutation probability	0.01	Workers attached to every main unit	10
Chromosome length	10–20	$Factor_1$	2–5
Population size	50–100	$Factor_2$	10–15
Max-number of generations	100-10,000	Number of migrants	
Selection type	Tournament	with size = 2	5
Crossover	Uniform	$Factor_3$	5

The parameter settings are shown in Table 14.3. Some of the values are given in ranges. It means that each benchmark has been evolved using an appropriate value from a range for each parameter. The purpose of the study was to find optimal circuits for the proposed benchmarks in a reasonable time. So the parameters were heuristically (experience-based) tuned to reach this purpose and the results are quite encouraging.

14.4.2 Experimentation and Results

Various experiments and tests have been performed on DRIMEP2 using a cluster of computers (LiDOng cluster of the TU Dortmund University, Germany, see [37] for more details about the total number of nodes, the CPU clock rate and other features. This support is here gladly acknowledged). The tests were divided in two parts depending on the literature published results. For the references [10, 16], the NCT library (Not, CNot, Toffoli) was used (see e.g. [4, 18, 23]) and the gate count was considered as an optimality measure. For [19, 36], the NCT library has been also used but the quantum cost was used as a factor of performance taking in account that some of the cases where a Toffoli gate is associated to a control not gate result in a Peres gate or in a negated Peres gate (Peres family). In this case, DRIMEP2 was performing using the NCT $\cup$ Peres family library.

We chose the maximum number of generations as a stopping criterion of the DRIMEP2 system for the following reason: for many benchmarks, the corresponding optimal circuits were found before this criterion was met, but keeping DRIMEP2 searching allows better exploration meanwhile preserving a reasonable response time. The results are shown in Figs. 14.6, 14.7, 14.8 and 14.9. Gate count is referred to as GC. Quantum cost is referred to as QC. Figure 14.5 shows an example of realization of the benchmark "$4b15_1$" from [19] to illustrate the performance of the DRIMPE2. This realization has the quantum cost of 35. The input-output specification of "$4b15_1$" is also given in Table 14.4. From the evolutionary point of view, this result was obtained under the following constellation: workers (W) with a population size of 50, main unit (MU) with a population 200 and $100,000$ generations, 44 min for 5 independent successful runs. A correct solution was obtained in each run, between $4,000$ and $40,000$ generations with an average of 18400 and a standard deviation of 12611. An additional experiment on this benchmark under stronger constraints was done: $10,000$ generations for the MU and $1/2$ of the population size. In 1.5 min execution time for 5 runs, two of which were successful, a very near optimal solution with a quantum cost of 36 was obtained.

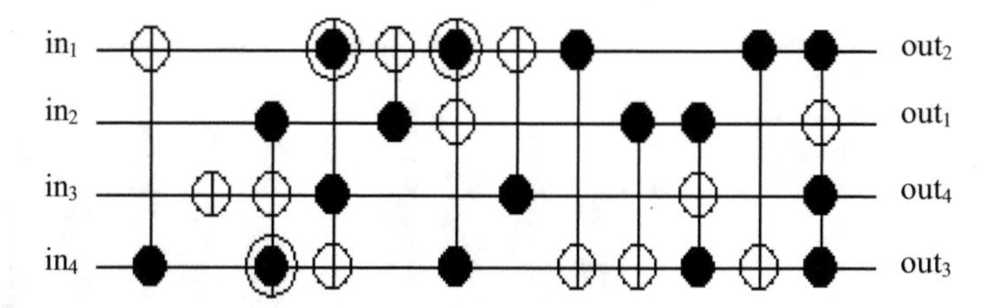

Fig. 14.5 DRIMEP2 Implementation of the benchmark “$4b15g_1$” at the gate level representation

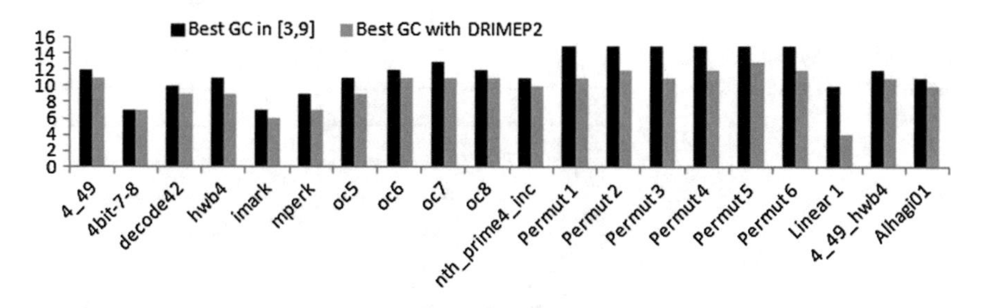

Fig. 14.6 DRIMEP2 results compared with reported benchmarks in [10, 16]. The gate count is considered as a performance measure

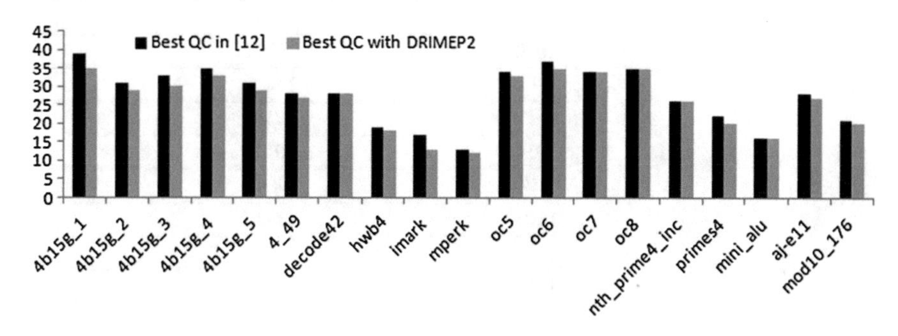

Fig. 14.7 DRIMEP2 results compared with reported benchmarks in [19] (part 1). The quantum Cost is considered

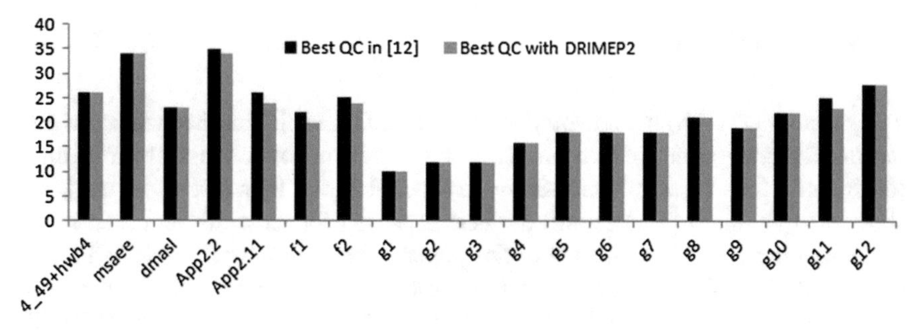

Fig. 14.8 DRIMEP2 results compared with reported benchmarks in [19] (part 2). The quantum Cost is considered

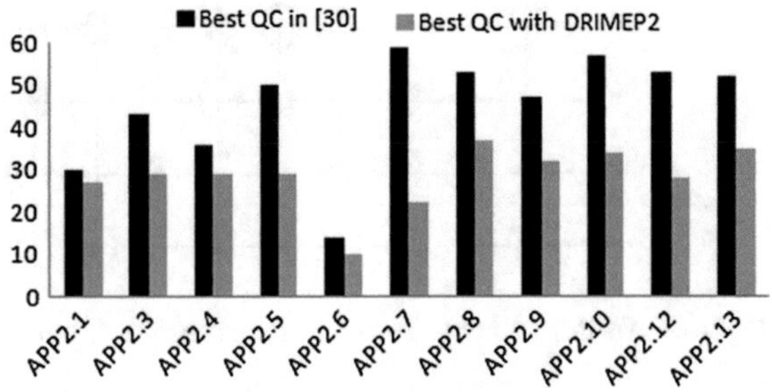

Fig. 14.9 DRIMEP2 results compared with reported benchmarks in [36]. The quantum Cost is considered

Table 14.5 The truth table of the 4-bit $4b15_1$ reversible problem

Inputs	Outputs
0000	0001
0010	0101
0010	0000
0011	1000
0100	1001
0101	1011
0110	1100
1101	0010
0111	1111
1000	0011
1001	1100
1010	0100
1011	0110
1100	1010
1101	1110
1110	1101
1111	0111

14.5 Conclusions

The hierarchical distributed evolutionary system DRIMEP2 has been tested with a reasonably large group of benchmark problems taken from a class of problems—(design of 4 bits reversible circuits)—considered in the literature to be hard, and showed to be able to outperform the best solutions for most of them (and match the remaining). The system used in this paper comprised only 2 levels and a total of at most 10 WUs. The computing time was not over 1 hour for the whole set of generations. In most cases, when processing the whole set of generations DRIMEP2 found several (different) solutions for a given circuit problem. Also, when obtaining a solution as good as the best known, DRIME2 found not necessarily the same circuit,

Fig. 14.10 Claudio Moraga and Fatima Zohra Hadjam at the *European Centre for Soft Computing* in Mieres (Asturias), Spain in 2014

but one with the same quantum cost or gate count (size). The fact that DRIMEP2 was able to obtain better solutions than (or as good as) the best known in the literature shows that it did an effective exploration together with a partial exploitation.

References

1. Benett, C.H.: Logical reversibility of computation, *IBM J. Res. Dev.* Vol. 17 (6), 1973, pp. 525–532.
2. Drechsler, R. and Wille, R: From truth tables to programming languages: progress in the design of reversible circuits, Proceedings of the 41st IEEE International Symposium on Multiple-Valued Logic, ISMVL '11, Washington, DC, USA. IEEE Computer Society, 2011, pp. 78–85.
3. Saeedi, S. and Markov, I. L.: Synthesis and optimization of reversible circuits – a survey, *ACM Computing Surveys*, Vol. 45 (2), February 2013.
4. Lukac, M., Kameyama, M., Miller, M. and Perkowski, M.: High Speed Genetic Algorithms in Quantum Logic Synthesis: Low Level Parallelization vs. Representation. *Multiple-Valued Logic & Soft Computing* Vol. 20, (1-2), 2012, pp. 89–120.
5. Lukac, M., Perkowski, M. and Kameyama, M.: Evolutionary Quantum Logic Synthesis of Boolean Reversible Logic Circuits Embedded in Ternary Quantum Space using Heuristics, 2011, CoRRabs/1107.3383.
6. Drechsler, R., Finder, A. and Wille, R.: Improving ESOP-Based Synthesis of Reversible Logic Using Evolutionary Algorithms, Proceedings of EvoApplications (2), Berlin, Heidelberg: Springer-Verlag (Lecture Notes in Computer Science, Vol. 6625), 2011, pp. 151-161.

7. Oltean, M.: Evolving Reversible Circuits for the Even-Parity Problem. Applications of Evolutionary Computing, Berlin, Heidelberg: Springer-Verlag (Lecture Notes in Computer Science Vol. 3449), 2005, pp. 225–234.
8. Hadjam, F. Z. and Moraga, C.: A new linear genetic programming based system for reversible digital circuit design: RIMEP2, Technical Report ECSC-2013-FSC-04, European Centre for Soft Computing, 2013, ISSN 2254 - 2736.
9. Spector L.: *Automatic Quantum Computer Programming: A Genetic Programming Approach.* Boston, MA, Kluwer Academic Publishers, 2006/2007.
10. Golubitsky, O. and Maslov, D.: A Study of Optimal 4-Bit Reversible Toffoli Circuits and their Synthesis, *IEEE Transactions on Computers*, vol. 61 (9), 2012, pp. 1341–1353.
11. Prasad, A. K., Shende V. V., Markov, I. L., Hayes, J. P. and Patel, K. N.: Data structures and algorithms for simplifying reversible circuits, *J. Emerg. Technol. Comput. Syst.* Vol. 2 (4), 2006, pp. 277–293.
12. Yang, G., Song, X., Hung, W. N. N. and Perkowski, M. A.: Fast synthesis of exact minimal reversible circuits using group theory, Proceedings of the ASP-DAC 2005. Asia and South Pacific Design Automation Conference ('ASP-DAC'), ACM Press, 2005, pp. 1002–1005.
13. [6] Yang, G., Song, X., Hung, W. N. N. and Perkowski, M. A.: Bi-Directional Synthesis for Reversible Circuits, Proceedings of the IEEE Computer Society Annual Symposium on VLSI: New Frontiers in VLSI Design (ISVLSI'05), 2005, pp. 14–19.
14. Yang, G., Song, X., Hung, W. N. N. and Perkowski, M. A.: Bi-Directional Synthesis of 4-Bit Reversible Circuits. *Comput. J.* vol. 51 (2), 2008, pp. 207–215.
15. Li, Z., Chen, H., Xu, B., Liu, W., Song, X. and Xue, X.: Fast algorithm for 4-qubit reversible logic circuits synthesis, Proceedings of the IEEE Congress on Evolutionary Computation, 2008, pp. 2202–2207.
16. Li, Z., Chen, H., Yang, G. and Liu, W.: Efficient Algorithms for Optimal 4-Bit Reversible Logic System Synthesis, *Journal of Applied Mathematics*, vol. 2013, Article ID 291410.
17. Szyprowski, M. and Kerntopf, P.: Reducing Quantum Cost in Reversible Toffoli Circuits, Proceedings of the Reed-Muller Workshop, Tuusula, Finland, 2011, pp. 127–136, also available at arXiv:1105.5831 [quant-ph].
18. Szyprowski, M. and Kerntopf, P.: An Approach to Quantum Cost Optimization in Reversible Circuits, Proceedings of the 11th IEEE Conference on Nanotechnology, 2011, pp. 1521–1526.
19. Szyprowski, M. and Kerntopf, P: A study of optimal 4-bit reversible circuit synthesis from mixed-polarity Toffoli gates, Proceedings of the 12th IEEE Conference on Nanotechnology (IEEE-NANO), 2012, Birmingham, pp. 1–6.
20. Szyprowski, M. and Kerntopf, P.: Optimal 4-bit Reversible Mixed-Polarity Toffoli Circuits, In: Reversible Computation, 4th International Workshop, RC 2012, Copenhagen, Denmark, July 2-3, 2012. Revised Papers, Berlin Heidelberg: Springer (Lecture Notes in Computer Science, Vol. 7581), 2013. pp. 138–151.
21. Soeken, M., Wille, R., Dueck, G. W. and Drechsler, R.: Window optimization of reversible and quantum circuits, Proceedings of the IEEE 13th International Symposium on Design and Diagnostics of Electronic Circuits and Systems, Vienna, 2010, pp. 341-345.
22. Hadjam, F. Z. and Moraga, C. Fatima: RIMEP2, Evolutionary Design of Reversible Digital Circuits. Submitted. ACM Journal on Emerging Technologies in Computing Systems. Special Issue on Computational Synthetic Biology and Regular Papers. Vol. 11 (3), December 2014 Article No. 27 ACM New York, NY, USA. http://jetc.acm.org/
23. Maslov, D.: Reversible logic synthesis benchmarks page, 2012. Available athttp://www.cs.uvic.ca/~dmaslov. Last accessed July 2012.
24. Barenco, A., Bennett, C.H., Cleve, Di Vincenzo, R., D.P., Margolus, N., Shor, P., Sleator, T., Smolin, J.A. and Weinfurter, H.:Elementary gates for quantum computation. Phys- Rev. A 52, 1995, pp. 3457–3467.
25. Maslov, D. and Dueck, G. W.: Reversible cascades with minimal garbage. *IEEE Trans. on CAD*, Vol. 23 (11), 2004, pp.1497–1509.
26. Alba, E. and Troya J. M.: A survey of parallel distributed genetic algorithms. *Complexity*, Vol. 4 (4), 1999, pp. 31–52.

27. Gustafson, S.: An Analysis of Diversity in Genetic Programming. Ph.D. Dissertation, School of Computer Science and Information Technology, University of Nottingham, Nottingham, U.K., 2004.
28. Munawar, A.; Wahib, M.; Munetomo, M. and Akama, K.: A Survey: Genetic Algorithms and the Fast Evolving World of Parallel Computing. Proceedings of the 10th IEEE International Conference on High Performance Computing and Communications, 2008, pp. 897–902.
29. Knysh D. S., Kureichik, V. M.: Parallel genetic algorithms: a survey and problem state of the art. *Journal of Computer and Systems Sciences International*, Vol. 49 (4), 2010, pp. 579–589.
30. Sudholt, D.: Parallel Evolutionary Algorithms, in: Kacprzyk, J. and Pedrycz, W. (eds.): *Handbook of Computational Intelligence*, Berlin Heidelberg: Springer-Verlag, 2015. pp. 929–959.
31. Cantu Paz, E.: A survey of parallel genetic algorithms, Technical Report, Illinois Genetic Algorithms Laboratory, University of Illinois at Urbana Champaign, Urbana, IL. 1997.
32. Nicoara, E. S.: Mechanisms to Avoid the Premature Convergence of Genetic Algorithms. *Petroleum - Gas University of Ploiesti, Bulletin Mathematics* - I. Vol. 61 (1), 2009, pp. 87–96.
33. Gupta, D. and Ghafir, S.: An Overview of methods maintaining Diversity in Genetic Algorithms, *International Journal of Emerging Technology and Advanced Engineering*, Vol. 2 (5), May 2012.
34. Den Heijer, E. and Eiben A. E.: Maintaining Population Diversity in Evolutionary Art using Structured Populations, Proceedings of the IEEE Congress on Evolutionary Computation (CEC), Cancun, 2013, pp. 529–536.
35. Hadjam, F. Z.: Tuning of Parameters of a Soft Computing System for the Synthesis of Reversible Circuits. MVLSC, *Multiple-Valued Logic & Soft Computing* 24.1-4, 2014, p. 341–368.
36. Younes, A.: Detection and Elimination of Non-trivial Reversible Identities. *International Journal of Computer Science, Engineering and Applications* (IJCSEA) Vol.2 (4), August 2012.
37. Lidong. Dortmund University, 2009, http://lidong.hrz.tu-dortmund.de/ldw/index.php/MainPage

Chapter 15
From Boolean to Multi-valued Bent Functions

Bernd Steinbach

15.1 Introduction

The growing utilization of the internet increases the requirements of cryptosystems. The aim of such systems is the transmission of a given plain message such that the transferred message contains the given message in an encrypted form which cannot be understood by anyone, see, e.g., Chunhui Wu and Bernd Steinbach in the chapter *Utilization of Boolean Functions in Cryptography* ([15], p. 220). Both the encryption and decryption are controlled by Boolean functions. The properties of these functions strongly influence the required efforts to decrypt a message without knowing the key.

The use of linear functions for the encryption preserves the probability that a certain character appears in the cipher-text. Hence, simple statistical methods can be successfully used in an attack against corresponding cryptosystems. Taking functions with the largest distance to all linear functions within a cryptosystem hedge these attacks. Such functions are called *Bent functions*. Bernd Steinbach and Christian Posthoff derived in *Classes of Bent Functions Identified by Specific Normal Forms and Generated Using Boolean Differential Equations* ([13], p. 359) the following definition of the Bent functions.

Definition 15.1.1 (*Bent Function*) Let $f(\mathbf{x})$ be a Boolean function of n variables, where n is even. $f(\mathbf{x})$ is a bent function if its non-linearity is as large as possible.

This contribution reflects some steps on the way of the specification of Boolean bent functions by O. S. Rothaus in ([8], p. 300) to the generalization to multi-valued bent function by Claudio Moraga et al. in *Multiple-valued Functions with Bent Reed-Muller Spectra* ([7], pp. 309–324).

B. Steinbach (✉)
Department of Computer Science, Freiberg University of Mining and Technology, Freiberg, Germany
e-mail: steinb@informatik.tu-freiberg.de

R. Seising and H. Allende-Cid (eds.), *Claudio Moraga: A Passion for Multi-Valued Logic and Soft Computing*, Studies in Fuzziness and Soft Computing 349, DOI 10.1007/978-3-319-48317-7_15

15.2 Fundamental Observations of Rothaus

O. S. Rothaus has published in 1976 the pioneering paper *On "Bent" Functions* ([8], pp. 300–305). He studied Boolean functions $P(\mathbf{x})$ from $GF(2^n)$ to $GF(2)$ and noticed that a small fraction of these functions has the property that all Fourier coefficients of $(-1)^{P(\mathbf{x})}$ are equal to ± 1. He called functions with this property *bent*.

On the first page of this paper Rothaus introduced the *Fourier coefficients* as follows:

Now let ω be the real number -1. Then $\omega^{P(\mathbf{x})}$ is a well-defined real function on V_n, and by the character theory for abelian groups we may write:

$$\omega^{P(\mathbf{x})} = \frac{1}{2^{n/2}} \sum_{\lambda \in V_n} c(\lambda) \omega^{<\lambda,\mathbf{x}>} \tag{15.1}$$

where the $c(\lambda)$, the *Fourier coefficients* of $\omega^{P(\mathbf{x})}$, are given by

$$c(\lambda) = \frac{1}{2^{n/2}} \sum_{\mathbf{x} \in V_n} \omega^{P(\mathbf{x})} \omega^{<\lambda,\mathbf{x}>} . \tag{15.2}$$

In words, $2^{n/2} c(\lambda)$ is the number of zeros minus the number of ones of the function $P(\mathbf{x}) + <\lambda, \mathbf{x}>$.

From Parseval's equation, we know that

$$\sum_{\lambda \in V_n} c^2(\lambda) = 2^n . \tag{15.3}$$

We call $P(\mathbf{x})$ a *bent* function if all Fourier coefficients of $\omega^{P(\mathbf{x})}$ are equal to ± 1."

Based on this definition O. S. Rothaus found some important propositions:

Proposition 15.2.1 ([8], p. 301) *The Fourier transform of a bent function is a bent function.*

Proposition 15.2.2 ([8], p. 301) *$P(\mathbf{x})$ is bent if and only if $\omega^{P(\mathbf{x}+\mathbf{y})}$ is a Hadamard matrix.*

Proposition 15.2.3 ([8], p. 301) *If $P(\mathbf{x})$ is a bent function on V_n, then n is even, $n = 2k$; the degree of $P(\mathbf{x})$ is at most k, except in the case $k = 1$.*

Proposition 15.2.4 ([8], p. 303) *If $P(\mathbf{x})$ is a bent function on V_{2k}, $k > 3$, of degree k, then $P(\mathbf{x})$ is irreducible.*

15.3 Boolean Differential Equations of Bent Functions

It is well-known that the solution of a Boolean equation is a set of Boolean vectors. Bernd Steinbach found in his Ph.D. Thesis that the solution of a Boolean differential equation is a *set of Boolean functions* ([10], p. 40). Methods to solve Boolean differential equations and many applications are recently published by Bernd Steinbach and Christian Posthoff in the book *Boolean Differential Equations* ([12], p. 1 ff.). Knowing these methods it is easy to solve a Boolean differential equation. However, it remains a difficult problem to find the Boolean differential equation that describes all Boolean functions that satisfy a certain property.

Bernd Steinbach and Christian Posthoff published in *Classes of Bent Functions Identified by Specific Normal Forms and Generated Using Boolean Differential Equations* a Boolean differential equation that specifies all bent functions of two variables ([13], p. 370) as well as a Boolean differential equation that specifies all bent functions of four variables ([13], p. 372).

All eight bent functions of two variables are specified by the Boolean differential equation

$$\frac{\partial^2 f(x_1, x_2)}{\partial x_1 \partial x_2} = 1 \ . \tag{15.4}$$

These are half of all $2^{2^2} = 16$ functions of two variables. The fraction of bent functions will be much smaller for Boolean functions of a larger number of variables. All 896 bent functions of four variables are specified by the Boolean differential equation

$$\frac{\partial^2 f(\mathbf{x})}{\partial x_1 \partial x_2} \cdot \frac{\partial^2 f(\mathbf{x})}{\partial x_3 \partial x_4} \oplus \frac{\partial^2 f(\mathbf{x})}{\partial x_1 \partial x_3} \cdot \frac{\partial^2 f(\mathbf{x})}{\partial x_2 \partial x_4} \oplus \frac{\partial^2 f(\mathbf{x})}{\partial x_1 \partial x_4} \cdot \frac{\partial^2 f(\mathbf{x})}{\partial x_2 \partial x_3} = 1 \ . \tag{15.5}$$

These are only 1.37 % of all $2^{2^4} = 65,536$ functions of four variables.

The approach to specify bent functions can be generalized for Boolean functions of more than four variables. A method of recursive extension of Boolean differential equations for bent functions was suggested by Bernd Steinbach and Christian Posthoff in ([13], p. 376), where a Boolean differential equation for bent functions of six variables was given

$$\begin{aligned}
&\frac{\partial^2 f(\mathbf{x})}{\partial x_1 \partial x_2} \cdot \left(\frac{\partial^2 f(\mathbf{x})}{\partial x_3 \partial x_4} \cdot \frac{\partial^2 f(\mathbf{x})}{\partial x_5 \partial x_6} \oplus \frac{\partial^2 f(\mathbf{x})}{\partial x_3 \partial x_5} \cdot \frac{\partial^2 f(\mathbf{x})}{\partial x_4 \partial x_6} \oplus \frac{\partial^2 f(\mathbf{x})}{\partial x_3 \partial x_6} \cdot \frac{\partial^2 f(\mathbf{x})}{\partial x_4 \partial x_5} \right) \oplus \\
&\frac{\partial^2 f(\mathbf{x})}{\partial x_1 \partial x_3} \cdot \left(\frac{\partial^2 f(\mathbf{x})}{\partial x_2 \partial x_4} \cdot \frac{\partial^2 f(\mathbf{x})}{\partial x_5 \partial x_6} \oplus \frac{\partial^2 f(\mathbf{x})}{\partial x_2 \partial x_5} \cdot \frac{\partial^2 f(\mathbf{x})}{\partial x_4 \partial x_6} \oplus \frac{\partial^2 f(\mathbf{x})}{\partial x_2 \partial x_6} \cdot \frac{\partial^2 f(\mathbf{x})}{\partial x_4 \partial x_5} \right) \oplus \\
&\frac{\partial^2 f(\mathbf{x})}{\partial x_1 \partial x_4} \cdot \left(\frac{\partial^2 f(\mathbf{x})}{\partial x_2 \partial x_3} \cdot \frac{\partial^2 f(\mathbf{x})}{\partial x_5 \partial x_6} \oplus \frac{\partial^2 f(\mathbf{x})}{\partial x_2 \partial x_5} \cdot \frac{\partial^2 f(\mathbf{x})}{\partial x_3 \partial x_6} \oplus \frac{\partial^2 f(\mathbf{x})}{\partial x_2 \partial x_6} \cdot \frac{\partial^2 f(\mathbf{x})}{\partial x_3 \partial x_5} \right) \oplus
\end{aligned}$$

$$\frac{\partial^2 f(\mathbf{x})}{\partial x_1 \partial x_5} \cdot \left(\frac{\partial^2 f(\mathbf{x})}{\partial x_2 \partial x_3} \cdot \frac{\partial^2 f(\mathbf{x})}{\partial x_4 \partial x_6} \oplus \frac{\partial^2 f(\mathbf{x})}{\partial x_2 \partial x_4} \cdot \frac{\partial^2 f(\mathbf{x})}{\partial x_3 \partial x_6} \oplus \frac{\partial^2 f(\mathbf{x})}{\partial x_2 \partial x_6} \cdot \frac{\partial^2 f(\mathbf{x})}{\partial x_3 \partial x_4} \right) \oplus$$
$$\frac{\partial^2 f(\mathbf{x})}{\partial x_1 \partial x_6} \cdot \left(\frac{\partial^2 f(\mathbf{x})}{\partial x_2 \partial x_3} \cdot \frac{\partial^2 f(\mathbf{x})}{\partial x_4 \partial x_5} \oplus \frac{\partial^2 f(\mathbf{x})}{\partial x_2 \partial x_4} \cdot \frac{\partial^2 f(\mathbf{x})}{\partial x_3 \partial x_5} \oplus \frac{\partial^2 f(\mathbf{x})}{\partial x_2 \partial x_5} \cdot \frac{\partial^2 f(\mathbf{x})}{\partial x_3 \partial x_4} \right) = 1 . \tag{15.6}$$

Unfortunately, these extended Boolean differential equations describe only a subset of all bent functions.

15.4 Enumeration of Bent Functions

It is a very hard problem to determine the numbers of bent functions for larger Boolean spaces. Natalia Tokareva uses in her paper *On the Number of Bent Functions from Iterative Constructions: Lower Bounds and Hypotheses* ([14], p. 609 ff.) an iterative approach to find tighter lower bounds for the number of bent functions. This approach has been based on

Theorem 15.4.1 ([14], p. 612) *Let the functions f_0, f_1, f_2 be bent functions in n variables. Then the function g defined by*

$$g(00, \mathbf{x}) = f_0(\mathbf{x}), g(01, \mathbf{x}) = f_1(\mathbf{x}), g(10, \mathbf{x}) = f_2(\mathbf{x}), g(11, \mathbf{x}) = f_3(\mathbf{x}), \tag{15.7}$$

is a bent function in $n + 2$ variables if and only if f_3 is a bent function in n variables and

$$\tilde{f}_0 \oplus \tilde{f}_1 \oplus \tilde{f}_2 \oplus \tilde{f}_3 = 1 . \tag{15.8}$$

Natalia Tokareva summarizes her results about the number of iterative bent functions $|\mathscr{BI}_n|$ in some propositions:

Proposition 15.4.1 ([14], p. 617)

$$|\mathscr{BI}_4| = 512, \quad |\mathscr{BI}_6| = 322,961,408 \approx 2^{28.3} . \tag{15.9}$$

Proposition 15.4.2 ([14], p. 618) *With probability 0.999 it holds that*

$$2^{87.36} < |\mathscr{BI}_8| < 2^{87.38} . \tag{15.10}$$

Proposition 15.4.3 ([14], p. 618)

$$|\mathscr{BI}_8| > 197,004,891,331,091,000,000,000,000 \approx 2^{87.35} . \tag{15.11}$$

Proposition 15.4.4 ([14], p. 618)

$$|\mathscr{B}_{10}| > |\mathscr{BI}_{10}| > 830,602,255,559,379 \cdot 10^{64} > 2^{262.16} . \tag{15.12}$$

Jon T. Butler tried in the paper *Bent Function Discovery by Reconfigurable Computer* ([1], pp. 1–19) to sieve bent functions from the set of all Boolean functions using the special SRC-6 reconfigurable computer. Despite of the low clock of 100 MHz of the reconfigurable computer, the acceleration of the programmed hardware reaches a speedup factor of 592.0 in comparison to a PC running on 2.8 GHz to find all 896 bent functions of four variables in less than one millisecond ([1], p. 5).

This speedup grows for increased numbers of variables. For example, the speedup factor reaches 6,805.9 for the task to sieve all bent functions of six variables. Despite this large speedup of the utilized hardware it will need 5,840 years to find all bent functions of six variables ([1], p. 5). The evaluated speedup grows to 62,111 in the case of Boolean functions of 10 variables; however, the expected time to calculate all bent functions of 10 variables on the reconfigurable computer SRC-6 is 3.67×10^{61} years ([1], p. 5).

15.5 Classes of Bent Functions

Due to the unique flat spectrum, defined by O. S. Rothaus to characterize the bent functions, it can be assumed that all bent functions have a similar structure. This assumption is not true. A first hint for this observation gives the approach *Class Separation* ([12], p. 64 ff.). The solution set of the Boolean differential equation for bent functions of four variables consists of 56 classes of 16 functions.

A detailed analysis of these classes was published by Bernd Steinbach and Christian Posthoff in *Classes of Bent Functions Identified by Specific Normal Forms and Generated Using Boolean Differential Equations* ([13], p. 362 ff.). The 56 classes can be divided into two subsets of classes. Each pair of associated classes of these subsets has the property that their functions are the complement of each other.

A possibility to distinguish these classes of bent functions consists in the evaluation of their complexity. Bernd Steinbach and Alan Mishchenko provided in the paper *SNF: A Special Normal Form for ESOPs* ([11], p. 69) the construction of a unique canonical *Exclusive Sum of Product (ESOP)* called SNF(f) among all 2^{3^n} ESOPs of n variables. Bernd Steinbach suggested in the journal paper *Most Complex Boolean Functions Detected by the Specialized Normal Form* ([9], p. 271) the number of cubes in the SNF(f) as a complexity measure of the Boolean function f.

Table 15.1, taken from ([13], p. 363), shows in the first column the number of cubes in the unique SNF(f) as a complexity measure and in the last column the number of bent functions belonging to these complexity classes. It can be seen that 48 bent functions of four variables belong to the complexity class of 30 cubes in their SNF and the 16 most complex bent functions of four variables generate 50 cubes in the SNF. Table 15.1 shows also that no bent function of four variables belongs to the set of the 24 most complex Boolean functions of four variables.

Representative bent functions of four variables for all classes are given in ([13], p. 378 ff.). An example of a simplest bent function is

Table 15.1 Distribution of 896 bent functions into classes of $SNF(f)$ for all 65536 boolean functions of 4 variables

Cubes in the		Number of	
SNF	Minimal ESOP	All Functions	Bent Functions
0	0	1	0
16	1	81	0
24	2	324	0
28	2	1296	0
30	2	648	48
32	3	648	0
34	3	3888	240
36	3	6624	0
36	4	108	0
38	3	7776	384
40	3	2592	0
40	4	6642	0
42	3	216	0
42	4	14256	192
44	4	12636	0
46	4	3888	0
46	5	1296	16
48	5	1944	0
50	5	648	16
54	6	24	0

$$f_{b1} = x_1\, x_2 \oplus x_3\, x_4\ , \tag{15.13}$$

and the positive polarity ESOP of one of the most complex bent functions of four variables is

$$f_{b28} = x_1\, x_2 \oplus x_1\, x_3 \oplus x_1\, x_4 \oplus x_2\, x_3 \oplus x_2\, x_4 \oplus x_3\, x_4 \oplus 1\ . \tag{15.14}$$

15.6 Bent Functions and Cryptography

Jon T. Butler and Tsutomu Sasao provide in their chapter *Boolean Functions and Cryptography* ([2], p. 33 ff.) a concise exposition of properties of Boolean functions utilized in cryptography. In addition to the to the bent functions they explore Boolean functions that satisfy other properties needed to resist attacks against the cipher. Bent functions have the benefit that their distance to all linear function is as large as possible; however, the drawback of bent functions is that they are not balanced. In order to eliminate this drawback other properties of Boolean functions are utilized in cryptography. Here are the definitions of such properties.

Definition 15.6.1 ([2], *p. 45*) A function f satisfies the *strict avalanche criterion (SAC)* if and only if complementing any single variable complements exactly half of the function values.

Definition 15.6.2 ([2], *p. 45*) An n-variable function f is *balanced* if and only if the number of function values 1 is equal to 2^{n-1}.

Definition 15.6.3 ([2], *p. 46*) An n-variable function f satisfies the *propagation criterion (PC(k))* if and only if for any assignment to k of n variables the resulting function satisfies SAC.

Definition 15.6.4 ([2], *p. 47*) An n-variable function f has *correlation immunity k* if and only if, for every fixed set S of k variables, $1 \leq k \leq n$, given the value of f, the probability that S takes on any of its 2^k assignments of values to the k variables is $\frac{1}{2^k}$.

Definition 15.6.5 ([2], *p. 47*) If an n-variable function f has correlation immunity k and is balanced then it has *resiliency* of order k.

A similar exploration *Utilization of Boolean Functions in Cryptography* was recently published by Chunhui Wu and Bernd Steinbach ([15], p. 220 ff.). In addition to the properties mentioned above, further relevant properties in cryptography are explored.

Definition 15.6.6 ([15], *p. 227*) The *nonlinearity* N_f describes the smallest distance between the Boolean function $f(\mathbf{x})$ and all linear (affine) functions $l(\mathbf{x}) \in L_n(\mathbf{x})$:

$$N_f = \min_{l(\mathbf{x}) \in L_n(\mathbf{x})} d(f(\mathbf{x}), l(\mathbf{x})) = \min_{l(\mathbf{x}) \in L_n(\mathbf{x})} w(f(\mathbf{x}) \oplus l(\mathbf{x})) \,. \tag{15.15}$$

There is the following relation between the nonlinearity and the Walsh spectrum according to their definitions:

$$N_f = \frac{2^n - \max_{\mathbf{w} \in GF(2)^n} |S_{(f)}(\mathbf{w})|}{2} \,. \tag{15.16}$$

Definition 15.6.7 ([15], *p. 228*) The *algebraic degree* $\deg(f(\mathbf{x}))$ of a function $f(\mathbf{x})$ is defined as the number of Boolean variables in the highest order monomials with a nonzero coefficient in its antivalence normal form.

There are trade-offs between correlation immunity CI_f and algebraic degree deg $(f(\mathbf{x}))$ as well as the correlation immunity CI_f and the nonlinearity N_f.

Definition 15.6.8 (*Algebraic Immunity,* [15], *p. 228*) Algebraic attacks have gained much concern since 2002. An over-defined system of high degree equations between the original status and the key stream is established and solved using linearization methods. In order to resist an algebraic attack, the Boolean function $f(\mathbf{x})$ should have the property that there is no non-zero Boolean function $g(\mathbf{x})$ such that $f(\mathbf{x})\, g(\mathbf{x}) = h(\mathbf{x})$ (or $\overline{f(\mathbf{x})}\, g(\mathbf{x}) = h(\mathbf{x})$) and $h(\mathbf{x})$ has a low algebraic degree. If $f(\mathbf{x})\, g(\mathbf{x}) = 0$ and $g(\mathbf{x}) \neq 0$, then $g(\mathbf{x})$ is called an *annihilator* of $f(\mathbf{x})$. It has been proven that the lowest degree of all the multiples of $f(\mathbf{x})$ and $\overline{f(\mathbf{x})}$ is equal to the lowest degree of all the annihilators of $f(\mathbf{x})$ and $\overline{f(\mathbf{x})}$. This lowest degree is defined as the algebraic immunity $\mathscr{A}\mathscr{I}$ of $f(\mathbf{x})$ (or $\overline{f(\mathbf{x})}$):

$$\begin{aligned}\mathscr{A}\mathscr{I}(f(\mathbf{x})) &= \mathscr{A}\mathscr{I}(\overline{f(\mathbf{x})}) \\ &= \min\{\deg(g(\mathbf{x}))| f(\mathbf{x})\, g(\mathbf{x}) = 0 \text{ or } \overline{f(\mathbf{x})} g(\mathbf{x}) = 0\}\,. \end{aligned} \qquad (15.17)$$

Chunhui Wu and Bernd Steinbach explored in *Utilization of Boolean Functions in Cryptography* ([15], p. 220 ff.) not only the theory of cryptography, but they also explained several practical applications. Very often the DES-algorithm is used for both encryption and decryption. Figure 15.1, taken from ([15], p. 234), shows the hardware structure that implements this algorithm.

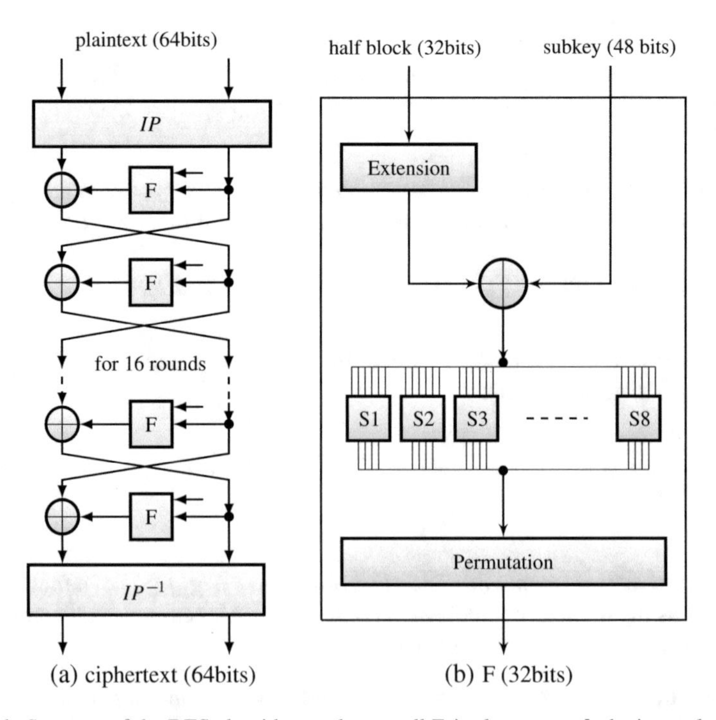

Fig. 15.1 Structure of the DES algorithm: **a** the overall Feistel structure, **b** the internal structure of the Feistel function F

Table 15.2 Properties of DES S-box and AES S-box

	bound ($n \times m$)	DES S-box (6×4)	AES S-box (8×8)
Orthogonality	-	Yes	Yes
Algebraic degree of every function	$\leq n-1$	5 (5)	7 (7)
Nonlinearity of every function	$\leq 2^{n-1} - 2^{\frac{n}{2}-1}$	$14 \sim 22$ (28)	112 (120)
Differential uniformity	$\geq 2^{n-m+1}$	16 (8)	4 (2)
Robustness	$\leq (2^{-1} + 2^{m-n-1}) \cdot (1 - 2^{-m+1})$	$0.316 \sim 0.469$ (0.546875)	0.984375 (0.9921875)

The security of block ciphers as DES mainly depends on the properties of the S-box, which is the only nonlinear part. In order to resist known attacks, the Boolean functions of the S-box require the properties of orthogonality (corresponding to the balance in the single-output Boolean function), a high algebraic degree, a high nonlinearity, and a high algebraic immunity. Additional properties for resisting differential attacks are the differential uniformity and robustness. All these requirements are completely satisfied by the four Boolean functions of six variable assigned to the eight S-boxes. These functions map $GF(2^6)$ to $GF(2)$.

Using the same basic structure of the DES implementation the advanced encryption standard (AES) was developed. The 6×4 S-box of the DES is replaced in the AES by an 8×8 S-box which is constructed based on a multiple-inverse operation on $GF(2^8)$. The analysis shows that the AES S-box is orthogonal, i.e., it is a Boolean permutation on $GF(2)^8$. Not only all eight Boolean functions, but the S-box as a whole has the maximum algebraic degree. The low differential uniformity and high robustness (close to its upper bound) guarantee the ability of AES to resist differential attacks. Both the eight Boolean functions and the S-box as a whole also have a high nonlinearity which are close to the totally nonlinear functions, i.e., the bent functions. The construction of the AES S-box is one step in cryptography from the Boolean into the multi-valued domain.

Table 15.2, taken from ([15], p. 237), compares the cryptographic properties of the S-boxes of DES and AES.

15.7 Bent Functions in the Multi-valued Domain

Claudio Moraga explored multi-valued bent function over a long period of time. Commonly with M. Luis he published already in 1989 the paper *On Functions with Flat Chrestenson Spectra* ([3], p. 406 ff.).

In the last years Claudio Moraga collaborated with Milena Stanković, Radomir S. Stanković, and Suzana Stojković of the University of Niś and published the new results in a series of papers and chapters of books.

In the chapter *Hyper-Bent Multiple-Valued Functions* ([4], p. 250 ff.) they have shown that multi-valued hyper-bent functions constitute a reduced subset of the multi-valued bent functions, have provided a simple lemma for their characterization, and have introduced the new concept of strict hyper-bent functions.

In the paper *Contribution to the Study of Multiple-Valued Bent Functions* ([5], p. 340 ff.) they used the Vilenkin-Chrestenson transform to determine the bent property of a multi-valued function.

Definition 15.7.1 (*Vilenkin-Chrestenson Transform:* [5], *p. 340*) The direct and inverse *Vilenkin-Chrestenson transform* of a function

$$f : GF(p)^n \rightarrow GF(p) , \tag{15.18}$$

are defined as follows:

$$S_f(w) = \sum_{z \in FG(p)^n} \chi_w^*(z) \cdot hf(z) , \tag{15.19}$$

$$f(z) = h^{-1}\left(p^{-1} \sum_{w \in FG(p)^n} chi_w(z) \cdot S(w)\right) , \tag{15.20}$$

where $\chi_w^*(z)$ denotes the complex conjugate of $\chi_w(z)$,

$$\chi_w(z) = \xi^{w^T z} \quad \text{and} \quad \chi_w^*(z) = \xi^{-w^T z} . \tag{15.21}$$

Using this definition, they generalized the definition of Rothaus regarding bent functions for the multi-valued case on two levels of requirements.

Definition 15.7.2 (*Multi-Valued Bent Function:* [5], *p. 341*) A function f (15.18), p prime, $p > 2$ is called a *multi-valued bent function* if all elements of its circular Vilenkin-Chrestenson Spectrum have the same magnitude. In this case the spectrum is said to be “flat”.

Definition 15.7.3 (*Strictly Multi-valued Bent Function:* [5], *p. 344*) Let f (15.18) be a bent function. This function will be called *strictly bent* if the elements of its spectrum may be expressed as rotations of $S_f(0)$ by powers of ξ.

Formally, f is strictly bent $\Leftrightarrow \exists g : GF(p)^n \rightarrow GF(p)$ such that

$$\forall w \in GF(p)^n, \; S_f(w) = S_f(0) \cdot \xi^{g(w)} . \tag{15.22}$$

The function g will be called the dual of f.

Based on these definitions they studied some of their properties and have proven, that there are 18 ternary bent functions with only one argument, 486 ternary bent functions with two arguments, and 100 five-valued bent functions with one argument. All of them are strictly bent.

The same authors presented their extended knowledge about this topic one year later in the paper *The Maiorana Method to Generate Multiple-Valued Bent Functions Revisited* ([6], p. 19 ff.). Reusing Definitions 15.7.2 and 15.7.3 they concluded after a detailed exploration that the representation of multi-valued functions based on matrices and the Vilenkin-Chrestenson transform allows a better understanding of both the power and the limitations of the Maiorana method. In the context of some proofs they noticed the interesting observation that the Maiorana method is more flexible due to permutations instead of the identity originally used by Rothaus. The new formal structure additionally allows an estimation of a lower bound on the number of multi-valued bent functions. For ternary bent functions of four arguments they found the lower bound of 708,588 functions.

Claudio Moraga, Milena Stanković, and Radomir S. Stanković continued this research and contributed recently in their chapter *Multi-Valued Functions with Bent Reed-Muller Spectra* an innovative exploration of bent functions in the multi-valued domain using the Reed-Muller transform ([7], p. 309 ff.).

Definition 15.7.4 (*Reed-Muller Transform,* [7], *p. 311*) The matrix **T** representing the basis of the Reed-Muller transform of p-valued functions is defined as follows:

$$\mathbf{T}(1) = \begin{bmatrix} 1 & x & x^2 & x^3 & \dots & x^{p-1} \end{bmatrix} \tag{15.23}$$
$$= \begin{bmatrix} 1 & 0 & 0 & 0 & \dots & 0 \\ 1 & 1 & 1 & 1 & \dots & 1 \\ 1 & 2 & 2^2 & 2^3 & \dots & 2^{p-1} \\ 1 & 3 & 3^2 & 3^3 & \dots & 3^{p-1} \\ \vdots & \vdots & \vdots & \vdots & \ddots & \vdots \\ 1 & p-1 & (p-1)^2 & (p-1)^3 & \dots & (p-1)^{p-1} \end{bmatrix} \bmod p\,,$$

$$\begin{aligned} \mathbf{T}(n) = {} & \begin{bmatrix} 1 & x_{n-1} & x_{n-1}^2 & \dots & x_{n-1}^{p-1} \end{bmatrix} \otimes \\ & \begin{bmatrix} 1 & x_{n-2} & x_{n-2}^2 & \dots & x_{n-2}^{p-1} \end{bmatrix} \otimes \dots \otimes \\ & \begin{bmatrix} 1 & x_1 & x_1^2 & \dots & x_1^{p-1} \end{bmatrix} \otimes \\ & \begin{bmatrix} 1 & x_0 & x_0^2 & \dots & x_0^{p-1} \end{bmatrix} \bmod p\,. \end{aligned} \tag{15.24}$$

Numerically:

$$\mathbf{T}(n) = (\mathbf{T}(1))^{\otimes n}\,, \tag{15.25}$$

where, $\otimes$, as the operation symbol, denotes the Kronecker product, and as the exponent, the n-fold Kronecker product of the argument with itself.

In detail they explored the Maiorana class which contains only n-place bent functions, where n is even.

Definition 15.7.5 (*Maiorana Class,* [7], *p. 317*) Let

$$\mathbf{M} = [m_{i,j}] \,, \tag{15.26}$$

with $m_{i,j} = i \cdot j \bmod p, i, j \in \mathbb{Z}_p$.

Lemma 15.7.1 *[Property of Maiorana Class, [7], p. 317] Extending the concept of modular weight to matrices,*

$$\varphi(\mathbf{M}) \equiv 0 \bmod p \,. \tag{15.27}$$

Lemma 15.7.2 *[Bent Reed-Muller Spectrum of the Maiorana Class, [7], p. 317] If* **R** *is a bent Reed-Muller spectrum of the Maiorana class, then*

$$\varphi(\mathbf{R}) \equiv 0 \bmod p \,. \tag{15.28}$$

In addition to the Maiorana class they defined the Γ-class of multi-valued bent functions.

Definition 15.7.6 (*γ-Class,* [7], *p. 316*) For a given p, let γ be the set of all one-place p-valued bent functions. Let Γ denote the set of all p-valued functions obtained as the tensor sum of one-place bent functions, including repetitions and reorderings.

Using the modular weight:

Definition 15.7.7 (*Modular Weight φ,* [7], *p. 313*) The modular weight of a p-valued vector is given by the sum mod p of all its components. The symbol φ will be used to denote the modular weight.

they have proven the lemma do determine bent functions within a Γ-class:

Lemma 15.7.3 *[Multi-valued Bent Functions, [7], p. 316] The elements of* Γ *are bent and if* **V** $\in \Gamma$ *then*

$$\varphi(\mathbf{V}) \equiv 0 \bmod p \,. \tag{15.29}$$

and concluded the consequences:

Consequence 15.7.1 *(*Bent Reed-Muller Spectrum, *[7],* p. 316*).*
A necessary *condition for a Reed-Muller spectrum* **R** *to be bent (and belong to* Γ*) is that*

$$\varphi(\mathbf{R}) \equiv 0 \bmod p \,. \tag{15.30}$$

Consequence 15.7.2 *(*Multi-valued Bent Function, *[7],* p. 316*).*
A necessary condition for an n-place p-valued function to have a bent Reed-Muller spectrum in Γ *is, that*

$$f(\pi_{n-1}) = 0 . \tag{15.31}$$

A remarkable observation of this work is that there are 18 ternary two-place bent functions such that their Reed-Muller spectra are also bent; these functions satisfy the necessary condition $f(4) = 0$ and their value vectors are palindromes.

Although a general characterization is still missing, they have found that for the Γ-class, the Maiorana class and their tensor sums, the fact that $f(\pi - n - 1) = 0$, (for an n-place function, $n > 1$), is a necessary condition for f to have a bent Reed-Muller spectrum, which will have a modular weight congruent to $(0 \mod p)$.

Acknowledgments We would like to thank Claudio Moraga for his constructive collaboration over many years and all his contributions to extend scientific insights.

References

1. J. T. Butler, *Bent Function Discovery by Reconfigurable Computer* in: Bernd Steinbach (ed.): Boolean Problems - 9th International Workshop, 2016, pp. 1–19.
2. J. T. Butler and T. Sasao, *Boolean Functions and Cryptography*, in: T. Sasao and J. T. Butler (eds.): *Progress in Applications of Boolean Functions*, Morgan & Claypool Publishers, San Rafael, California, USA, 2010, pp. 33–53.
3. M. Luis, C. Moraga: *On Functions with Flat Chrestenson Spectra*, in: Proceedings of the IEEE 19th International Symposium on Multiple-Valued Logic, IEEE-CS-Press, 1989, Fukuoka, 1989, pp. 406–413.
4. C. Moraga and M. Stanković and R. S. Stanković and S. Stojković: *Hyper-Bent Multiple-Valued Functions* in: R. Moreno-Diaz, F. Pichler, A. Quesada-Arencibia (Editors), Lecture Notes in Computer Science, Volume 8112, Berlin Heidelberg: Springer, 2013, pp. 250–257.
5. C. Moraga and M. Stanković and R. S. Stanković and S. Stojković: *Contribution to the Study of Multiple-Valued Bent Functions*, in: Proceedings of the IEEE 43th International Symposium on Multiple-Valued Logic, IEEE-CS-Press, 2014, Toyama, 2014, pp. 19–24.
6. C. Moraga and M. Stanković and R. S. Stanković and S. Stojković: *The Maiorana Method to Generate Multiple-Valued Bent Functions Revisited*, in: Proceedings of the IEEE 44th International Symposium on Multiple-Valued Logic, IEEE-CS-Press, 2014, Bremen, 2014, pp. 340–345.
7. C. Moraga and M. Stanković and R. S. Stanković: *Multiple-valued Functions with Bent Reed-Muller Spectra*, in: Steinbach (Editor), Problems and New Solutions in the Boolean Domain, Newcastle upon Tyne: Cambridge Scholars Publishing, 2016, pp. 309–324.
8. O. S. Rothaus: *On "Bent" Functions*, *Journal of Combinatorial Theory*, (A) 20, 1976, pp. 300–305.
9. B. Steinbach: *Most Complex Boolean Functions Detected by the Specialized Normal Form*, *FACTA UNIVERSITATIS* (NIŠ), Series: Electronics And Energetics, Vol. 20 (3), 2007, pp. 259–279.
10. B. Steinbach: Lösung binärer Differentialgleichungen und ihre Anwendung auf binäre Systeme. Ph.D. Thesis, TH Karl-Marx-Stadt, 1981.
11. B. Steinbach and A. Mishchenko: *SNF: A Special Normal Form for ESOPs*, in: Proceedings of the 5th International Workshop on Application of the Reed-Muller Expansion in Circuit Design (RM 2001), Mississipi State University, Starkville (Mississipi), 2001, pp. 66–81.

12. B. Steinbach and C. Posthoff: *Boolean Differential Equations*, Morgan & Claypool Publishers, San Rafael, California, 2013.
13. B. Steinbach and C. Posthoff: *Classes of Bent Functions Identified by Specific Normal Forms and Generated Using Boolean Differential Equations, FACTA UNIVERSITATIS* (NIŠ), Series: Electronics And Energetics, Vol. 24 (3), December 2011, pp. 357–383.
14. N. Tokareva, *On the Number of Bent Functions from Iterative Constructions: Lower Bounds and Hypotheses, Advances in Mathematics of Communications*, Vol. 5 (4), 2011, pp. 609–621.
15. C. Wu and B. Steinbach: *Utilization of Boolean Functions in Cryptography*, in: B. Steinbach (ed.): *Problems and New Solutions in the Boolean Domain*, Newcastle upon Tyne: Cambridge Scholars Publishing, 2016, pp. 220–240.

Chapter 16
Claudio Moraga and the University of Santiago de Compostela: Many Years of Collaboration

Senén Barro, Alberto Bugarín and Alejandro Sobrino

16.1 A Long-Term Collaboration with the University of Santiago de Compostela

The relationship of Professor Claudio Moraga with the University of Santiago de Compostela has been very extensive in time. One of his first visits (may be the first one?) was to deliver one of the lectures at the University summer course on "Fuzzy logic: theory and applications" we organized in July 1992. For our Intelligent Systems Group of the *University of Santiago de Compostela* it was a very remarkable occasion, since this course was the first event that we organized[1] and it gave us the opportunity to host and to present to the Galician students' and research community some researchers of the highest relevance, as Prof. Moraga. His talk dealt about the "The *ORBE* (Fuzzy Experimental Computer) project", which gathered together and aligned research efforts at that time of most of the relevant research groups of the Spanish fuzzy community in the areas of fuzzy theory, fuzzy hardware and fuzzy control. Selected talks of the course were edited in a book in which Prof. Moraga kindly agreed to contribute with a chapter [17].

This participation in research dissemination and training events was not the single contribution of Prof. Moraga to our University summer courses' program. Many years after, following the organization of the course "Information and Communications Technologies in 2000", we edited the volume "Frontiers of Computation", in

[1]Many other summer courses, training activities as well as other scientific events followed this first one in the forthcoming years.

S. Barro · A. Bugarín (✉)
Centro Singular de Investigación
en Tecnoloxías da Información (CiTIUS), Santiago, Spain
e-mail: alberto.bugarin.diz@usc.es

A. Sobrino
Facultade de Filosofía, Universidade de Santiago de Compostela, Santiago, Spain

R. Seising and H. Allende-Cid (eds.), *Claudio Moraga: A Passion for Multi-Valued Logic and Soft Computing*, Studies in Fuzziness and Soft Computing 349, DOI 10.1007/978-3-319-48317-7_16

which he again accepted to contribute (this time co authoring with Prof. Enric Trillas, Senén Barro and Alberto Bugarín) with the chapter entitled "Fuzzy Computation" [2] (Fig. 16.1). In this chapter some of the basics of fuzzy sets and fuzzy reasoning were introduced as a general framework for Mamdani's fuzzy control. This perspective was maintained for years in other papers by Prof. Moraga and Prof. Trillas and could be considered a far antecedent of Professor Moraga's later interest about the formal aspects of fuzzy controllers reasoning and design we will comment about in Sect. 16.3. The chapter was an important contribution to the book, where other emerging (by those years) computing paradigms were presented (among them, Optical, Biomolecular, Quantum or Ubiquitous Computing). Some of these paradigms are still now frontier fields for research, whilst others have consolidated as technological developments and are a part of our daily life. In this regard, it is anecdotic to recover one of the statements of the starting paragraph of Prof. Moraga's et al. chapter, regarding the authors concerns about having to develop the topic of "Fuzzy Computing" in the little space a single book chapter provides, given that by that days "... the Handbook of Fuzzy Computation has more than 800 pages ..."). Such worries should have exponentially increased by now, if we recall that Springer's 2015

Fig. 16.1 Covers of the books "Studies of fuzzy logic and its applications" [17] and "Frontiers of Computation" [2] in which Prof. Moraga contributed to. Both books were the result of two Summer Courses respectively held at the University of Santiago de Compostela in 1992 and 2000

"Handbook of Fuzzy Computation" [10] comprises 81 chapters and more than 1500 pages!

These are just a few samples of some of the first collaborations of Prof. Moraga with research dissemination activities in the *University of Santiago de Compostela*. The collaboration became much more intense when it came to young researchers training, through his participation in Ph.D. assessment committees or as a lecturer in the doctoral programs of the Departments of Electronics and Computer Science (now the *Research Centre on Information Technologies*, CiTIUS) and Logic and Philosophy of Science of our university. Throughout the years, Prof. Moraga came to Santiago in a number of occasions to teach our researchers in a number of different emerging fields and hot topics, such as Neural Networks, Fuzzy Computability Theory, Fuzzy Formal Languages or Multi-nets.

Once again with Prof. Enric Trillas, Prof. Claudio Moraga greatly contributed to give international visibility of our University as one of the key persons that facilitated the organization in 1996 of the *26th IEEE International Symposium on Multiple-Valued Logic* (ISMVL) in Santiago de Compostela from 29th to 31st May 1996 (Fig. 16.2).

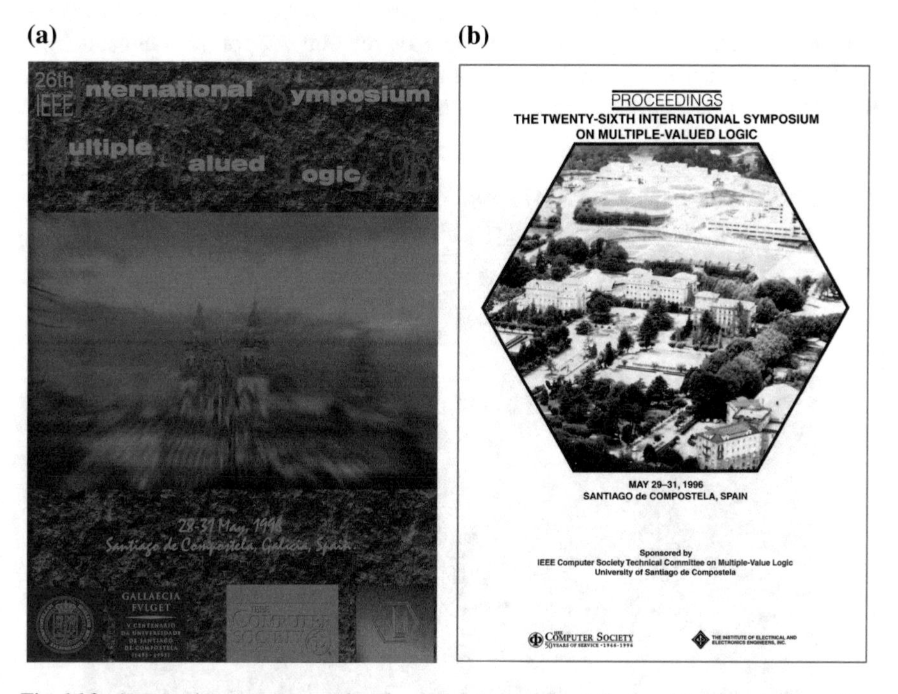

Fig. 16.2 Announcing poster greeted to the attendants and Front Cover of the Proceedings of the *26th International Symposium on Multiple-Valued Logic*, held in Santiago de Compostela. Prof. Moraga was one of the key persons on the event organization and served as one its Programme co-Chairs

This event was organized within the framework of the 5th Centennial of the University of Santiago de Compostela, which programmed and supported a number of selected scientific events for that solemn celebration. ISMVL'96 was the first relevant international scientific event we organized in the Intelligent Systems Research Group and that gave us the opportunity of welcoming for the first time in Santiago the father of fuzzy logic, Prof. Lotfi A. Zadeh (University of California at Berkeley), who delivered one of the keynote plenary lectures, about "Inference in Fuzzy Logic via Generalized Constraint Propagation". The other plenary speaker was Prof. Claudi Alsina, from the Open University of Catalonia with the talk "As you like them: connectives in fuzzy logic". Prof. Moraga was one of the promoters of Santiago's candidature to host the event and also served as one of the Program Co-Chairs of the Conference. The Program Committee of ISMVL'96 selected 46 papers by 92 authors working in 17 different countries that were refereed by reviewers from 14 countries. These are good indicators of the world-wide international dimension this event had. The topics the conference dealt with ranged from Logic, Algebra, AI, Soft Computing to Logic Design and Decision Diagrams. The conference also hosted a special session dedicated to honor the memory of the late Polish mathematician Helena Rasiowa (Fig. 16.3).

Apart of this scientific angle of the event we will always have memories about the relevance that Prof. Moraga always gave to providing support to attendants from developing countries in order to facilitate their participation in the event, following the best scientific tradition. He always kept in mind that we as organizers should

Fig. 16.3 Gala dinner of the *26th International Symposium on Multiple-Valued Logic*, held in Santiago de Compostela (May 31st, 1996). Prof. Moraga can be guessed in front of the column at the right, mostly hidden behind the gray-haired head, near his late wife. At the dinner the Symposium organizers (also the authors of this chapter) are being acknowledged by the MVL Committee

help as much as possible in this regard, and we are pretty sure that some researchers would not have been able to attend the symposium without the support achieved through Prof. Moraga's mediation. This concern, worry and care for this kind of non-scientific, but also equally important responsibilities towards others colleagues and communities, have always been a constant in Prof. Moraga's life attitude and work ethic. Twenty years later, it is a fact that ISMVL still keeps in a very good scientific healthy state, since its 46th edition has been recently held in Sapporo (Japan). Prof. Moraga kept actively participating in this event through the years, since, for example, three papers authored/co-authored by him were included in the final program of 2016 edition.

16.2 Professor Moraga's Relationship with the Spanish Fuzzy Logic and Soft Computing Community

The links between professor Moraga and the Spanish community of researchers in Fuzzy Computational Intelligence has a long and rich tradition. The beginning of the relationship perhaps took place in June, 1980, when Professor Enric Trillas met him at the *International Symposium of Multiple-Valued Logic*, held in Evanston, Illinois, USA, was kept alive for decades and was consolidated when, in 1980, he joined the 'Fundamentals of Soft Computing' Research Unit, headed in the *European Centre for Soft Computing* (ECSC), Mieres, Spain, by Prof. Trillas. So, from the previous paragraph it can be inferred that, to a good extent, the relationships between Prof. Moraga and the fuzzy community in Spain is largely due to the friendship and professional relationship cultivated by both scientists.

Since Prof. Trillas is acknowledged as the father of Soft Computing in Spain, many of his colleagues and disciples had also the opportunity to meet and collaborate with Prof. Moraga. Those that did so can certify the commitment, rigor, talent and generosity decorating him. Prof. Moraga is a paramount scientist and all of us who attended his lessons recognize many more merits than those announced by the mere reading of his curriculum (already outstanding by itself).

Moraga is an extraordinary worker, a tireless man in the pursuit of (many-valued) truth and perfection. His lectures and presentations show clairvoyance, intensity and supreme accuracy. From his presentations is also possible to infer his artistic sense and his taste for beauty and transcendence. His friends know that he is an accomplished photographer and a remarkable guitar player. But perhaps his great passion is computer science, initiated in his native Chile, extended in the USA and developed most of his life in Dortmund, Germany.

In the pursuit of the Computational Intelligence challenges, Prof. Moraga shows an unusual expertise in four areas: soft computing, multiple-valued logic, spectral techniques and reversible computing. The papers he wrote in collaboration with the Spanish fuzzy community are mostly focused to the soft computing and multiple-valued logic topics, those strictly related to the computer implications from considering more than two logical states or values. The coauthored Moraga's contributions with Spanish researchers can be organized around people and institutions as follows:

(a) Contributions with E. Trillas, Trillas' disciples and people linked to the ECSC, (Mieres, Spain), that covered many relevant topics about soft computing and many-valued logic, including computing with words [6, 20], fuzzy control [14], fuzzy logic and fuzzy sets [1, 2, 12, 16, 18, 19, 23, 24] and many valued logic [21].
(b) Contributions with M. Delgado, F. Herrera, and people linked to DECSAI (*Dept. of Computer Science and Artificial Intelligence*, University of Granada, Spain), focused in the fields of multilayer perceptrons [4] and genetic algorithms [8, 9]
(c) Contribution with IIIA (*Research Institute on Artificial Intelligence*, CSIC, Barcelona, Spain), researchers, about the role of fuzzy logic in the design of a VLSI chip-architecture [5].

The above papers are a small part of the research conducted by Professor Moraga and a no residual part of the Spanish inquiries in fuzzy and multiple-valued logic applications to computational intelligence. His suggestions and findings contributed to increase the interest and knowledge in those areas, but more importantly, his advices helped to motivate and encourage many Spanish researchers for pursuing the highest goals in his or her investigations. We hope and wish that Professor Moraga will continue to work with the Spanish fuzzy community for many years.

16.3 His Research Within the Context of the Intelligent Systems Group

Some of the previously indicated research topics Prof. Moraga has developed in his long career have had contact points with part of the research lines we have conducted in the Intelligent Systems Research Group at the University of Santiago de Compostela. We will comment in this section about some of these research topics, without the intention of being exhaustive in its description and presentation, since our intention is just to provide a short sample of the wide range of research fields Prof. Moraga has dealt with through the years.

The early research of our group in this area had to do with computationally efficient ways of performing the Mamdani fuzzy reasoning process, by following a hardware approach (through the proposal of systolic architectures for the reasoning execution) as well as algorithmic approaches (through moving off-line part of the operations involved in the reasoning procedure).

In this first topic, the research by Prof. Moraga was pioneering, together with the first proposals in fuzzy hardware by H. Watanabe, M. Togai or T. Yamakawa. Claudio Moraga's pioneering work in the field of (fuzzy) hardware was initiated with the proposal of systolic algorithms and systems for related operations, such as generalized transitive closure [11] and followed by the proposal of CMOS multiple valued circuits [3] It is very remarkable his proposal in the late 90s for using Soft Computing approaches (in particular, different types of Evolutionary Computation) as an innovative tool for the design of reversible digital circuits [7, 15]. This field of research was preceded by his proposals of hybrid paradigms in soft computing, with papers where he described fuzzy knowledge-based genetic algorithms or genetic engineering for Artificial Neural Networks, among others.

Apart of this, there were also other contact points between our group's and Prof. Moraga's research, such as the proposal and use of "First-Aggregate-Then-Infer" (FATI) approaches in fuzzy reasoning, the relation between some Fuzzy Formal Languages and Fuzzy Petri Nets, or the use of the concept of linguistic variable as a formal support for computing with words in natural language [13]. Nevertheless, what we find closest contact point is a common concern rather than the use or proposal of a particular research paradigm. Such concern is related to the need of providing useful guidelines or criteria that help designers or users of fuzzy (rule based) systems to correctly address the non-that-easy task of designing fuzzy systems. After having ourselves arrived to the field of fuzzy logic from an applied and modeler perspective, we always found that the gap between the theoretical and formal results of fuzzy logic and the practical needs of designers should be needed to be filled somehow. Both the mathematical and the logical expressions and language were not always that understandable for other scientist or engineers that confronted the task of providing comprehensible models for implementing successful systems that solve practical problems. For us, this need became clear when researching in the field of fuzzy quantification, where a number of models had been proposed in the literature. It was well-known that some of them exhibited non-plausible behaviors and we found an actual need for characterizing them in terms of understandable (high-level) properties that allow designers to anticipate (unwanted) effects that some of these models could provoke in systems.

In a similar way, the process of building fuzzy rule-based (Mamdani) systems, and the design decisions modelers have to take, such as the proper definition of fuzzy terms and partitions, the choice of the conjunction, conditional or aggregation operators to be implemented, as well as the selection of the most appropriate defuzzification operators, was traditionally referred as a trial-and-error task. In this regard, the work by Prof. Moraga, together with Prof. Enric Trillas and Dr. Sergio Guadarrama [14, 18], helped much, thanks to their consideration of this design process as a task that should be accomplished with "care". They provide indications for the different stages that make the design process up, such as conditionals, rule of inference and reasoning schemes, and established some criteria for selection a particular fuzzification or defuzzification operator. Also plausible reasons for performing such careful design are described, not only for the membership functions of fuzzy sets, but also for linguistic connectives, modifier or quantifiers. Using their own authors words [18], the papers are related to this research are "*…a reflection on modeling… starting by asking what happens when the meaning of a predicate P is interpreted by the membership function of a fuzzy set*" After extending this reflection to the other knowledge representation and reasoning elements in fuzzy-rule based systems, the final conclusion is that "*the involved operations in rules and systems should be at* *least carefully chosen, if not specially designed"*. This allows Prof. Moraga and his co-authors *"... to improve the famous Zadeh's statement 'In fuzzy logic everything is a matter of degree,'* by extending it to the new form *"In fuzzy logic everything is not only a matter of degree, but also of design,"* that can be labeled as their final summary (and brilliantly brief) statement for their point of view in this realm.

16.4 Epilogue

The involvement of Professor Claudio Moraga in many doctoral meetings, summer courses, conferences and many other activities held at the University of Santiago de Compostela helped our Intelligent Systems Group (and University in general) to increase our knowledge about fuzzy logic and its applications, as well as to the improve the training of our doctoral students and researchers in this area.

Furthermore, and together with Prof. Enric Trillas, Prof. Moraga participated from the very beginning in some of the initial and most relevant scientific events organized by the Intelligent Systems Group in Santiago de Compostela, and we will be always in debt to him for his support in these early stages. Prof. Moraga's role as a supporter for scientific dissemination in many research fields such as multiple-valued and fuzzy logics applied to computer science was also very successfully played by him in many other groups world-wide (and the list of contributors to this book is just a short sample of the impact he achieved).

Last, but not least, his friendship and politeness, sometimes with a German touch, make him a lovely person who never forgets to compliment Christmas to his colleagues and friends. In a certain sense, we can say that Christmas does not actually begin until the email brings us Claudio's trilingual[2] Christmas Greeting Card. We sincerely hope to keep receiving Prof. Moraga's Card promptly every December 22nd morning for many and many years.

Acknowledgments This work was supported by the Spanish Ministry for Economy and Innovation and by the European Regional Development Fund (ERDF/FEDER) under grant TIN2014-56633-C3-1-R.

References

1. C. Alsina, E. Trillas, C. Moraga: Combining degrees of impairment: the case of the index of Balthazard, *Mathware and Soft Computing*, 2003, vol. 10 (1), pp. 23–41.
2. S. Barro, A. Bugarín, C. Moraga, E. Trillas: Computación fuzzy. In: S. Barro, A. Bugarín (eds.): *Fronteras de la computación*, Ed. Díaz de Santos-Fundación Dintel, 2002, vol. 8, pp. 313–368.
3. X. Chen, C. Moraga: Design of multivalued circuits based on an algebra for current-mode CMOS multivalued circuits. *J. Comput. Sci. Technol*, vol. 10 (6), 1995, pp. 564–568.

[2]May be in more than three idioms? Who can tell ...?.

4. M. Delgado, C. J. Mantas, C. Moraga: A fuzzy rule based backpropagation method for training binary multilayer perceptrons, *Information Sciences*, 1999, vol. 113, (1), pp. 1–17.
5. R. Felix, A. Höffmann, C. Moraga, L. Godo, C. Sierra: VLSI chip-architecture selection using reasoning based on fuzzy logic, 19th International Symposium on Multiple Valued Logic ISMVL, Guangzhou, 1989, pp. 165–171.
6. L. Garmendia, A. Salvador, E. Trillas, C. Moraga: On measuring μ-T-inconditionality of fuzzy relations, *Soft Computing*, vol. 9 (3), 2005, pp. 164–171.
7. F. Z. Hadjam, C. Moraga: Evolutionary design of reversible digital circuits using IMEP the case of the even parity problem, Proceedings IEEE Congress on Evolutionary Computation, 2010, pp. 1–6.
8. F. Herrera, M. Lozano, C. Moraga: Hierarchical distributed genetic algorithms, *International Journal of Intelligent Systems*, vol 14 (11), 1999, pp. 1099–1121.
9. F. Herrera, M. Lozano, C. Moraga: Hybrid distributed real-coded genetic algorithms, in A. Eiban et al. (eds.): *Parallel Problem Solving from Nature—PPSN IV*, Springer (LNCS, vol. 1498), 1998, pp. 603–612.
10. J. Kacprzyk, W. Pedrycz (eds.): *Handbook of Computational Intelligence*, Springer, 2015.
11. C. Moraga: A Systolic Algorithm for the Generalized Transitive Closure, Proc. Architecture of Computing Systems (ARCS), 1988, pp. 70–79.
12. C. Moraga, M. Sugeno, E. Trillas: Optimizations of fuzzy if-then rules by evolutionary tuning operations, Proceedings of the 39th International Symposium on Multiple-Valued Logic, IEEE-CS Press, 2009, pp. 221–226.
13. C. Moraga, E. Trillas: A Computing with Words path from fuzzy logic to natural language. Proc. 2009 EUSFLAT Conference, pp. 687–692.
14. C. Moraga, E. Trillas, S. Guadarrama: Multiple-Valued Logic and AI fundamentals of fuzzy control revisited, *Artificial Intelligence Review*, vol. 20, (3-4), 2003, pp. 169–197.
15. C. Moraga, W. Wang, Evolutionary Methods in the Design of Quaternery Digital Circuits, Proceedings of the 28th International Symposium on Multiple-Valued Logic Conference (ISMVL 1998), pp. 89–94.
16. A. Pradera, E. Trillas, C. Moraga: Clarifying Elkan's theoretical result, Proceedings EUSFLAT Conference, Zittau, 2003, pp. 604–608.
17. A. Sobrino, S. Barro (eds.): Estudios de lógica borrosa y sus aplicaciones. Serv. Publicacións Universidade de Santiago de Compostela, 1993.
18. E. Trillas, C. Moraga: Reasons for a careful design of fuzzy sets, Proceedings of the EUSFLAT Conference, Atlantis Press, 2013.
19. E. Trillas, C. Moraga, S. Guadarrama: A (naïve) glance at Soft Computing, *International Journal of Computational Intelligence Systems*, vol. 3, (2), 2010, pp. 197–201.
20. E. Trillas, C. Moraga, S. Guadarrama, S. Cubillo, E. Castiñeira: Computing with antonyms, in: Nikravesh, M. et al. (eds.): *Forging New Frontiers: Fuzzy Pioneers I*, Springer, 2007, pp. 133–153.
21. E. Trillas, C. Moraga, E. Renedo: On Aristotle's NC and EM Principles in three-valued logics, Proceedings IFSA/EUSFLAT Conference 2009, pp. 879–884.
22. E. Trillas, C. Moraga, A. Sobrino: On 'Family Resemblance' with fuzzy sets, Proceedings IFSA/EUSFLAT Conference, 2009, pp. 306–311.
23. E. Trillas, C. Moraga, G. Triviño: Weighting the support conjectures inherit from premises, *IEEE Transactions on Fuzzy Systems*, 2014, vol 23, (4), pp. 1299–1305.
24. E. Trillas, S. Termini, C. Moraga: A naïve way of looking at fuzzy sets, *Fuzzy Sets and Systems*, vol. 292(C), 2016, pp. 380–395.

Chapter 17
Using Background Knowledge for AGM Belief Revision

Christian Eichhorn, Gabriele Kern-Isberner and Katharina Behring

17.1 Introduction

Possible worlds and their syntactical representations as complete conjunctions of literals are vital building blocks of various logical techniques, for instance, to bridge the gap between syntax and semantics, to define the state of a logical system, or to define the core of a set of formulas believed by a reasoning agent. We here define semantical distance measures between possible worlds, using syntactical operations on semantical relations between possible worlds.

Classical distance measures on possible worlds, like, for instance, the Hamming distance, are strictly syntactical. That means the distance between two worlds depends on the different instantiations of the variables only, and does not take the "meaning" of the variables into account. This makes syntactical measures sensible to (semantically irrelevant) operations like renaming of variables as well as extension or contraction of the logical language with (semantically irrelevant) variables.

However, syntactical measures are easily implementable, and the distance between two possible worlds can be determined in polynomial time that depends on the cardinality of the logical alphabet. Semantical operations, however, usually have a computational complexity depending on the cardinality of the set of possible worlds, which is exponential in the cardinality of the alphabet. Distances between worlds are needed in different areas, for instance when changing the belief of an agent. Belief revision is the process of adjusting the belief of a reasoning agent to new information about the world, where the new information may contradict the actual belief of the agent, and thus to incorporate the new information, formulas that are believed a priori to the revision may be retracted from the belief set. The seminal

C. Eichhorn (✉) · G. Kern-Isberner
Lehrstuhl Informatik 1, Technische Universität Dortmund, Dortmund, Germany
e-mail: christian.eichhorn@tu-dortmund.de

K. Behring
Accenture PLM GmbH, Leinfelden-Echterdingen, Germany

R. Seising and H. Allende-Cid (eds.), *Claudio Moraga: A Passion for Multi-Valued Logic and Soft Computing*, Studies in Fuzziness and Soft Computing 349, DOI 10.1007/978-3-319-48317-7_17

article of Alchourrón et al. (AGM) [2] describes postulates to characterise rational belief revision. It has been shown that an AGM revision can be instantiated using possible worlds and relations that order worlds with respect to their dissimilarity to the original belief [12]. Distance measures can be used to define such orderings.

In this paper, based on and extending the ideas from the Master's thesis of Diekmann [5], we combine the advantages of syntactical and semantical approaches by defining semantical distance measures between possible worlds that have the implementational merits of syntactical distances while incorporating the semantical information of satisfying or violating given background knowledge. We use conditional structures [14] to connect each possible world to its individual evaluation by the conditionals of a belief base, and define distances on the syntax of these semantical structures. We show that semantical distance measures defined in this paper can be used to instantiate AGM belief revision operations. We compare the resulting operations with Dalal's revision [3], which is based on strictly syntactical distance, and show that incorporating the semantical relationships from background knowledge in this way leads to more intuitive results without losing the implementational advantages of syntactical measures.

This paper is organised as follows: This introduction is followed by the presentation of the used syntax and semantics of propositional logic and conditionals as trivalent logical entities and building blocks of knowledge bases which formalize rule-based knowledge about the world in Sect. 17.2. After than, we present the basic techniques used in this paper, which consist of the approach of conditional structures that assign the evaluation of conditionals to possible worlds using abstract symbols of verification and falsification in Sect. 17.3.1, the formal background for distance measures in general, and with the Hamming distance a concrete but strictly syntactical distance measure between possible worlds in Sect. 17.3.2, and a short introduction to AGM belief revision in general, and Dalal's belief revision with minimal models as concrete revision operation in Sect. 17.3.3. In Sect. 17.4 we define semantical distance measures on the syntax of conditional structures and show basic properties of these measures, which are plugged into the general revision operator to define semantical belief revision operators in Sect. 17.5. We conclude in Sect. 17.6.

17.2 Preliminaries

Let $\Sigma = \{V_1, \ldots, V_p\}$ be a set of propositional atoms and a *literal* a positive or negative atom representing variables in their positive (negated) form. We write $\dot{v}_i$ to indicate an arbitrary but fixed outcome of V_i. The set of formulas $\mathfrak{L}$ over Σ joined with the symbols for tautology ($\top$) and contradiction ($\bot$) and closed under the application of the connectives $\wedge$ (*and*), $\vee$ (*or*), and $\neg$ (*not*) shall be defined in the usual way. For $\varphi, \psi \in \mathfrak{L}$, we omit the connective $\wedge$ and write $\varphi\psi$ instead of $\varphi \wedge \psi$ as well as indicate negation by overlining, that is, $\overline{\varphi}$ means $\neg\varphi$.

Interpretations, or *possible worlds*, are also defined in the usual way; the set of all possible worlds is denoted by Ω. We often use the 1-1 association between worlds and *complete conjunctions*, that is, conjunctions of literals where every variable $V_i \in \Sigma$ appears exactly once. A *model* ω of a propositional formula $\varphi \in \mathfrak{L}$ is a possible world that satisfies φ, written as $\omega \models \varphi$. The set of all models $\omega \models \varphi$ is denoted by $Mod(\varphi)$. For formulas $\varphi, \psi \in \mathfrak{L}$, φ *entails* ψ, written as $\varphi \models \psi$, if and only if $Mod(\varphi) \subseteq Mod(\psi)$, that is, if and only if for all $\omega \in \Omega$, $\omega \models \varphi$ implies $\omega \models \psi$. For sets of formulas $\Phi \subseteq \mathfrak{L}$ we have $Mod(\Phi) = \bigcap_{\varphi \in \Phi} Mod(\varphi)$.

The set of *consequences* of a set $\varphi \subseteq \mathfrak{L}$ is the deductively closed set of classical entailments of φ, formally

$$Cn(\varphi) = \{\psi | \varphi \models \psi\} = \{\psi | Mod(\varphi) \subseteq Mod(\psi)\} \tag{17.1}$$

The set of *theories* of a set $\Omega' \subseteq \Omega$ is the set of formulas for which the worlds in Ω' are models, formally

$$Th(\Omega') = \{\varphi | \Omega' \subseteq Mod(\varphi)\} = Cn(\bigvee_{\omega \in \Omega'} \omega) \tag{17.2}$$

Note that Th is antitone, i.e., $\Omega'' \subseteq \Omega'$ implies $Th(\Omega') \subseteq Th(\Omega'')$.

A *belief set* $\mathscr{K}$ is the deductively closed set over a *core* φ, $\mathscr{K} = Cn(\varphi) = Th(Mod(\varphi))$. A *conditional* $(\psi|\varphi)$ with $\varphi, \psi \in \mathfrak{L}$ encodes a defeasible rule "if φ then *usually* ψ" with the trivalent evaluation [4, 14]

$$[(\psi|\varphi)]_\omega = \begin{cases} \textit{true} & \text{iff} \quad \omega \models \varphi\psi \quad \text{(verification)} \\ \textit{false} & \text{iff} \quad \omega \models \varphi\overline{\psi} \quad \text{(falsification)} \\ \textit{undefined} & \text{iff} \quad \omega \models \overline{\varphi} \quad \text{(non-applicability)} \end{cases} \tag{17.3}$$

We denote by Δ a set of conditionals $\Delta = \{(\psi_1|\varphi_1), \ldots, (\psi_n|\varphi_n)\}$ and call Δ a *knowledge base* that represents the (rule-based) knowledge an agent holds about relationships in the world and is to be used as a base for the agent's reasoning. We assume that this (background) knowledge has been built up from experiences the agent made in its everyday life. So possible sources of this knowledge include, but are not limited to, experiences and experiments made by the agent itself (that is, abductive knowledge collected/learned by the agent on its own), rules provided by an expert, and methods that mine data sets for rules like, for instance, the Apriori algorithm [1].

A conditional $(B|A)$ is *tolerated* by Δ if and only if there is a world $\omega \in \Omega$ such that $\omega \models AB$ and $\omega \models A_i \Rightarrow B_i$ for every $1 \leq i \leq n$. Δ is *consistent* if and only if for every non-empty subset $\Delta' \subseteq \Delta$ there is a conditional $(B|A) \in \Delta'$ that is tolerated by Δ' [10].

Example 17.2.1 For our running example, we use the alphabet $\Sigma = \{R, S, D\}$ where R indicates whether it is raining (r) or not ($\overline{r}$), S indicates whether the sun is shining (s) or not ($\overline{s}$), and D indicates whether the ground is dry (d) or not ($\overline{d}$). The set of possible worlds for this alphabet is $\Omega = \{r\,s\,d, r\,s\,\overline{d}, r\,\overline{s}\,d, r\,\overline{s}\,\overline{d}, \overline{r}\,s\,d, \overline{r}\,s\,\overline{d}, \overline{r}\,\overline{s}\,d, \overline{r}\,\overline{s}\,\overline{d}\}$. As example background knowledge we use the conditionals

$\delta_1 : (\overline{d}|r)$ "If it rains, the ground is usually not dry."
$\delta_2 : (d|\overline{r})$ "If it does not rain, the ground is usually dry."
$\delta_3 : (d|s)$ "If the sun is shining, the ground is usually dry."
$\delta_4 : (\overline{r}|s)$ "If the sun is shining, it usually is not raining."
$\delta_5 : (\overline{s}|r)$ "If it is raining, usually the sun is not shining."

which we combine to the knowledge base $\Delta = \{\delta_1, \delta_2, \delta_3, \delta_4, \delta_5\}$.

17.3 Basic Techniques

In this section we recall distance measures (Sect. 17.3.2), the strictly syntactical Dalal distance between worlds, and (AGM) belief revision (Sect. 17.3.3) as basic groundwork for this paper.

17.3.1 Conditional Structures

In the logic preliminaries, we recalled conditionals to be three-valued logical entities that are evaluated to *true*, *false*, or *undefined* in a given possible world. In this section we recall the approach of *conditional structures* [14] which uses the evaluation function (17.3) to connect the set of possible worlds with a knowledge base.

Let $\mathfrak{F}_\Delta = \{\mathbf{a}_1^-, \mathbf{a}_1^+, \ldots, \mathbf{a}_n^-, \mathbf{a}_n^+\}$ be a set of abstract symbols such that each symbol $\mathbf{a}_i^+$ ($\mathbf{a}_i^-$) indicates the verification (falsification) of the conditional $(\psi_i|\varphi_i)$ of a knowledge base $\Delta = \{(\psi_1|\varphi_1), \ldots, (\psi_n|\varphi_n)\}$. We define by $\sigma_{\Delta,i}$ a function that assigns to each world $\omega \in \Omega$ the evaluation of the conditional $(\psi_i|\varphi_i) \in \Delta$ in the following way [14, 15]:

$$\sigma_{\Delta,i}(\omega) = \begin{cases} \mathbf{a}_i^+ & \text{iff } [(\psi_i|\varphi_i)]_\omega = \textit{true} \quad \text{iff } \omega \models \psi_i\varphi_i \\ \mathbf{a}_i^- & \text{iff } [(\psi_i|\varphi_i)]_\omega = \textit{false} \quad \text{iff } \omega \models \psi_i\overline{\varphi}_i \\ 1 & \text{iff } [(\psi_i|\varphi_i)]_\omega = \textit{undefined} \quad \text{iff } \omega \models \overline{\psi}. \end{cases} \tag{17.4}$$

With these indicators we define the free abelian group $\mathfrak{A}_\Delta = (\mathfrak{F}_\Delta, \cdot, 1)$ on Δ with generators $\mathfrak{F}_\Delta$, the commutative operation $\cdot$ and neutral element 1. This group consists of all products $(\mathbf{a}_1^+)^{\alpha_1} \cdot (\mathbf{a}_1^-)^{\beta_1} \cdot \cdots \cdot (\mathbf{a}_n^+)^{\alpha_n} \cdot (\mathbf{a}_n^-)^{\beta_n}$ with $\alpha_i, \beta_i \in \mathbb{Z}$ for all $1 \leq i \leq n$.

With these individual algebraic indicators we define the conditional structure of a world as follows:

Definition 17.3.1 (*Conditional structure* (σ_Δ) [14, 15])
Let $\Delta = \{(\psi_1|\varphi_1), \ldots, (\psi_n\ |\varphi_n)\}$ be a conditional knowledge base. The *conditional structure* $\sigma_\Delta(\omega)$ of a possible world $\omega \in \Omega$ in the context of Δ is the product (concatenation) of the individual abstract indicators of all conditionals in Δ, formally

Table 17.1 Conditional structures for the running example

ω	$r s d$	$r s \overline{d}$	$r \overline{s} d$	$r \overline{s} \overline{d}$	$\overline{r} s d$	$\overline{r} s \overline{d}$	$\overline{r} \overline{s} d$	$\overline{r} \overline{s} \overline{d}$
$\sigma_\Delta(\omega)$	$\mathbf{a}_1^- \mathbf{a}_3^+ \mathbf{a}_4^- \mathbf{a}_5^-$	$\mathbf{a}_1^+ \mathbf{a}_3^- \mathbf{a}_4^- \mathbf{a}_5^-$	$\mathbf{a}_1^- \mathbf{a}_5^+$	$\mathbf{a}_1^+ \mathbf{a}_5^+$	$\mathbf{a}_2^+ \mathbf{a}_3^+ \mathbf{a}_4^+$	$\mathbf{a}_2^- \mathbf{a}_3^- \mathbf{a}_4^+$	$\mathbf{a}_2^+$	$\mathbf{a}_2^-$

$$\sigma_\Delta(\omega) = \prod_{i=1}^{n} \sigma_{\Delta,i}(\omega). \tag{17.5}$$

Note that the conditional structures of possible worlds from Definition 17.3.1 only cover a subset of the group $\mathfrak{A}_\Delta$; the conditional structure $\sigma_\Delta(\omega)$ of a world $\omega \in \Omega$ is the juxtaposition of its respective verification/falsification symbols of all conditionals in the knowledge base Δ or 1 if ω neither verifies nor falsifies any conditional in Δ.

Example 17.3.1 We illustrate conditional structures with the running example. Here, for instance, the world $rs\overline{d}$ verifies δ_1, falsifies δ_3, δ_4, and δ_5, and δ_2 is not applicable, hence $\sigma_\Delta(rs\overline{d}) = \mathbf{a}_1^+ \cdot 1 \cdot \mathbf{a}_3^- \cdot \mathbf{a}_4^- \cdot \mathbf{a}_5^- = \mathbf{a}_1^+\mathbf{a}_3^-\mathbf{a}_4^-\mathbf{a}_5^-$. Table 17.1 shows the conditional structures for every possible world in the running example.

The function σ_Δ is not injective, it is possible for different worlds to have the same conditional structure, as we show in the following example.

Example 17.3.2 Let $\Sigma = \{A, B\}$ and $\Delta = \{(b|a)\}$. Then the worlds $\overline{a}b$ and $\overline{a}\overline{b}$ have the same conditional structure, $\sigma_\Delta(\overline{a}b) = \sigma_\Delta(\overline{a}\overline{b}) = 1$.

This means that two worlds can be *structurally* equivalent even if they are not identical; not withstanding two worlds that are identical always have the same conditional structure, since this structure is calculated by a semantical evaluation of the conditionals with these worlds.

17.3.2 Distance Measures

A distance between two worlds is one option to express similarity or dissimilarity between the worlds: the shorter the distance, the more similar both worlds are. In this section we recall the basics of distances between worlds and sets of worlds and metrics. We also recall the Hamming resp. Dalal distance [3, 11] as a strictly syntactical distance measure between possible worlds.

In general, a distance measure is a function $\mathrm{dist} : \Omega \times \Omega \to \mathbb{N}_0$ that assigns an integer (the distance) to each pair ω, ω' of worlds in Ω. Let $\mathscr{A}, \mathscr{B} \subseteq \Omega$ be sets of worlds. We define the distance between $\mathscr{A}$ and ω as

$$\mathrm{dist}(\mathscr{A}, \omega) = \min\{\mathrm{dist}(\omega', \omega) | \omega' \in \mathscr{A}\} \tag{17.6}$$

and the distance between $\mathscr{A}$ and $\mathscr{B}$ as

$$\text{dist}(\mathscr{A}, \mathscr{B}) = \min\{\text{dist}(\omega, \omega') | \omega \in \mathscr{A}, \omega' \in \mathscr{B}\}. \tag{17.7}$$

With this extension of distance measures we define the set of *minimal models* in $\mathscr{B}$ with respect to $\mathscr{A}$ and dist as the set

$$\text{Min}^{\text{dist}}_{\mathscr{A}}(\mathscr{B}) = \{\omega | \omega \in \mathscr{B}, \nexists\, \omega' \in \mathscr{B} \quad \text{such that} \quad \text{dist}(\mathscr{A}, \omega') < \text{dist}(\mathscr{A}, \omega)\}. \tag{17.8}$$

For formulas $\varphi, \psi \in \mathfrak{L}$, we overload this notion and define the minimal models of ψ with respect to a set of worlds $\mathscr{A}$, or a formula φ, and dist as

$$\text{Min}^{\text{dist}}_{\mathscr{A}}(\psi) = \text{Min}^{\text{dist}}_{\mathscr{A}}(Mod(\psi)) \tag{17.9}$$

$$\text{Min}^{\text{dist}}_{\varphi}(\psi) = \text{Min}^{\text{dist}}_{Mod(\varphi)}(Mod(\psi)). \tag{17.10}$$

Definition 17.3.2 (*Metric* [13]) A *metric* is a distance measure $\text{dist}(\omega, \omega')$ between possible worlds if and only if it satisfies the following properties:

Symmetry	$d(\omega, \omega') = d(\omega', \omega)$
Identity	$d(\omega, \omega) = 0$
Positivity	$d(\omega, \omega') \geq 0$ and $d(\omega, \omega') = 0$ only if $\omega = \omega'$
Triangle inequality	$d(\omega, \omega'') \leq d(\omega, \omega') + d(\omega', \omega'')$

A metric assigns a total, transitive, antisymmetric, and reflexive ordering $\leq_\omega$ to each $\omega \in \Omega$. The respective strict ordering $<_\omega$ is defined as usual, we have $\omega <_\omega \omega'$ if and only if $\omega \leq_\omega \omega'$ and $\omega' \nleq_\omega \omega$.

To relate distance measures to belief sets, we use $\mathscr{K}$-persistent relations which order the set of possible worlds with respect to their distance to a belief set $\mathscr{K}$.

Definition 17.3.3 ($\mathscr{K}$*-persistent relation* [12]) Let $\mathscr{K} = Cn(\varphi)$ be a belief set. The relation $\leq_{\mathscr{K}} \subseteq \Omega \times \Omega$ is a $\mathscr{K}$*-persistent relation* if and only if we have $\omega \in Mod(\varphi)$ implies $\omega \leq_{\mathscr{K}} \omega'$ for all $\omega' \in \Omega$ and $\omega \in Mod(\varphi)$ implies $\omega <_{\mathscr{K}} \omega'$ for all $\omega' \notin Mod(\varphi)$.

Proposition 17.3.1 *Let $\mathscr{K} = Cn(\varphi)$ be a belief set. The relation*

$$\omega \leq_{\mathscr{K}} \omega' \qquad \textit{iff} \qquad dist(Mod(\varphi), \omega) \leq dist(Mod(\varphi), \omega'). \tag{17.11}$$

is a $\mathscr{K}$-persistent relation if dist is a metric.

Proof If dist is a metric, then by positivity we have $\text{dist}(\omega, \omega') = 0$ iff $\omega = \omega'$ and $\text{dist}(\omega, \omega') > 0$, otherwise. By (17.8) we have

$$\text{dist}(Mod(\varphi), \omega) = \min\{\text{dist}(\omega, \omega') | \omega' \in Mod(\varphi)\}$$

which gives us $\text{dist}(Mod(\varphi), \omega) = 0$ iff $\omega \in Mod(\varphi)$ and $\text{dist}(Mod(\varphi), \omega) > 0$, otherwise. So we have $\text{dist}(Mod(\varphi), \omega) = \text{dist}(Mod(\varphi), \omega')$ if ω and $\omega' \in Mod(\varphi)$ and $\text{dist}(Mod(\varphi), \omega) < \text{dist}(Mod(\varphi), \omega')$ if $\omega \in Mod(\varphi)$ and $\omega' \notin Mod(\varphi)$. Therefore (17.11) gives us $\omega \leq_{\mathcal{K}} \omega'$ if $\omega, \omega' \in Mod(\varphi)$ and $\omega <_{\mathcal{K}} \omega'$ if $\omega \in Mod(\varphi)$ and $\omega' \notin Mod(\varphi)$, according to the definition of $\mathcal{K}$-persistent relations in Definition 17.3.3.

As baseline for the semantical measures of the following sections, we define a syntactical edit distance between worlds as Hamming distance [11] between the literals of the worlds:

Definition 17.3.4 (*Dalal distance* [3]) Let ω, ω' be a pair of worlds from Ω. The *Dalal distance* $\text{dist}^D(\omega, \omega')$ between ω and ω' is the number of literals in which the worlds differ, formally

$$\text{dist}^D(\omega, \omega') = \sum_{\substack{V \in \Sigma \\ \omega \models \dot{v}, \omega' \not\models \dot{v}}} 1. \tag{17.12}$$

We illustrate the Dalal distance with the running example as follows.

Example 17.3.3 The Dalal distance between $rs\overline{d}$ and $\overline{r}sd$ is 2 because the worlds differ in the literals of R and D; Table 17.2 shows all Dalal distances between the worlds of the running example. Figure 17.1 shows the $\mathcal{K}$-persistent relation $\leq^D_{\mathcal{K}}$ for $\mathcal{K} = Cn(r\overline{s}\overline{d})$ and dist^D.

Proposition 17.3.2 *The Dalal distance is a metric.*

Proof The Dalal distance counts differences of the literals of two possible worlds, therefore it cannot be negative and is symmetric, and only worlds where all literals are identical (and hence worlds that are identical) have a distance of 0, hence symmetry, positivity and identity follow directly from the definition. We show that the Dalal

Table 17.2 Dalal distance between every pair of worlds in the running example

	rsd	$rs\overline{d}$	$r\overline{s}d$	$r\overline{s}\overline{d}$	$\overline{r}sd$	$\overline{r}s\overline{d}$	$\overline{r}\overline{s}d$	$\overline{r}\overline{s}\overline{d}$
rsd	0	1	1	2	1	2	2	3
$rs\overline{d}$	1	0	2	1	2	1	3	2
$r\overline{s}d$	1	2	0	1	2	3	1	2
$r\overline{s}\overline{d}$	2	1	1	0	3	2	2	1
$\overline{r}sd$	1	2	2	3	0	1	1	2
$\overline{r}s\overline{d}$	2	1	3	2	1	0	2	1
$\overline{r}\overline{s}d$	2	3	1	2	1	2	0	1
$\overline{r}\overline{s}\overline{d}$	3	2	2	1	2	1	1	0

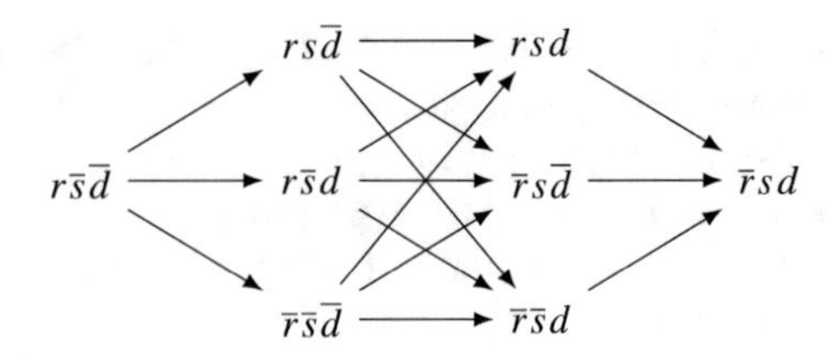

Fig. 17.1 Strict $\mathscr{K}$-persistent relation $<^D_{\mathscr{K}}$ with $\mathscr{K} = Cn(r\overline{s}\overline{d})$ and the knowledge base from the running example based on the Dalal distance for conditional structures. Note that $\omega \rightarrow \omega'$ means $\omega <^D_{\mathscr{K}} \omega'$

distance also satisfies the triangle inequality: Let $\Sigma = \{V_i, ..., V_p\}$ and let ω, ω' and $\omega'' \in \Omega$, then the definition of dist^D can be rewritten to

$$\text{dist}^D(\omega, \omega') = \sum_{i=1}^{p} \begin{cases} 0 \text{ iff } \omega \models \dot{v}_i \text{ and } \omega' \models \dot{v}_i \\ 1 \text{ otherwise.} \end{cases} \tag{17.13}$$

The difference and therefore distance between the literals of individual variables is independent from all other variables. We define by $H_i(\omega, \omega')$ the Dalal distance between the literals of V_i in ω and ω' such that $H_i(\omega, \omega') = 0$ if and only if $\omega \models \dot{v}_i$ and $\omega' \models \dot{v}_i$, and $H_i(\omega, \omega') = 1$ otherwise, that is, $\omega \models \dot{v}_i$ and $\omega' \not\models \dot{v}_i$, which is equivalent to $\omega' \models \overline{\dot{v}}_i$. With this distance on literals we rewrite the definition to

$$\text{dist}^D(\omega, \omega') = \sum_{i=1}^{p} H_i(\omega, \omega'). \tag{17.14}$$

As first step of the proof that the Dalal distance satisfies the triangle inequality, we show that the triangle inequality holds for H_i and each $1 \leq i \leq p$, that is, we show that $H_i(\omega, \omega'') \leq H_i(\omega, \omega') + H_i(\omega', \omega'')$ for all $1 \leq i \leq p$ by case analysis:

1. If $\omega \models \dot{v}_i, \omega' \models \dot{v}_i, \omega'' \models \dot{v}_i$ then $H_i(\omega, \omega'') = 0 = H_i(\omega, \omega') + H_i(\omega', \omega'')$.
2. If $\omega \models \dot{v}_i$, $\omega' \models \dot{v}_i$, $\omega'' \not\models \dot{v}_i$ then $H_i(\omega, \omega'') = 1 = 0 + 1 = H_i(\omega, \omega') + H_i(\omega', \omega'')$.
3. If $\omega \models \dot{v}_i$, $\omega' \not\models \dot{v}_i$, $\omega'' \models \dot{v}_i$ then $H_i(\omega, \omega'') = 0 \leq 0 + 1 = H_i(\omega, \omega') + H_i(\omega', \omega'')$.
4. If $\omega \models \dot{v}_i$, $\omega' \not\models \dot{v}_i$, $\omega'' \not\models \dot{v}_i$ then $H_i(\omega, \omega'') = 1 = 1 + 0 = H_i(\omega, \omega') + H_i(\omega', \omega'')$.

By this case analysis we obtain that each individual distance satisfies the triangle inequality. We recall that aligned inequalities can be added, that is, for numbers a, b, c, d, e, f we have $a \leq b + c$ and $d \leq e + f$ implies $a + d \leq b + e + c + f$, therefore, since we have shown that the triangle inequality holds for each individual H_i, it also holds for the sum of all H_i, $1 \leq i \leq p$, which is dist^D. This completes the proof.

Since dist^D is a metric we can instantiate a $\mathscr{K}$-persistent relation with dist^D according to (17.11) and use this relation to define sets of minimal models with respect to a belief set:

Definition 17.3.5 (*Minimal models*) Let $\mathscr{K}$ be a belief set and let $<_{\mathscr{K}}$ be a $\mathscr{K}$-persistent relation. Let ψ be a formula. The set of *minimal models* of ψ with respect to $<_{\mathscr{K}}$ using $\mathscr{K}$-persistent relations as defined in Definition 17.3.3 is the set

$$\text{Min}_{<_{\mathscr{K}}}(\psi) = \{\omega \in Mod(\psi) | \nexists\, \omega' \in Mod(\psi) \text{s.t.} \omega' <_{\mathscr{K}} \omega\} \tag{17.15}$$

We instantiate this definition of minimal models with the Dalal distance and the respective $\mathscr{K}$-persistent relation as follows:

$$\text{Min}^D_{<_{\mathscr{K}}}(\psi) = \{\omega \in Mod(\psi) | \nexists\, \omega' \in Mod(\psi) \text{s.t.} \omega' <^D_{\mathscr{K}} \omega\} \tag{17.16}$$

Example 17.3.4 In the running example, the set of minimal models of $\overline{r}\,\overline{s}$ with respect to $\mathscr{K} = Cn(r\overline{s}\overline{d})$ is the set $\text{Min}^D_{Cn(r\overline{s}\overline{d})}(\overline{r}\,\overline{s}) = \{\overline{r}\,\overline{s}\overline{d}\}$, as can be seen in Fig. 17.1.

17.3.3 Belief Revision

An operation of *belief revision* changes a deductively closed belief set $\mathscr{K}$ with a formula ψ to obtain a revised belief $\mathscr{K} * \psi$. In this section we recall the basics of AGM belief revision.

Definition 17.3.6 (*AGM Belief revision*) A belief revision is an *AGM belief revision* if it satisfies the following postulates [2]:

(∗ 1) $\mathscr{K} * \varphi$ is a belief state
(∗ 2) $\varphi \in \mathscr{K} * \varphi$
(∗ 3) $\mathscr{K} * \varphi \subseteq \mathscr{K} + \varphi$
(∗ 4) $\overline{\varphi} \notin \mathscr{K}$ implies $\mathscr{K} + \varphi \subseteq \mathscr{K} * \varphi$
(∗ 5) $\mathscr{K} * \varphi \equiv \perp$ if and and only if $\varphi \equiv \perp$
(∗ 6) $\varphi \equiv \psi$ implies $\mathscr{K} * \varphi = \mathscr{K} * \psi$
(∗ 7) $\mathscr{K} * (\varphi\psi) \subseteq (\mathscr{K} * \varphi) + \psi$
(∗ 8) $\psi \notin \mathscr{K} * \varphi$ implies $(\mathscr{K} * \varphi) + \psi \subseteq \mathscr{K} * (\varphi\psi)$

Here $+$ is the AGM expansion operator that satisfies the six postulates of expansion (see [2]) and hence is characterised by the deductive closure of $\mathscr{K}$ extended with φ, that is, $\mathscr{K} + \varphi = Cn(\mathscr{K} \cup \{\varphi\})$ [8].

Proposition 17.3.3 [12] *A belief revision operator $*$ for belief sets $\mathscr{K}$ and formulas $\varphi \in \mathfrak{L}$ is an AGM belief revision operator if and only if for each belief set $\mathscr{K}$ there is a $\mathscr{K}$-persistent relation $\leq_{\mathscr{K}}$ on Ω such that*

$$\mathscr{K} * \varphi = Th\big(Min_{<_{\mathscr{K}}}(\varphi)\big) \tag{17.17}$$

A special case of the belief revision operator (17.17) is the belief revision operator of Dalal:

Definition 17.3.7 (*Dalal revision* [3]) Let $\mathcal{K} = Cn(\varphi)$ be a belief set with a core $\varphi \in \mathfrak{L}$. The Dalal revision of $\mathcal{K}$ with a formula $\psi \in \mathfrak{L}$ is the set of theories of the minimal models of ψ with respect to $\mathcal{K}$, formally

$$Cn(\varphi) *^D \psi = Th\big(\mathrm{Min}^D_{<_{Cn(\varphi)}}(\psi)\big). \tag{17.18}$$

Example 17.3.5 Let $\mathcal{K} = Cn(r\overline{s}\overline{d})$, that is, we believe that it is raining, the sun is not shining and the ground is not dry, and we get to know that it is not raining, that is, we revise with $\overline{r}$. Using the Dalal revision we calculate the set of minimal models to be $\mathrm{Min}^D_{\mathcal{K}}(\overline{r}) = \{\overline{r}\,\overline{s}d\}$ (see Fig. 17.1) and thus as result of the revision we have $Cn(r\overline{s}\overline{d}) * \overline{r} = Th(\overline{r}\,\overline{s}d)$.

This example illustrates that a strictly syntactical distance measure, like the Dalal distance, results in a revision operation that may return counter intuitive results: In the running example we know that the facts "it is raining" and "the ground is not dry" are connected, but the Dalal distance cannot incorporate this knowledge. In the following section we will define semantical measures that are based on the background knowledge to overcome this problem.

17.4 Conditional Distance Measures

We defined the conditional structure of a world using the verification/falsification behaviour with respect to individual conditionals encoded in the abstract symbols $\mathbf{a}_i^+$ and $\mathbf{a}_i^-$. In this section we use these building blocks to define the building blocks for the semantical distances.

As first semantical distance measure, we define a measure in parallel to the Dalal distance (Definition 17.3.4) but on conditional structures.

Definition 17.4.1 (*Conditional distance*) Let $\Delta = \{(\psi_1|\varphi_1), \ldots, (\psi_n|\varphi_n)\}$ be a conditional knowledge base. We define by S the distance between two algebraic symbols according to Sect. 17.3.1 such that for two worlds ω, ω' and each $1 \leq i \leq n$

$$S(\sigma_{\Delta,i}(\omega), \sigma_{\Delta,i}(\omega')) = \begin{cases} 0 \text{ iff } \sigma_{\Delta,i}(\omega) = \sigma_{\Delta,i}(\omega') \\ 1 \text{ otherwise.} \end{cases}$$

and define the *conditional distance* between ω and ω' as

$$\mathrm{dist}^C_\Delta(\omega, \omega') = \begin{cases} 0 & \text{iff } \omega = \omega' \\ 1 + \sum_{i=1}^{n} S(\sigma_{\Delta,i}(\omega), \sigma_{\Delta,i}(\omega')) & \text{otherwise.} \end{cases}$$

Table 17.3 Conditional distance in the running example

	rsd	$rs\overline{d}$	$r\overline{s}d$	$r\overline{s}\overline{d}$	$\overline{r}sd$	$\overline{r}s\overline{d}$	$\overline{r}\overline{s}d$	$\overline{r}\overline{s}\overline{d}$
rsd	0	3	4	5	5	6	6	6
$rs\overline{d}$	3	0	5	4	6	5	6	6
$r\overline{s}d$	4	5	0	2	6	6	4	4
$r\overline{s}\overline{d}$	5	4	2	0	6	6	4	4
$\overline{r}sd$	5	6	6	6	0	3	3	4
$\overline{r}s\overline{d}$	6	5	6	6	3	0	4	3
$\overline{r}\overline{s}d$	6	6	4	4	3	4	0	2
$\overline{r}\overline{s}\overline{d}$	6	6	4	4	4	3	2	0

Example 17.4.1 Table 17.3 gives the conditional distance between any two worlds in the running example; we have, for example, $\text{dist}^C_\Delta(rsd, r\,\overline{s}\,\overline{d}) = 5$ because $\sigma_\Delta(rsd) = \mathbf{a}_1^-\mathbf{a}_3^+\mathbf{a}_4^-\mathbf{a}_5^-$ and also $\sigma_\Delta(r\,\overline{s}\,\overline{d}) = \mathbf{a}_1^+\mathbf{a}_5^+$, so both worlds differ in the evaluation of δ_1, δ_3, δ_4, and δ_5.

Proposition 17.4.1 *The conditional distance is a metric.*

Proof By definition, the conditional distance between worlds can only be 0 if the worlds are identical, so we obtain directly that the conditional distance is symmetric, positive, and satisfies identity. We show that the conditional distance satisfies the triangle inequality, that is, for three worlds ω, ω' and ω'' we have $\text{dist}^C_\Delta(\omega, \omega'') \leq \text{dist}^C_\Delta(\omega, \omega') + \text{dist}^C_\Delta(\omega', \omega'')$. By $cd(\omega, \omega')$ we abbreviate the expression $1 + \sum_{i=1}^n S(\sigma_{\Delta,i}(\omega), \sigma_{\Delta,i}(\omega'))$ and differentiate between the following cases:

1. If $\omega = \omega' = \omega''$ then $\text{dist}^C_\Delta(\omega, \omega'') = 0 = \text{dist}^C_\Delta(\omega, \omega') = \text{dist}^C_\Delta(\omega', \omega'')$.
2. If $\omega \neq \omega' = \omega''$ then $\text{dist}^C_\Delta(\omega, \omega'') = cd(\omega, \omega'') = cd(\omega, \omega') = \text{dist}^C_\Delta(\omega, \omega')$ and $\text{dist}^C_\Delta(\omega', \omega'') = 0$, which gives us $cd(\omega, \omega'') \leq cd(\omega, \omega'') + 0$ if inserted into the triangle inequality.
3. If $\omega = \omega' \neq \omega''$ then $\text{dist}^C_\Delta(\omega, \omega'') = cd(\omega, \omega'') = cd(\omega', \omega'') = \text{dist}^C_\Delta(\omega', \omega'')$ and $\text{dist}^C_\Delta(\omega, \omega') = 0$, which gives us $cd(\omega, \omega'') \leq 0 + cd(\omega, \omega'')$ when inserted into the triangle inequality.
4. If $\omega \neq \omega' \neq \omega''$ and $\omega = \omega''$ then $\text{dist}^C_\Delta(\omega, \omega') = cd(\omega, \omega') = cd(\omega'', \omega') = \text{dist}^C_\Delta(\omega', \omega'')$ and $\text{dist}^C_\Delta(\omega, \omega'') = 0$, which gives us $0 \leq cd(\omega, \omega'') + cd(\omega', \omega'')$ for the triangle inequality.

All these cases satisfy the triangle inequality, directly. We finally have to examine the case $\omega \neq \omega' \neq \omega''$ and $\omega \neq \omega''$. Like for the proof of Proposition 17.3.2, the summands of the conditional distance are independent from one another, so it suffices to show that the triangle inequality holds for every $1 \leq i \leq n$, which gives us that the inequality holds for the sum of the values, that is, dist^C_Δ, as well. Let $\sigma_{\Delta,i}(\omega) = \alpha_i$, $\sigma_{\Delta,i}(\omega') = \beta_i$ and $\sigma_{\Delta,i}(\omega'') = \gamma_i$. We show that $S(\alpha_i, \gamma_i) \leq S(\alpha_i, \beta_i) + S(\beta_i, \gamma_i)$ by case differentiation.

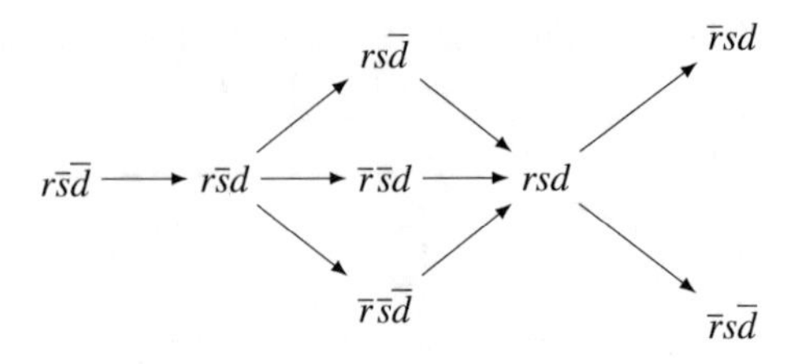

Fig. 17.2 $\mathcal{K}$-persistent relation $<^C_{\mathcal{K},\Delta}$ with $\mathcal{K} = Cn(r\overline{s}\overline{d})$ and the knowledge base from the running example based on the conditional distance

1. If $\alpha_i = \beta_i = \gamma_i$ then $S(\alpha_i, \gamma_i) = 0 \leq 0 + 0 = S(\alpha_i, \beta_i) + S(\beta_i, \gamma_i)$.
2. If $\alpha_i = \beta_i \neq \gamma_i$ then $S(\alpha_i, \gamma_i) = 1 \leq 0 + 1 = S(\alpha_i, \beta_i) + S(\beta_i, \gamma_i)$.
3. If $\alpha_i \neq \beta_i = \gamma_i$ then $S(\alpha_i, \gamma_i) = 1 \leq 1 + 0 = S(\alpha_i, \beta_i) + S(\beta_i, \gamma_i)$.
4. If $\alpha_i \neq \beta_i \neq \gamma_i$ and $\alpha_i = \gamma_i$ then $S(\alpha_i, \gamma_i) = 0 \leq 1 + 1 = S(\alpha_i, \beta_i) + S(\beta_i, \gamma_i)$.
5. If $\alpha_i \neq \beta_i \neq \gamma_i$ and $\alpha_i \neq \gamma_i$ then $S(\alpha_i, \gamma_i) = 1 \leq 1 + 1 = S(\alpha_i, \beta_i) + S(\beta_i, \gamma_i)$.

So the triangle inequality holds for each $1 \leq i \leq n$, which gives us that dist^C_Δ satisfies the triangle inequality, by which we conclude that the conditional distance is a metric.

The conditional distance being a metric allows us to define a total $\mathcal{K}$-persistent preorder according to (17.11), which we illustrate with the following example.

Example 17.4.2 Figure 17.2 shows the $\mathcal{K}$-persistent relation $\leq^D_{\mathcal{K},\Delta}$ for $\mathcal{K} = Cn(r\overline{s}\overline{d})$ and the knowledge base Δ from the running example which can be derived from Table 17.3.

The conditional distance does not take into account that conditional structures, like conditionals, are trivalent. The following distances differentiate between the different changes between verification (falsification) and not-applicability, and between verification and falsification.

Definition 17.4.2 (*Weighted conditional distance*) Let $\Delta = \{(\psi_1|\varphi_1), \ldots, (\psi_n|\varphi_n)\}$ be a conditional knowledge base. We define a distance function $W(\cdot, \cdot) : \mathfrak{F}_\Delta \times \mathfrak{F}_\Delta \to \{0, 1, 2\}$ that measures the distance for a switch between indicator symbols as follows:

- $W(\mathbf{a}_i^+, \mathbf{a}_i^+) = W(1, 1) = W(\mathbf{a}_i^-, \mathbf{a}_i^-) = 0$
- $W(\mathbf{a}_i^+, 1) = W(1, \mathbf{a}_i^+) = 1$
- $W(\mathbf{a}_i^-, 1) = W(1, \mathbf{a}_i^-) = 1$
- $W(\mathbf{a}_i^+, \mathbf{a}_i^-) = W(\mathbf{a}_i^-, \mathbf{a}_i^+) = 2$,

and with this the *weighted conditional distance* dist^W_Δ of Δ as

$$\mathrm{dist}^W_\Delta(\omega, \omega') = \begin{cases} 0 & \text{iff } \omega = \omega \\ 1 + \sum\limits_{i=1}^{n} W(\sigma_{\Delta,i}(\omega), \sigma_{\Delta,i}(\omega')) & \text{otherwise.} \end{cases} \tag{17.19}$$

Example 17.4.3 Table 17.4 shows the weighted conditional distances between all worlds of the running example, we have, for example, $\mathrm{dist}^W_\Delta(rsd, r\,\overline{s}\,\overline{d}) = 1 + 2 + 1 + 1 + 2 = 7$.

Table 17.4 Weighted conditional distance in the running example

	$r\,s\,d$	$r\,s\,\overline{d}$	$r\,\overline{s}\,d$	$r\,\overline{s}\,\overline{d}$	$\overline{r}\,s\,d$	$\overline{r}\,s\,\overline{d}$	$\overline{r}\,\overline{s}\,d$	$\overline{r}\,\overline{s}\,\overline{d}$
$r\,s\,d$	0	5	5	7	6	8	6	6
$r\,s\,\overline{d}$	5	0	7	5	8	6	6	6
$r\,\overline{s}\,d$	5	7	0	3	6	6	4	4
$r\,\overline{s}\,\overline{d}$	7	5	3	0	6	6	4	4
$\overline{r}\,s\,d$	6	8	6	6	0	5	3	5
$\overline{r}\,s\,\overline{d}$	8	6	6	6	5	0	5	3
$\overline{r}\,\overline{s}\,d$	6	6	4	4	3	5	0	3
$\overline{r}\,\overline{s}\,\overline{d}$	6	6	4	4	5	3	3	0

Proposition 17.4.2 *The weighted conditional distance is a metric.*

Proof From the definition we obtain directly that the weighted conditional distance is symmetric, positive, and satisfies identity. For being a metric, finally we have to show that the triangle inequality holds for the weighted conditional distance. Following the argumentation from the proof of Proposition 17.4.1, it suffices to show that the triangle inequality is satisfied on W for each individual distance $1 \leq i \leq n$. Let ω, ω' and $\omega'' \in \Omega$, we show that for $\sigma_{\Delta,i}(\omega) = \alpha_i$, $\sigma_{\Delta,i}(\omega') = \beta_i$ and $\sigma_{\Delta,i}(\omega'') = \gamma_i$ we have $W(\alpha_i, \gamma_i) \leq W(\alpha_i, \beta_i) + W(\beta_i, \gamma_i)$ for all configurations $\alpha_i, \beta_i, \gamma_i \in \{\mathbf{a}_i^+, \mathbf{a}_i^-, 1\}$ by case analysis.

1. $\alpha_i = \beta_i = \gamma_i$. In this case $W(\mathbf{a}, \gamma_i) = W(\mathbf{a}, \beta_i) + W(\beta_i, \gamma_i) = 0$.
2. $\alpha_i \neq \beta_i$ and $\beta_i = \gamma_i$. In this case either $W(\alpha_i, \gamma_i) = 1 = 1 + 0 = W(\alpha_i, \beta_i) + W(\alpha_i, \gamma_i)$ or $W(\mathbf{a}, \gamma_i) = 2 = 2 + 0 = W(\alpha_i, \beta_i) + W(\beta_i, \gamma_i)$
3. $\alpha_i = \beta_i$ and $\beta_i \neq \gamma_i$. In this case either $W(\alpha_i, \gamma_i) = 1 = 0 + 1 = W(\alpha_i, \beta_i) + W(\beta_i, \gamma_i)$ or $W(\alpha_i, \gamma_i) = 2 = 0 + 2 = W(\mathbf{a}, \beta_i) + W(\beta_i, \gamma_i)$
4. $\alpha_i \neq \beta_i$, $\beta_i \neq \gamma_i$ and $\alpha = \gamma$. In this case $W(\alpha_i, \gamma_i) = 0 \leq 1 + 1 = W(\alpha_i, \beta_i) + W(\beta_i, \gamma_i)$ or $W(\alpha_i, \gamma_i) = 0 \leq 2 + 2 = W(\alpha_i, \beta_i) + C(\beta_i, \gamma_i)$.
5. $\alpha_i \neq \beta_i$, $\beta_i \neq \gamma_i$, and $\alpha \neq \gamma$. Here we have to differentiate between the different configurations of α_i, β_i and γ_i:

   - $\alpha_i = \mathbf{a}_i^+$, $\beta_i = \mathbf{a}_i^-$ and $\gamma_i = 1$
     In this case $W(\alpha_i, \gamma_i) = 2 = 1 + 1 = W(\alpha_i, \beta_i) + W(\beta_i, \gamma_i)$.
   - $\alpha_i = \mathbf{a}_i^+$, $\beta_i = 1$ and $\gamma_i = \mathbf{a}_i^-$
     In this case $W(\alpha_i, \gamma_i) = 2 = 1 + 1 = W(\alpha_i, \beta_i) + W(\beta_i, \gamma_i)$.
   - $\alpha_i = 1$, $\beta_i = \mathbf{a}_i^+$ and $\gamma_i = \mathbf{a}_i^-$
     In this case $W(\alpha_i, \gamma_i) = 1 \leq 1 + 1 = W(\alpha_i, \beta_i) + W(\beta_i, \gamma_i)$.
   - $\alpha_i = 1$, $\beta_i = \mathbf{a}_i^-$ and $\gamma_i = \mathbf{a}_i^+$
     In this case $W(\alpha_i, \gamma_i) = 1 \leq 1 + 1 = W(\alpha_i, \beta_i) + W(\beta_i, \gamma_i)$.
   - $\alpha_i = \mathbf{a}_i^-$, $\beta_i = \mathbf{a}_i^+$ and $\gamma_i = 1$
     In this case $W(\alpha_i, \gamma_i) = 1 \leq 2 + 1 = W(\alpha_i, \beta_i) + W(\beta_i, \gamma_i)$.
   - $\alpha_i = \mathbf{a}_i^-$, $\beta_i = 1$ and $\gamma_i = \mathbf{a}_i^+$
     In this case $W(\alpha_i, \gamma_i) = 2 = 1 + 1 = W(\alpha_i, \beta_i) + W(\beta_i, \gamma_i)$.

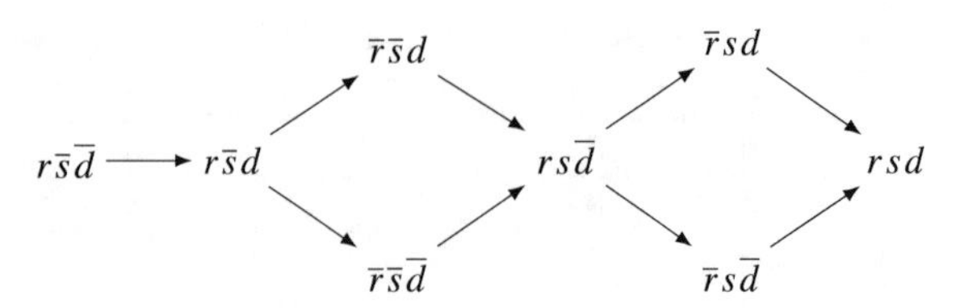

Fig. 17.3 $\mathscr{K}$-persistent relation $<^W_{\mathscr{K},\Delta}$ for $\mathscr{K} = Cn(r\overline{s}\overline{d})$ in the running example

We obtain that the triangle inequality holds for all configurations of α_i, β_i and γ_i and hence for the whole distance measure, and conclude that the weighted conditional distance is a metric.

After having shown that the weighted conditional distance is a metric, we can define a total $\mathscr{K}$-persistent relation according to (17.11). We illustrate this relation with the running example as follows.

Example 17.4.4 We use Table 17.4 to derive the $\mathscr{K}$-persistent relation $\leq^W_{\mathscr{K},\Delta}$ for $\mathscr{K} = Cn(r\overline{s}\overline{d})$ and the knowledge base from the running example (Fig. 17.3).

Differentiating between the verification and the non-applicability of a conditional may be too rigorous. With the following distance we introduce a measure that differentiates only between falsifying and not falsifying a conditional.

Definition 17.4.3 (*Conditional penalty distance*) Let Δ be a knowledge base. We define a distance function $P(\cdot,\cdot) : \mathfrak{F}_\Delta \times \mathfrak{F}_\Delta \to \{0, 1\}$ that measures the distance between indicator symbols as:

- $P(\mathbf{a}_i^+, \mathbf{a}_i^+) = P(1, 1) = P(\mathbf{a}_i^-, \mathbf{a}_i^-) = P(\mathbf{a}_i^+, 1) = P(1, \mathbf{a}_i^+) = 0$
- $P(\mathbf{a}_i^-, 1) = P(1, \mathbf{a}_i^-) = 1$
- $P(\mathbf{a}_i^+, \mathbf{a}_i^-) = P(\mathbf{a}_i^-, \mathbf{a}_i^+) = 1$

With this function we define the *conditional penalty distance* dist^P_Δ of Δ as

$$\mathrm{dist}^P_\Delta(\omega, \omega') = \begin{cases} 0 & \text{iff } \omega = \omega' \\ 1 + \sum_{i=1}^{n} P(\sigma_{\Delta,i}(\omega), \sigma_{\Delta,i}(\omega')) & \text{otherwise.} \end{cases} \tag{17.20}$$

Example 17.4.5 In the running example we have, for example, $\mathrm{dist}^P_\Delta(rsd, r\overline{s}\overline{d}) = 1 + 1 + 0 + 1 + 1 = 4$; Table 17.5 shows all conditional penalty distances between the worlds in this setting.

Proposition 17.4.3 *The conditional penalty distance is a metric.*

This can be shown with a case analysis similar to the proof of Proposition 17.4.2, and hence we can define a total $\mathscr{K}$-persistent preorder according to (17.11), which we illustrate in the following example.

Table 17.5 Conditional penalty distance for the running example

	rsd	$rs\overline{d}$	$r\overline{s}d$	$r\overline{s}\overline{d}$	$\overline{r}sd$	$\overline{r}s\overline{d}$	$\overline{r}\overline{s}d$	$\overline{r}\overline{s}\overline{d}$
rsd	0	3	3	4	4	6	4	5
$rs\overline{d}$	3	0	5	4	4	4	4	5
$r\overline{s}d$	3	5	0	2	2	4	2	3
$r\overline{s}\overline{d}$	4	4	2	0	1	3	1	2
$\overline{r}sd$	4	4	2	1	0	3	1	2
$\overline{r}s\overline{d}$	6	4	4	3	3	0	3	2
$\overline{r}\overline{s}d$	4	4	2	1	1	3	0	2
$\overline{r}\overline{s}\overline{d}$	5	5	3	2	2	2	2	0

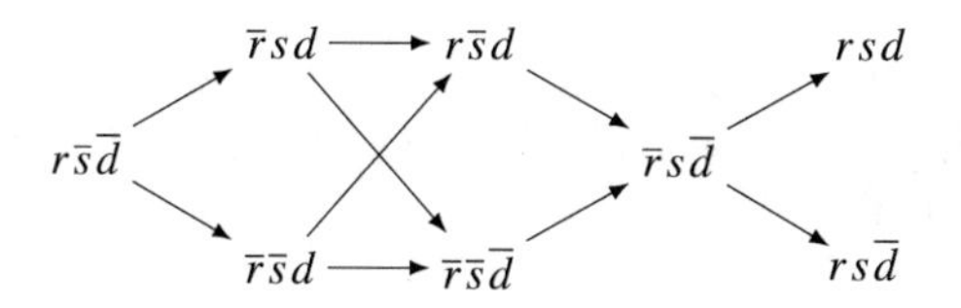

Fig. 17.4 $\mathscr{K}$-persistent relation $<^P_{\mathscr{K},\Delta}$ for $\mathscr{K} = Cn(r\overline{s}\overline{d})$ in the running example

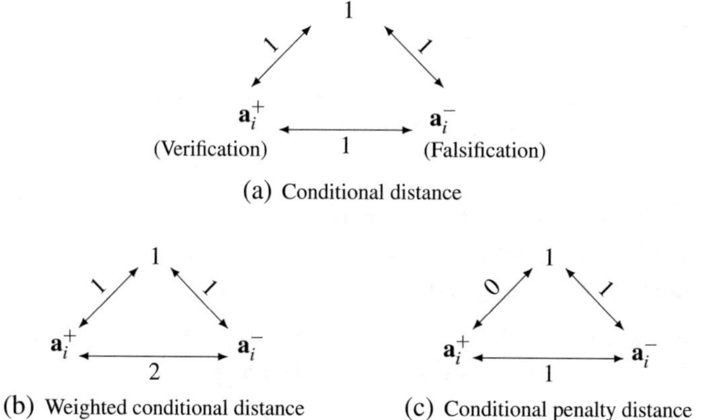

Fig. 17.5 Illustration of the three conditional distance measures on a single indicator symbol, where the weight on the edges indicate the costs for a switch between these instantiations

Example 17.4.6 Figure 17.4 shows the $\mathscr{K}$-persistent relation $<^P_{\mathscr{K},\Delta}$ for $\mathscr{K} = Cn(r\overline{s}\overline{d})$ and the knowledge base from the running example derived from Table 17.5.

All distance measures defined in this section are semantic distances in that they make use of the semantical relationships encoded in Δ. We close this section with an illustration of all three distance measures in Fig. 17.5.

17.5 Revision with Conditional Distance Measures

In this section we instantiate AGM belief revision with the defined conditional distances. This is obtained by instantiating minimal models (Definition 17.3.5) with the $\mathcal{K}$-persistent relations based on conditional, weighted conditional and conditional penalty distance, which then is used to instantiate the generic belief revision operator from Proposition 17.3.3. We illustrate the resulting revision operations with examples, discuss the performance of the revision operations in comparison to the Dalal revision and the implementational properties of the conditional revision operations.

We illustrate the minimal models generated by conditional distances with the running example and include the minimal models generated with the Dalal distance for comparison:

Example 17.5.1 Let $\mathcal{K} = Cn(r\overline{s}\overline{d})$ and let Δ be the knowledge base from the running example. Using Definition 17.3.5 and Figs. 17.1, 17.2, 17.3 and 17.4 we get the following $\mathcal{K}$-minimal models for $\overline{r}\,\overline{s}$:

$$\begin{aligned}
\mathrm{Min}^{D}_{Cn(r\overline{s}\overline{d})}(\overline{r}\,\overline{s}) &= \{\overline{r}\,\overline{s}\overline{d}\}\\
\mathrm{Min}^{C}_{Cn(r\overline{s}\overline{d}),\Delta}(\overline{r}\,\overline{s}) &= \{\overline{r}\,\overline{s}d, \overline{r}\,\overline{s}\overline{d}\}\\
\mathrm{Min}^{W}_{Cn(r\overline{s}\overline{d}),\Delta}(\overline{r}\,\overline{s}) &= \{\overline{r}\,\overline{s}d, \overline{r}\,\overline{s}\overline{d}\}\\
\mathrm{Min}^{P}_{Cn(r\overline{s}\overline{d}),\Delta}(\overline{r}\,\overline{s}) &= \{\overline{r}\,\overline{s}d\}
\end{aligned}$$

We use the generic Definition (17.18) to instantiate conditional revision operators that are based on background knowledge represented as conditional knowledge base as follows:

Conditional revision $\quad Cn(\varphi) *^{C}_{\Delta} \psi = Th(\mathrm{Min}^{C}_{\varphi,\Delta}(\psi))$
Weighted conditional revision $\quad Cn(\varphi) *^{W}_{\Delta} \psi = Th(\mathrm{Min}^{W}_{\varphi,\Delta}(\psi))$
Conditional Penalty revision $\quad Cn(\varphi) *^{P}_{\Delta} \psi = Th(\mathrm{Min}^{P}_{\varphi,\Delta}(\psi))$

We have constructed all relations $\leq^{x}_{\Delta}$ with $x \in \{C, W, P\}$ to be $\mathcal{K}$-persistent relations, therefore with Proposition 17.3.3 we directly obtain

Corollary 17.5.1 *The conditional revision, weighted conditional revision, and conditional penalty revision are AGM revision operations.*

In the following, we give three examples to illustrate the revision processes using the conditional, weighted conditional and penalty revision and compare the results to the Dalal revision. We list the instantiations of minimal models for an easier understanding of the examples, beforehand:

$$\mathrm{Min}^{D}_{\mathscr{K}}(\psi) = \{\omega \in Mod(\psi) | \nexists \omega' \in Mod(\psi) \text{s.t.} \omega' <^{D}_{\mathscr{K}} \omega\}$$
$$\mathrm{Min}^{C}_{\mathscr{K},\Delta}(\psi) = \{\omega \in Mod(\psi) | \nexists \omega' \in Mod(\psi) \text{s.t.} \omega' <^{C}_{\mathscr{K},\Delta} \omega\}$$
$$\mathrm{Min}^{W}_{\mathscr{K},\Delta}(\psi) = \{\omega \in Mod(\psi) | \nexists \omega' \in Mod(\psi) \text{s.t.} \omega' <^{W}_{\mathscr{K},\Delta} \omega\}$$
$$\mathrm{Min}^{P}_{\mathscr{K},\Delta}(\psi) = \{\omega \in Mod(\psi) | \nexists \omega' \in Mod(\psi) \text{s.t.} \omega' <^{P}_{\mathscr{K},\Delta} \omega\}$$

Example 17.5.2 We recall Example 17.3.5 to show the defined conditional revisions in comparison with the Dalal revision, so we believe that it rains, the sun is not shining and the ground is wet, that is, $\mathscr{K} = Cn(r\overline{s}\overline{d})$. In this situation we get to know that it is not raining, that is, we revise with $\overline{r}$, and ask ourselves whether the ground would be dry or not. We already determined the sets of minimal models in Example 17.5.1, so we obtain:

- With the Dalal distance, the closest $\overline{r}$-satisfying world to $\mathscr{K}$ is $\overline{r}\,\overline{s}\overline{d}$. Here we have $\overline{d} \in Th(\overline{r}\,\overline{s}\overline{d}) = Cn(r\overline{s}\overline{d}) *^{D} \overline{r}$, that is, after the revision we conclude that the ground is not dry.
- Using the conditional distance, the closest worlds that satisfy $\overline{r}$ are the worlds $\overline{r}\,\overline{s}d$ and $\overline{r}\,\overline{s}\overline{d}$, thus $Cn(r\overline{s}\overline{d}) *^{C}_{\Delta} \overline{r} = Th(\{\overline{r}\,\overline{s}d, \overline{r}\,\overline{s}\overline{d}\}) = Cn(\overline{r}\,\overline{s})$. Since neither $d \in Cn(\overline{r}\,\overline{s})$ nor $\overline{d} \in Cn(\overline{r}\,\overline{s})$, it cannot be concluded whether the ground is dry or not.
- The closest $\overline{r}$-satisfying worlds under weighted conditional distance are $\overline{r}\,\overline{s}d$ and $\overline{r}\,\overline{s}\,\overline{d}$. It is neither $d \in Th(\overline{r}\,\overline{s}d, \overline{r}\,\overline{s}\,\overline{d}) = Cn(r\overline{s}\overline{d}) *^{W}_{\Delta} \overline{r}$, nor $\overline{d} \in Th(\overline{r}\,\overline{s}d, \overline{r}\,\overline{s}\,\overline{d}) = Cn(r\overline{s}\overline{d}) *^{W}_{\Delta} \overline{r}$ that is, after the revision it cannot be concluded whether the ground is dry or not.
- With the conditional penalty distance, the closest worlds to $r\overline{s}\overline{d}$ that satisfy $\overline{r}$ are $\overline{r}sd$ and $\overline{r}\,\overline{s}d$, and so $d \in Cn(r\overline{s}\overline{d}) *^{P}_{\Delta} \overline{r}$, and after the revision we conclude that the ground is dry.

In this example, we see that using the (syntactical) Dalal distance we get the counter-intuitive result to believe to be in a world where the ground is not dry when it is not raining, while our background knowledge states that normally the ground should be dry if it is not raining. In line with the background knowledge, the conditional distances withdraw the belief that the ground is not dry, and penalty distance even allows to conclude that the ground is dry.

Example 17.5.3 Let $\mathscr{K} = Cn(\overline{r}\overline{d})$, that is, we believe that the ground is not dry and it is not raining. In this situation we get to know that it is raining, that is, we revise with r. We use Tables 17.2, 17.3, 17.4 and 17.5 to determine the respective minimal worlds and obtain:

- $Cn(\overline{r}\overline{d}) *^{D} r = Th(\{rs\overline{d}, r\overline{s}\overline{d}\}) = Cn(r\overline{d})$
- $Cn(\overline{r}\overline{d}) *^{C}_{\Delta} r = Th(\{r\overline{s}d, r\overline{s}\overline{d}\}) = Cn(r\overline{s})$
- $Cn(\overline{r}\overline{d}) *^{W}_{\Delta} r = Th(\{r\overline{s}d, r\overline{s}\overline{d}\}) = Cn(r\overline{s})$
- $Cn(\overline{r}\overline{d}) *^{P}_{\Delta} r = Th(\{r\overline{s}\overline{d}\}) = Cn(r\overline{s}\overline{d})$

Again we see that the Dalal revision preserves the knowledge about the ground, but also the ignorance about whether the sun is shining or not. Using conditional operations, we also believe that the sun is not shining with respect to δ_5 "If it is raining, usually the sun is not shining". Also, both the Dalal and penalty revision preserve the information about the ground not being dry, whereas the conditional and weighted conditional revisions contract this belief since the justification r for $\overline{d}$ is contracted.

Example 17.5.4 Finally we assume to believe that the ground is dry and it is either raining or the sun is shining, which we formalize as $\mathscr{K} = Cn(d \wedge (r \Leftrightarrow \overline{s}))$, and get to know that the ground is not dry, that is, we revise with $\overline{d}$. We use Tables 17.2, 17.3, 17.4 and 17.5 to determine the respective minimal worlds and obtain:

- $Cn(d \wedge (r \Leftrightarrow \overline{s})) *^D \overline{d} = Th(\{r\overline{s}\overline{d}, \overline{r}s\overline{d}\}) = Cn(\overline{d})$
- $Cn(d \wedge (r \Leftrightarrow \overline{s})) *^C_\Delta \overline{d} = Th(\{r\overline{s}\overline{d}\}) = Cn(r\overline{s}\overline{d})$
- $Cn(d \wedge (r \Leftrightarrow \overline{s})) *^W_\Delta \overline{d} = Th(\{r\overline{s}\overline{d}\}) = Cn(r\overline{s}\overline{d})$
- $Cn(d \wedge (r \Leftrightarrow \overline{s})) *^P_\Delta \overline{d} = Th(\{r\overline{s}\overline{d}\}) = Cn(r\overline{s}\overline{d})$

In this example, revising with the Dalal revision relinquishes the belief in sunshine and rain being mutually exclusive. This belief is preserved in the conditional revisions, and additionally the world that incorporates this belief. The world satisfying $\overline{d}$ (the new knowledge) is selected as the world believed to be true from the worlds in accordance with this belief.

The three Examples 17.5.2 through 17.5.4 illustrate that including background knowledge in the form of a conditional knowledge base via our conditional measures in the process of revising an agent's belief leads to more intuitive results with respect to the represented background knowledge.

We will finish this section with a discussion about the implementational properties of the defined conditional measures. The defined conditional distance measures are defined as a sum of distances, each being a distance between the evaluation of individual conditionals in the knowledge base. Using the abstract evaluation symbols of conditional structures allowed us to therefore generate the distance between two worlds as a sum of distances between the building blocks of elements of the free abelian group $\mathfrak{A}_\Delta$, that is, we defined conditional distance measures on the syntax of the elements of $\mathfrak{A}_\Delta$. These elements can be padded with the neutral element to generate strings of identical length, and the distance between two of these elements can therefore be calculated by reading each symbol in a pair of strings once. Other than using strictly syntactical distances like Dalal (Levenshtein, Hamming) distance, the length of each element does not depend on the size of the alphabet Σ but on the size of the knowledge base Δ. We expect that the number of variables exceeds the number of conditionals in honest-to-life examples, so the calculation of the distance of two conditional structures should be less or equally computationally complex than the calculation of the syntactical distance of two possible worlds. To obtain the conditional structure of a world, however, all conditionals in Δ have to be evaluated. Applying an interpretation to a formula and checking whether it is evaluated to *true* or *false* can be done in linear time, but this has to be repeated at most twice for every conditional, and hence no more than four times per pair of worlds. Overall, the computational complexity of calculating a conditional distance between two worlds can be done in polynomial time with respect to the size of the knowledge base comparison, compared to calculating the Dalal distance in polynomial time with respect to the size of the alphabet. For the general case we neither know the cardinality of the alphabet nor the cardinality of the knowledge base. Using conditional structures partitions the possible worlds into $3^{|\Delta|} - 1$ equivalence classes equivalent to Gilio's constituents (see, e.g., [9]). If the proportion of empty partitions is high, this leads to an overhead in computational time. But by using approaches to decompose knowledge into sets of mutually relevant information (confer, for example, [6, 7]) we can ensure that there is no unnecessary blow-up in either.

17.6 Conclusion

In this paper we defined semantical distance measures that calculate the distance between worlds based on the evaluation of worlds by conditionals from a knowledge base. This distance was defined on the individual evaluation of the conditionals stored as abstract symbols of conditional structures. By using this data structure, we defined distance measures as syntactical distances on the semantical relations generated from the conditional knowledge base; we discuss that this allows for the semantical distance between two worlds to be calculated in polynomial time mainly dependent on the size of the knowledge base. We used these distance measures to define persistent relations on the worlds with respect to the belief set. With these relations we derived semantical revision operators. The belief revision operations change the belief set of an agent, but not the background knowledge. Since the conditional structure relies only on the latter, the structures and distances that have been calculated between two worlds are not changed by a revision and therefore the distances can be used for repeated revisions.

All of the distances have been set up as symmetrical measures. A part of our ongoing research, we examine the impact of abandoning this symmetry, such that, for instance, the distance from verification to falsification may be larger than the distance from falsification to verification, thus favoring transitions from falsification of conditionals to verification of conditionals over transitions from verification to falsification. In this line of work, we plan to compare the results of the resulting measures with the results of [16] regarding belief revision with pseudo-distances.

We implemented belief revision with the semantical metrics as defined in this paper using the *Tweety* collection of Java libraries for logical aspects of artificial intelligence and knowledge representations [17] as a free web service called "SEDIMENT" (**SE**mantical **DI**stance **ME**asureme**NT**s).[1]

---

[1] http://airconditionals.cs.tu-dortmund.de:8080/SDPlugin/.

Acknowledgments This work was supported by DFG-Grant KI1413/5-1 of Prof. Dr. Gabriele Kern-Isberner as part of the Priority Program "New Frameworks of Rationality" (SPP 1516). Christian Eichhorn is supported by this grant. Thanks to Richard Niland for his work on the implementation and for carefully proof-reading this paper.

References

1. Agrawal, Rakesh; Srikant, Ramakrishnan: Fast Algorithms for Mining Association Rules in Large Databases, in: Proceedings of the 20th International Conference on Very Large Data Bases VLDB '94, 1994, pp. 487–499.
2. Alchourrón, Carlos E.; Gärdenfors, Peter; Makinson, David: On the logic of theory change: Partial meet contraction and revision functions, *Journal of Symbolic Logic*, vol 50, pp. 510–530.
3. Dalal, Mukesh: Investigations Into a Theory of Knowledge Base Revision: Preliminary Report, Mitchell, Tom M.; Smith, Reid G. (eds.): Proceedings of the Seventh National Conference on Artificial Intelligence (AAAI-88), Cambridge, MA: MIT Press, 1988, pp. 475–479.
4. de Finetti, Bruno: *Theory of Probability*, New York: John Wiley & Sons, vols. 1, 2, 1974.
5. Diekmann, Katharina: Ähnlichkeitsbasierte Inferenzen für kontrafaktische Konditionale, Master's Thesis Technische Universiät Dortmund, 2013.
6. Eichhorn, Christian; Kern-Isberner, Gabriele: LEG Networks for Ranking Functions, in: Fermé, Eduardo; Leite, João (eds.): Logics in Artificial Intelligence (Proceedings of the 14th European Conference on Logics in Artificial Intelligence (JELIA'14)), Springer International (Lecture Notes in Computer Science, vol. 8761) 2014, pp. 210–223.
7. Eichhorn, Christian; Kern-Isberner, Gabriele: Using inductive reasoning for completing OCF-networks, *Journal of Applied Logic*, vol. 13 (4, Part 2): Special Issue dedicated to Uncertain Reasoning at FLAIRS, 2015, pp. 605–627.
8. Gärdenfors, Peter; Rott, Hans: Belief Revision, in: Gabbay, Dov M.; Hogger, C. J.; Robinson, J. A. (eds.): *Handbook of Logic in Artificial Intelligence and Logic Programming*, vol. 4, New York: Oxford University Press, 1994, pp. 35–132.
9. Gilio, Angelo: Probabilistic reasoning under coherence in system P*, *Annals of Mathematics and Artificial Intelligence* vol. 34, 2002, pp. 5–34.
10. Goldszmidt, Moisés; Pearl, Judea: Qualitative probabilities for default reasoning, belief revision, and causal modeling, *Artificial Intelligence*, vol. 84 (1–2), 1996, pp. 57–112.
11. Hamming, R. W.: Error detecting and error correcting codes, *The Bell System Technical Journal*, vol. 29 (2) 1950, pp. 147–160.
12. Katsuno, Hirofumi; Mendelzon, Alberto O.: Propositional knowledge base revision and minimal change *Artificial Intelligence*, vol. 52 (3), 1991, pp. 263–294.
13. Kelley, J. L.: *General Topology*, New York: Van Nostrand, 1955.
14. Kern-Isberner, Gabriele: *Conditionals in Nonmonotonic Reasoning and Belief Revision – Considering Conditionals as Agents*, Berlin: Springer Science+Business Media (Lecture Notes in Computer Science, Nr. 2087), 2001.
15. Kern-Isberner, Gabriele: A Thorough Axiomatization of a Principle of Conditional Preservation in Belief Revision, *Annals of Mathematics and Artificial Intelligence*, vol. 40 (1–2), 2004, pp. 127–164.
16. Lehmann, Daniel; Magidor, Menachem; Schlechta, Karl: Distance Semantics for Belief Revision, *The Journal of Symbolic Logic*, vol. 66 (1), 2001, pp. 295–317.
17. Thimm, Matthias: Tweety – A Comprehensive Collection of Java Libraries for Logical Aspects of Artificial Intelligence and Knowledge Representation, Proceedings of the 14th International Conference on Principles of Knowledge Representation and Reasoning (KR'14), 2014.

Chapter 18
Associative Globally Monotone Extended Aggregation Functions

Tomasa Calvo, Gaspar Mayor and Jaume Suñer

18.1 Introduction

Aggregation is the process of combining several input values into a single representative output value, and the functions that carried out this process are called aggregation functions. Perhaps the oldest example of aggregation function is the arithmetic mean, which has been used during all the history of physics and all experimental sciences.

It is easy to understand that aggregation functions play an important role in many fields: pure and applied mathematics, computer and engineering sciences, economics and finance, social sciences as well as many other applied fields of physics and natural sciences.

The problems of aggregation are, in general, very broad and heterogeneous and the task of defining or choosing the right class of aggregation functions for an specific problem is often difficult, considering the huge variety of potential aggregation functions. Here we restrict ourselves in this contribution to the specific topic of the aggregation of a finite number of real inputs only. In this spirit, if the number of input values is fixed, say n, an aggregation function is a real function of n variables and, if not explicity stated, we will assume throughout that both inputs and outputs are from the unit interval $[0, 1]$, and hence an n-ary aggregation function is a mapping from $[0, 1]^n$ into $[0, 1]$. Evidently, not all these functions are candidates to be an aggregation function and, in this sense, two requirements are commonly accepted

T. Calvo (✉)
University of Alcalá, Alcalá de Henares, Spain
e-mail: tomasa.calvo@uah.es

G. Mayor · J. Suñer
University of the Balearic Islands, Palma de Mallorca, Spain
e-mail: gmayor@uib.es

J. Suñer
e-mail: sunyer@uib.es

R. Seising and H. Allende-Cid (eds.), *Claudio Moraga: A Passion for Multi-Valued Logic and Soft Computing*, Studies in Fuzziness and Soft Computing 349, DOI 10.1007/978-3-319-48317-7_18

in the field: boundary conditions and increasing monotonicity, that we adopt as the basic definition of an n-ary aggregation function.

Moreover, it is often the case that aggregation of inputs of various sizes has to be considered in the same framework. For instance, in some applications, input vectors may have a varying number of components and, in this case, it is appropriate to consider a family of functions of $n = 1, 2, 3, \ldots$ arguments with the same underlying properties (boundary conditions and increasing monotonicity).The concept of extended aggregation function [1–7] allows us to work with such families of aggregation functions of any number of arguments. It is clear that two members of such families need be related somehow in order to give consistency to the process of aggregation. This can be done in several ways: to compute each member of the family using one generic formula (arithmetic mean, geometric mean); requiring some grouping property (decomposability, associativity); some kind of stability (self-identity, duplication); and others [8, 9].

The increasing number of research papers appeared in the last decades that either make use of aggregation functions or contribute to its theoretical study stands for the great interest in this subject from a theoretical and applied point of view. At this point, it is worth saying that the publication of several monographs on the subject in question has contributed to the presentation of a general framework where new concepts can still appear and develop new techniques in the field of aggregation [10–13].

Our aim in this contribution is to continue dealing with a type of global monotonicity as a minimum requirement of consistency for an extended aggregation function [14]. We are here interested in those associative aggregation functions that define global monotone extended aggregation functions.
After some preliminaries, in Sect. 3 we recall the concept of global monotonicity, and give two basic examples of extended aggregation functions that satisfy this requirement. The Sect. 4 is devoted to analyze the relationship between associativity and global monotonicity.

18.2 Preliminaries

All the concepts and results in this section can be found in [10, 12, 13] and references therein.

Definition 18.2.1 An (n–ary) aggregation function is a function $F : [0, 1]^n \to [0, 1]$ with the properties:

(i) $F(0, \ldots, 0) = 0, \; F(1, \ldots, 1) = 1$.
(ii) $F(\mathbf{x}) \leq F(\mathbf{y})$ whenever $\mathbf{x} \leq \mathbf{y}$, for all $\mathbf{x} = (x_1, \ldots, x_n)$, $\mathbf{y} = (y_1, \ldots, y_n)$ in $[0, 1]^n$.
(iii) In case $n = 1$, $F(x) = x \; \forall x \in [0, 1]$.

Proposition 18.2.1 *For an aggregation function* $F : [0,1]^n \to [0,1]$, *the following properties are equivalent:*

(i) $F(\overbrace{a, \ldots, a}^{n}) = a$ *for all* $a \in [0,1]$ *(*F *is idempotent).*
(ii) $\min(x_1, \ldots, x_n) \le F(x_1, \ldots, x_n) \le \max(x_1, \ldots, x_n)$ *(*F *is compensative).*

If an aggregation function is compensative, we also say that it is a mean or an average.

Consider a weighting vector $\mathbf{w} = (w_1, \ldots, w_n)$ where $w_i \ge 0,\ i = 1, \ldots, n$ and $\sum_{i=1}^{n} w_i = 1$. Two classic types of means associated to $\mathbf{w}$ can be considered.

Definition 18.2.2 Given a weighting vector $\mathbf{w}$, the weighted arithmetic mean defined by $\mathbf{w}$ is the function $F(x_1, \ldots, x_n) = \sum_{i=1}^{n} w_i x_i$.

The only weighted mean which is symmetric is the arithmetic mean

$$F(x_1, \ldots, x_n) = \frac{1}{n} \sum_{i=1}^{n} x_i .$$

Definition 18.2.3 Given a weighting vector $\mathbf{w}$, the ordered weighted arithmetic mean (OWA operator, [15]) defined by $\mathbf{w}$ is

$$F(x_1, \ldots, x_n) = \sum_{i=1}^{n} w_i x_{(i)},$$

where $(x_{(1)}, \ldots, x_{(n)})$ denotes the vector obtained from $(x_1, \ldots, x_n)$ by arranging its components in decreasing order $x_{(1)} \ge x_{(2)} \ge \ldots \ge x_{(n)}$.

Definition 18.2.4 An extended aggregation function is a mapping $F : \bigcup_{n \ge 1} [0,1]^n \to [0,1]$ such that the restriction of this mapping to each $[0,1]^n$ is an n–ary aggregation function.

We say that an *extended aggregation function* is idempotent (compensative) if its restriction to each $[0,1]^n$ is idempotent (compensative). In this contribution we are interested in extended aggregation functions which are idempotent.

18.3 Extended Monotonicity

Let us consider the following binary relation on $\bigcup_{n\geq 1}[0,1]^n$:

Definition 18.3.1 Let $\mathbf{x} = (x_1, \ldots, x_n)$ and $\mathbf{y} = (y_1, \ldots, y_m)$. Then $\mathbf{x} \leq \mathbf{y}$ means:

- If $n = m$, $x_i \leq y_i$ for all $i = 1, \ldots, n$
- If $n < m$, $x_i \leq y_i$ for all $i = 1, \ldots, n$ and $\max(x_1, \ldots, x_n) \leq \min(y_{n+1}, \ldots, y_m)$
- If $n > m$, $x_i \leq y_i$ for all $i = 1, \ldots, m$ and $\max(x_{m+1}, \ldots, x_n) \leq \min(y_1, \ldots, y_m)$

The binary relation given in the definition above is a pre-order[1] on the set $\bigcup_{n\geq 1}[0,1]^n$. For any $n \geq 1$, the restriction of this pre–order to the set of n–dimensional lists $[0, 1]^n$ coincides with the usual product order.

Remark 1 From Definition 18.3.1, we have immediately that $\mathbf{x} \leq \mathbf{y}$ and $\mathbf{y} \leq \mathbf{x}$ if, and only if, $\mathbf{x} = (\overbrace{a, \ldots, a}^{p})$, $\mathbf{y} = (\overbrace{a, \ldots, a}^{q})$, where $a \in [0, 1]$, $p, q \geq 1$.

Definition 18.3.2 A function $F : \bigcup_{n\geq 1}[0,1]^n \to [0, 1]$ is globally monotone (increasing) if $F(\mathbf{x}) \leq F(\mathbf{y})$ whenever $\mathbf{x} \leq \mathbf{y}$.

In the following we will denote $\min(x_1, \ldots, x_n) = \wedge x_i$ and $\max(x_1, \ldots, x_n) = \vee x_i$.

Proposition 18.3.1 *An extended aggregation function* $F : \bigcup_{n\geq 1}[0,1]^n \to [0, 1]$ *is globally monotone if, and only if,*

$$F(x_1, \ldots, x_n, \wedge x_i) \leq F(x_1, \ldots, x_n) \leq F(x_1, \ldots, x_n, \vee x_i) \tag{18.1}$$

for all $(x_1, \ldots, x_n) \in [0, 1]^n$, $n \geq 1$.

Proof Let us suppose first that F is globally monotone, then we have (18.1) because

$$(x_1, \ldots, x_n, \wedge x_i) \leq (x_1, \ldots, x_n) \leq (x_1, \ldots, x_n, \vee x_i)\,, \text{ for all } (x_1, \ldots, x_n),\ n \geq 1.$$

Reciprocally, let us consider F satisfying the condition (18.1) and $\mathbf{x} = (x_1, \ldots, x_n)$, $\mathbf{y} = (y_1, \ldots, y_m)$ such that $\mathbf{x} \leq \mathbf{y}$. If $n = m$ then $F(\mathbf{x}) \leq F(\mathbf{y})$ because the restriction of F to $[0, 1]^n$ is increasing in each variable. In case, $n < m$, from (18.1) we can write

$$\begin{aligned} F(\mathbf{x}) \;=\; F(x_1, \ldots, x_n) &\leq F(x_1, \ldots, x_n, \overbrace{\vee x_i, \ldots, \vee x_i}^{m-n}) \leq \\ &\leq F(y_1, \ldots, y_n, y_{n+1}, \ldots, y_m) \;=\; F(\mathbf{y}). \end{aligned}$$

[1] Reflexive and transitive.

Finally, if $n > m$, we have

$$\begin{aligned} F(\mathbf{x}) &= F(x_1, \ldots, x_m, x_{m+1}, \ldots, x_n) \leq F(y_1, \ldots, y_m, \overbrace{\wedge y_i, \ldots, \wedge y_i}^{n-m}) \leq \\ &\leq F(y_1, \ldots, y_m) = F(\mathbf{y}). \end{aligned}$$

Thus F is globally monotone. □

Proposition 18.3.2 *If an extended aggregation function F is globally monotone then it is idempotent and, therefore, compensative.*

Proof Let $a \in [0, 1]$ and $n \geq 1$. Since $(\overbrace{a, \ldots, a}^{n}) \leq (a) \leq (\overbrace{a, \ldots, a}^{n})$ we have that

$$F(\overbrace{a, \ldots, a}^{n}) \leq F(a) \leq F(\overbrace{a, \ldots, a}^{n})$$

and then $F(\overbrace{a, \ldots, a}^{n}) = a$ since $F(a) = a$. Thus F is idempotent. Proposition 18.2.1 proves the compensativeness. □

Remark 2 Note that the family of extended aggregation functions which are globally monotone is closed under convex linear combinations: if F and G are in this family, then $H = (1 - k)F + kG$, $0 \leq k \leq 1$, is also a member of that family. It is also closed by duality: if F is a globally monotone extended aggregation function, so is $F^*(x_1, \ldots, x_n) = 1 - F(1 - x_1, \ldots, 1 - x_n)$.

The set of weighting vectors w^n, $n = 1, 2, \ldots$, arranged as indicated below is called a weighting triangle. We denote it by $\triangle w_i^n$.

$$\begin{array}{c} 1 \\ w_1^2 \quad w_2^2 \\ w_1^3 \quad w_2^3 \quad w_3^3 \\ w_1^4 \quad w_2^4 \quad w_3^4 \quad w_4^4 \\ \cdots \end{array}$$

The following two propositions give conditions to determine whether an extended weighted arithmetic mean and an extended ordered weighted arithmetic mean are globally monotone in terms of their corresponding associated weighting triangles. More details can be seen in [6].

Proposition 18.3.3 *An extended weighted arithmetic mean* $F(x_1, \ldots, x_n) = \sum_{i=1}^{n} w_i^n x_i$ *is globally monotone if and only if its weighting triangle* $\triangle w_i^n$ *satisfies:*

$$w_i^{n+1} \leq w_i^n$$

for each $n > 1$ and $i = 1, \ldots, n$. In this case we say that the weighting triangle is left descending.

Similarly, we state the following.

Proposition 18.3.4 *An extended ordered weighted arithmetic mean $F(x_1, \ldots, x_n) = \sum_{i=1}^{n} w_i^n x_{(i)}$ is globally monotone if and only if its weighting triangle Δw_i^n satisfies the following conditions for each n and all $p = 1, \ldots, n$*

$$\sum_{i=1}^{p} w_i^{n+1} \leq \sum_{i=1}^{p} w_i^n \leq \sum_{i=1}^{p+1} w_i^{n+1}$$

In this case, we say that the weighting triangle is left regular.

In [16] the problem of global monotonicity for the class of maximum entropy extended OWA functions is stated. Some partial results have been obtained but the complete solution of the problem remains open.

A method for constructing weighting triangles can be obtained from sequences of non-negative real numbers $\lambda_1, \lambda_2, \lambda_3, \ldots$ with $\lambda_1 > 0$. Thus, defining $w_i^n = \frac{\lambda_i}{\sum_{j=1}^{n} \lambda_j}$ for any $n \geq 1$ and $i = 1, \ldots, n$ we obtain a weighting triangle that we call generated by the sequence $\{\lambda_n\}$. Obviously not every weighting triangle can be constructed from such a type of sequence.

In the following proposition we characterize those sequences which define left descending and left regular weighting triangles.

Proposition 18.3.5 *If Δw_i^n is the weighting triangle generated by the sequence $\{\lambda_n\}$, then:*

(i) It is left descending.
(ii) It is left regular if, and only if, the sequence $\left\{\frac{\lambda_{n+1}}{\lambda_1+\ldots+\lambda_n}\right\}$ is decreasing.

Example 1 The sequence $1, 2, 4, \ldots, 2^{n-1}, \ldots$ generates the weighting triangle Δw_i^n given by $w_i^n = \frac{2^{i-1}}{2^n-1}$. According to Proposition 18.3.5, it is left descending and left regular. Thus, from Propositions 18.3.3 and 18.3.4 we can say that the extended generalized weighted arithmetic mean and the extended generalized ordered weighted arithmetic mean defined by Δw_i^n are globally monotone.

There exist other methods for constructing weighting triangles based on quantifiers, negations, and fractals (see [4]).

18.4 Associativity and global monotonicity

In this section, we deal with associative extended aggregation functions and combinations of them in order to study their consistency (global monotonicity).

It is important to remark that the only definition we know of "consistency" of an extended aggregation function is given to be used in the context of economic analysis [8] and it is rather restrictive in the sense that it is achieved only when the extended aggregation function is symmetric and associative.

Definition 18.4.1 An extended aggregation function $F : \bigcup_{n\geq 1} [0, 1]^n \to [0, 1]$ is **associative** if for all $m, n \geq 1$ and for all $(x_1, \ldots, x_n) \in [0, 1]^n$, $(y_1, \ldots, y_m) \in [0, 1]^m$:

$$F(x_1, \ldots, x_n, y_1, \ldots, y_m) = F(F(x_1, \ldots, x_n), F(y_1, \ldots, y_m)).$$

Associativity is a well-known algebraic property which allows us to omit "parentheses" in an aggregation of at least three elements. Implicit in the assumption of associativity is a consistent way of going unambiguously from the aggregation of n elements to $n + 1$ elements, which implies that any associative extended aggregation function F is completely determined by its (binary) restriction to $[0, 1]^2$: $F(x_1, \ldots, x_{n+1}) = F(F(x_1, \ldots, x_n), x_{n+1})$.

A complete description of the extended aggregation functions which are continuous, idempotent and associative is given in [12].

The consistency of associative extended aggregation functions is characterized as follows.

Proposition 18.4.1 *Let F be an associative extended aggregation function. Then F is consistent (globally monotone) if and only if it is idempotent.*

Proof Suppose first that F is globally monotone. Then from Proposition 18.3.2 we obtain that F is idempotent. Reciprocally, let us assume now that F is idempotent and let us prove that F is globally monotone. For this purpose, we write for all $(x_1, \ldots, x_n) \in [0, 1]^n$, $n \geq 1$:

$$\begin{gathered} F(x_1, \ldots, x_n, \wedge x_i) = F(F(x_1, \ldots, x_n), \wedge x_i) \leq max(F(x_1, \ldots, x_n), \wedge x_i) = \\ F(x_1, \ldots, x_n) = min(F(x_1, \ldots, x_n), \vee x_i) \leq F(F(x_1, \ldots, x_n), \vee x_i) = \\ F(x_1, \ldots, x_n, \vee x_i). \end{gathered}$$

Consequently, we have $F(x_1, \ldots, x_n, \wedge x_i) \leq F(x_1, \ldots, x_n) \leq F(x_1, \ldots, x_n, \vee x_i)$ that, according to Proposition 18.3.1, is equivalent to say that F is globally monotone. □

Remark 3 Because of their associativity, t-norms[2] are defined for any number of arguments $n \geq 1$ (with the usual convention $T(x) = x$), hence they are extended aggregation functions. Also observe that extended t-norms are increasing with respect to argument cardinality: $T(x_1, \ldots, x_n, x_{n+1}) \leq T(x_1, \ldots, x_n)$, therefore they satisfy the left hand side of condition (18.1) in Proposition 18.3.1. An extended t-norm T is globally monotone if and only if it is the minimum.[3]

Dually, we have a similar result for t-conorms.[4] They satisfy the right hand side of condition (18.1) in Proposition 18.3.1. An extended t-conorm S is globally monotone if and only if it is the maximum.[5]

Two well known families of associative mixed aggregation functions[6] are uninorms[7] and nullnorms,[8] which are strongly related to triangular norms and conorms. Details on these aggregation functions can be found in gleb07. Next we are going to describe under which conditions they are globally monotone.

Proposition 18.4.2 *(i) An extended uninorm is globally monotone if and only if it is idempotent.*

(ii) An extended nullnorm is globally monotone if and only if it is idempotent.

Proof The result comes directly from Proposition 18.4.1, given associativity of uninorms and nullnorms. □

Remark 4 (i) There are different kinds of idempotent uninorms (once fixed the neutral e in $]0, 1[$). See, for instance, [17].

(ii) Once fixed the absorbent a in $[0, 1]$ there is only one idempotent nullnorm.

18.5 Conclusions

The global monotonicity condition is presented as a minimum requirement for an extended aggregation function to be considered consistent. Here, we focus our interest in studying this property for associative aggregation functions.

Acknowledgments The authors have written this contribution in tribute to Prof. Claudio Moraga in recognition of his important and extensive research in many areas of Soft Computing.
This contribution has been partially supported by the Spanish Grant TIN2013-42795-P.

---

[2] A t-norm is a two-variable function $T : [0, 1]^2 \longrightarrow [0, 1]$ which is associative, symmetric, increasing in each variable and has neutral 1.

[3] The minimum t-norm ($T(x_1, x_2) = min(x_1, x_2)$) is the only idempotent t-norm.

[4] A t-conorm is a two-variable function $S : [0, 1]^2 \longrightarrow [0, 1]$ which is associative, symmetric, increasing in each variable and has neutral 0.

[5] The maximum t-conorm ($S(x_1, x_2) = max(x_1, x_2)$) is the only idempotent t-conorm.

[6] A mixed aggregation function exhibits different types of behavior on different parts of the domain.

[7] A uninorm is a two-variable function $U : [0, 1]^2 \longrightarrow [0, 1]$ which is associative, symmetric, increasing in each variable and has neutral e belonging to $[0, 1]$.

[8] A nullnorm is a two-variable function $V : [0, 1]^2 \longrightarrow [0, 1]$ which is associative, symmetric, increasing in each variable and has absorbent a belonging to $[0, 1]$.

References

1. Calvo, T., Carbonell, M., Canet, P.: Medias Casilineales Ponderadas Extendidas. Proc. Estylf'97, Tarragona pp. 33–38 (1997).
2. Calvo, T.: Two ways of generating Extended Aggregation Functions. Proc. IPMU'98, Paris, pp. 825–831 (1998).
3. Calvo, T., Mayor, G.: Remarks on two types of Extended Aggregation Functions. Tatra Mount. Math. Publ. 16–1, 235–255 (1999).
4. Calvo, T., Mayor, G., Torrens, J., Suñer, Mas, M., Carbonell, M.: Generation of Weighting Triangles Associated with Aggregation Functions. International Journal of Uncertainty, Fuzziness and Knowledge-Based Systems 8 (4), 417–452 (2000).
5. Mas, M., Mayor, G., Suñer, J., Torrens, J., Generación de funciones de Agregación Extendidas. Proc. Estylf'97, Tarragona, pp. 39–44 (1997).
6. Mayor, G., Calvo, T.: On Extended Aggregation Functions. Proc. Seventh IFSA Congress Vol. 1, Prague, pp. 281–285 (1997).
7. Suñer, J.: Funcions d'Agregació Multidimensionals. Ph.D. (in Catalan) (2002).
8. Pursiainen, H.: Consistency in aggregation, quasilinear means and index numbers. Discussion paper No. 244, University of Helsinki and HECER (Helsinki Center of Economic Research) (2008).
9. Rojas, K., Gómez, D., Montero, J., Tinguaro, J.: Strictly stable families of aggregation operators. Fuzzy Sets and Systems 228, 44–63 (2013).
10. Beliakov, G., Pradera, A., Calvo, T.: Aggregation Functions A Guide for Practitioners. Beliakov, G., Pradera, A., Calvo, T. (eds.) Springer, New-York, (2007).
11. Calvo, T., Mayor, G. and Mesiar, R. (eds.): Aggregation operators, New Trends and Applications, Physica-Verlag, Heidelberg, (2002).
12. Grabisch, M., Marichal, J.-L., Mesiar, R. and Pap, E.: Aggregation Functions. Cambridge University Press, (2009).
13. Torra, V., Narukawa, Y.: Modeling decisions information fusion and aggregation operators. Springer, New York (2007).
14. Calvo, T., Mayor, G. and Suñer, J.: Monotone Extended Aggregation Functions. In Enric Trillas: A Passion for Fuzzy Sets. Springer, New York, pp. 49–66, (2015).
15. Yager, R.R.: On Ordered Weighted Averaging Operators in Multicriteria Decisionmaking. IEEE Transactions on Systems, Man and Cybernetics 18, 183–190 (1988).
16. Carbonell, M., Mas, M., Mayor, G.: On a class of Monotonic multidimensional OWA Operators. Proc. Sixth IEEE Int. Conference on Fuzzy Systems, Barcelona, pp. 1695–1700, (1997).
17. Ruiz-Aguilera, D., Torrens, J., De Baets, B., Fodor, J.: Some Remarks on the Characterization of Idempotent Uninorms. IPMU 2010, LNAI 6178, 425–434 (2010).

Chapter 19
Distributed Machine Learning with Context-Awareness for the Regression Task

Héctor Allende-Cid

19.1 Introduction

In the last decades, there has been an increasing development of high throughput data acquisition technologies in a large number of very different domains, such as Biological Sciences, Environmental Sciences, Astronomy, Meteorological Sciences, etc. At the same time there have been big advances in communication technologies, digital storage and computing architectures. Both of these advances in different areas present unprecedented opportunities for scientists to seize the wealth of the distributed information by means of automatic learning and decision making. With the development of related technologies and the increase of Cloud storage services available, the total amount of potential data to be used is almost incalculable. A rough estimation of data stored in the entire cloud in 2012 was in the order of 300 exabytes [1]). Most of the times, the information of interest is distributed in different geographical locations, making the problem even harder.

Because of these characteristics, automatic analysis and learning for distributed datasets has become of great importance over the last years. The rapid growth of the distributed data available presents new opportunities for applications of Machine Learning and Automatic Data Analysis in order to generate insight from this data. Human reasoning is limited regarding the ability of handling large data sets, so the need of automatic data analyzers is nowadays a necessity. Thereby, the scalability and efficiency of these learning algorithms have become of central importance for many researchers and data scientists.

Although there are many reasons why it is necessary to use the wealth of the distributed information, there are several challenges that need to be addressed [2]:

H. Allende-Cid (✉)
Escuela de IngenieríaInformática, Pontificia Universidad Católica de Valparaíso, Valparaíso, Chile
e-mail: hallende@gmail.com; hector.allende@pucv.cl

R. Seising and H. Allende-Cid (eds.), *Claudio Moraga: A Passion for Multi-Valued Logic and Soft Computing*, Studies in Fuzziness and Soft Computing 349, DOI 10.1007/978-3-319-48317-7_19

- Data repositories are large in size, dynamic and physically distributed. Consequently, it is neither desirable nor feasible to gather all the data in a centralized location for analysis, due to the high storing, processing and communication costs.
- Data sources are autonomously owned and operated. This leads to the problem that the range of operations that can be performed on the data source (e.g. types of queries allowed), and the precise mode of allowed interactions may be quite diverse.
- Data sources are heterogeneous in structure (e.g. relational databases, flat files, etc.) and content (names and types of attributes used to represent the same data).
- Data sources may have the same attributes, but contextual heterogeneity. From a statistical point of view, the underlying law of probability of each of the distributed data sources may be different.
- There is also a problem of privacy among different data sources. Distributed data sources may have different "owners", who may not be willing or may not be allowed to share "raw" data.

The challenges previously described, are nowadays present in a variety of real problems, so it is necessary to keep working in new modelling paradigms that are able to face them. In this sense, Distributed Data Mining (DDM) is a fast growing research field that deals with most of the problems described above. More generally DDM deals with the problem of finding data patterns or relations between inputs and outputs, in environments with distributed data and computation, but taking into account the previously mentioned challenges. Although today most data analysis systems require centralized storage, the increasing merger of computation with communication nowadays is likely to demand data mining environments that can exploit the full benefit of distributed computation. Most classic Machine Learning approaches demand monolithic datasets, an approach not compatible with today's needs.

Classic machine learning algorithms usually work with the entire data set loaded into main memory. When the size of data is large, this is impossible in practical terms, since the algorithms are not be able to load the whole data set in the memory (e.g., for training purposes), or it is impractical due to computational restrictions (data size and computation time). Thus, the centralization of the data for analytical purposes has indeed become a restriction. Moreover, if the amount of information that is distributed or the number of distributed sources is too large, this could lead to problems related to Big Data. In order to overcome these problems, Parallel and Distributed approaches are having a lot of attention from the Machine Learning community these days. Distributed Machine Learning (DML) is often mentioned together with Parallel Machine Learning (PML) in the literature. These two approaches attempt to improve the performance of traditional Data Analysis systems, but is necessary to clarify that they are defined by different system architectures and use different approaches. In DML, computers and data are distributed and communicate through message passing and are defined to specific master/node architectures. In PML, a parallel system is assumed with processors sharing memory and/or storage. This difference in architecture greatly influences algorithm design, cost model, and performance measure in Distributed and Parallel Machine Learning.

Since it was proposed, the area of Distributed Machine Learning (DML) has been very active and is enjoying a growing amount of attention from the Data Science community. There are many real world applications where the data is distributed naturally and stored over distributed sources. In most cases, the amount of data distributed is so large, that is unfeasible to send it to a centralized node, so there is no alternative other than to treat the problem with a Distributed Learning approach. Most of the current DML techniques treat the distributed data sets as a single virtual table and assume that there is a global model which could be generated if the data were combined or centralized, completely neglecting the different semantic contexts that this distributed data sets may have [3]. In other words, generally DML algorithms aim to infer a global model and try to approximate the results one would get from a single joint data set. Because of this, there are deeper implications, that make the problem more complicated, since distribution itself may have a meaning and unexpected side effects. If we see this as a statistical learning problem, we deal with samples of data that follow different underlying laws of probability. As was explained above, there is a mismatch between the architecture of most classic data mining systems and the needs of mining systems for distributed applications. This mismatch may cause a fundamental bottleneck in many emerging distributed applications. The traditional data-warehouse architecture for data mining works by regularly uploading mission critical data in the warehouse for subsequent centralized data mining application. This centralized approach is fundamentally inappropriate for most of the distributed and ubiquitous data mining applications. The problems are the long response time, lack of proper use of distributed resources, and the fundamental characteristics of centralized data mining algorithms. We need data mining architectures that pay careful attention to the distributed resources of data, computing, and communication in order to consume them in a near optimal fashion. DDM considers data mining in this broader context. The objective of DDM is to perform the data mining operations based on the type and availability of the distributed resources.

In this chapter we discuss some classic distributed machine learning algorithms and approaches that consider different contexts, thus preventing communication overhead and not exchanging raw data which is a requirement for applications with privacy concerns.

Most of the works that are present in the literature assume that the underlying laws of probability of the distributed sources are the same (see e.g. [4, 5]). This is an assumption that is inherited from the ensemble-based approaches in classic Machine Learning [6]. When the distributed sources are re-sampled, it is assumed that the re-sampled data follow the same underlying law of probability as the original set. In real-world distributed problems, it is impossible to assure that, because the real underlying law of probability is unknown. The majority of learning algorithms that are found in the state-of-the-art focus on combining the predictions of a set of classifiers [7]. Combining predictions avoids potential problems with concept descriptions and knowledge representation. Most of these works do not consider that the underlying law of probability could change from site to site, making the task of mapping the feature vectors to the independent variable more difficult. If the underlying probability distribution of all sites is the same, we could call this sub-category as homogeneous distributed sites, or with contextual homogeneity. If the distribution changes along the sites, the problem is called heterogeneous distributed sites (contextual heterogeneity). If we see this as a statistical learning problem, we deal with sources of data that might not follow the same underlying laws of probability. We may have two different subproblems depending on how we define this. The first one corresponds to the case when data come from the same type of law of probability but with different parameters, and the second one, when data follow different types of underlying laws of probability (e.g. normal, gamma, exponential, etc.). In most cases the true underlying law of probability is unknown, that is why we should not make strong assumptions. Instead we should find ways to represent it, using only the available data. The way the data sources are distributed can be divided into two main categories: with homogeneous and heterogeneous attributes. In the former case the databases located at different sites have the same attributes in the same format (same data dictionary). In the latter case, each local site may collect different data, thus having different number of attributes, types of attributes, format of attributes, etc. In the literature this categorization is also referred to as horizontal and vertical fragmentation: Horizontal fragmentation, wherein (possibly overlapping) subsets of data tuples are stored at different sites (examples are distributed accross the different sources); and Vertical fragmentation, wherein (possible overlapping) sub-tuples of data tuples are stored at different sites (attributes are distributed accross the data sources). This proposal will deal mainly with the first case (data sources with homogeneous attributes or horizontally fragmented).

The other category deals with the underlying law of probability that governs each distributed dataset. Related works in DDM often assume that the underlying law of probability that each of the sources follows is the same [4, 5, 8, 9].

The organization of this chapter is the following: In Sect. 19.2 a review of the work done in the field of Distributed Data Mining and Machine Learning is presented. In Sect. 19.3 a context-aware distributed machine learning models is presented for the task of regression. In the last Section some concluding remarks are presented and future work is discussed.

19.2 State of the Art

There is a fair amount of research done in the field of Distributed Data Mining. Most of the proposed algorithms that we can find in the literature fall in one of these three categories:

- Data combination: This category deals with algorithms that combine distributed data sources in a single large centralized dataset during the learning process and that allow communication of the intermediate results. The learned concept represents the distributed data and can be applied to incoming data instances. The number of examples should ideally be less than the union of the entire dataset to reduce communication costs.

- Local Model Combination: Learn at individual local sites independently, and combine the locally learned models to form a global concept. For example, the locally learned model can be some kind of rules or decision trees. The global concept is the final rule set or decision tree made by the combination of the local models. The combination at this level is strictly related with the type of local models. Black box models can not be combined at this level since, it is not possible to find interpretability in its topology.
- Predictive Model Combination: When receiving a new instance or example, local models are used to predict the output and then the locally learned concept in each local data source is used to form consensus. The final output is then estimated generally with a voting strategy.

The first approach that combines the distributed datasets (or fractions of the datasets) is either ineffective or infeasible, because of storage costs, communication costs, computational costs and privacy issues. The other two approaches (combining local models, and combining predictive models) seem more appealing to researchers for two main reasons. First, local models are normally much smaller than the raw data; sending the learnt model instead of the raw data reduces the traffic load in the network as well as the network bandwidth requirement. Second, sharing only the model, instead of the raw data, would give reasonable security since it overcomes partially the issues of privacy and security of the raw data.

Several algorithms have been introduced in the literature for combining local models. In [10, 11] a method to convert decision trees from distributed sites into a single set of rules is presented. In [12] the authors proposed an alternative method that learns a single rule for each class at each local site. The rule of each class is shortened by an order of importance with respect to confidence, support and deviation. The final rule set contains a set of the first n rules in the list. In [2], authors develop a framework for learning from distributed data based on two steps: extraction of sufficient statistics from every data set and an hypothesis generation step. To generate the hypothesis, the framework uses a statistics gathering scheme from every distributed source. Authors claim that this approach can be provably exact meaning that the decision tree constructed from distributed data is identical to that obtained in the centralized setting.

A larger fraction of DDM algorithms focuses on combining predictive models. This approach has emerged from empirical experimentation due to a requirement for higher prediction accuracy. Recently, several researchers treat distributed learning systems as a centralized ensemble-based method [13]. Several learning algorithms are applied at each local site, using separate training data to mine local knowledge. A new data point is then classified/predicted from the predictions of all local sites using ensemble methods such as stacking, bagging, boosting with different output combination schemes like majority voting, simple average, or winner-takes-all [14]. In general, DDM approaches apply ensemble methods to minimize the communication costs and to enhance the system predictions. Most of the approaches assume that the distributed data comes from the same underlying law of probability or have the same semantic meaning. Classic ensemble learning models work under the assumption, that the each of the models that are trained with re-sampling of the data, work with the same probability density function. In real distributed cases this is not straightforward. Therefore more care should be put into this.

Most DDM algorithms are designed upon the potential parallelism they can apply over the given distributed data. Typically the same algorithm operates on each distributed data site concurrently, producing one local model per site. Subsequently all local models are aggregated to produce the final model. In essence, the success of DDM algorithms lies in the aggregation. Each local model represents locally coherent patterns, but lacks details that may be required to induce globally meaningful knowledge. For this reason, many DDM algorithms require a centralization of a subset of local data to compensate it. Therefore minimum data transfer is another key attribute of the successful DDM algorithm.

In the following subsection, we present a literature review on DDM algorithms on both regression and classification tasks.

19.2.1 Distributed Classifier Learning

Most distributed classifiers are based in classical ensemble learning models [15–17]. This approach has been used in various applications and domains in order to improve the classification accuracy of predictive models. The result is a set of multiple models (base classifiers or learners), typically from horizontally fragmented data subsets, that are combined in order to improve accuracy. Typically, in order to aggregate them, voting schemes (weighted or un-weighted) are employed.

The ensemble approach is most of the times directly applicable to the distributed scenario (only in cases where the underlying laws of probability are the same in all distributed sources). Different models can be generated at different sources and then aggregated using classical ensemble combining strategies. In [18] an AdaBoost-based ensemble approach for distributed scenarios is discussed. Breiman [19] considered Arcing as a mean to aggregate multiple blocks of data, especially in on-line settings. Homogeneous distributed classifiers are the most studied type of algorithms in DDM. One notable ensemble approach to learn distributed classifier is the meta-learning framework [20–22]. It offers a way to mine classifiers from homogeneous, distributed data. In this approach, supervised learning techniques are first used to learn classifiers at local data sites; then meta-level classifiers are learned from a dataset generated using the locally learned concepts. The meta-level learning may be applied recursively, producing hierarchy of meta-classifiers. Learning at a meta-level can work in many different ways. For example, we may generate a new dataset using the locally learned classifiers. We may also move some of the original training data from the local sites, combine it with the data artificially generated by the local classifiers, and then run any learning algorithm to train the meta-level classifiers. The output of the meta-classifier can be also decided by counting votes cast by different base classifiers. Other examples of homogeneous distributed classifiers can be found in [9, 23]. Approaches for vertically fragmented distributed learning, that will not be addressed in the proposal, can be found in [2, 24–26].

For heterogeneous distributed classifiers, the ensemble approach is not straightforward to apply to this kind of distributed problems. In heterogeneous distributed data, we observe the incomplete knowledge about the complete dataset. Different local models represent disjoint regions of the problem and DDM has to develop a global data model, associations, and other patterns with only limited access to the features observed at non-local sites. For this reason, it is generally believed that mining of heterogeneous data is more challenging. These issues in mining from heterogeneous data are discussed in [25, 26] from the perspective of inductive bias. This work notes that such heterogeneous partitioning of the feature space can be addressed by decomposing the problem into smaller sub-problems when the problem is site-wise decomposable. However, this approach is too restrictive to handle problems that involve inter-site correlations.

In [24] the authors note that any inter-site pattern cannot be captured by the aggregation of heterogeneous local classifiers. To detect such patterns, they first identify a subset of data that any local classifier cannot classify with a high confidence. Identified subset is merged in a central site and another classifier (central classifier) is constructed from it. When a combination of local classifiers cannot classify an unseen data with a high confidence, the central classifier is used instead. This approach exhibits a better performance than a simple aggregation of local models. However, its performance is sensitive to the sample size (or confidence threshold).

In [2] a solution to deal with the semantical heterogeneity problem is proposed, introducing ontology-extended data sources and define a user perspective consisting of an ontology and a set of interoperation constraints between data source ontologies and the used ontology. She shows how these constraints can be used to define mappings and conversion functions needed to answer statistical queries from semantically heterogeneous data viewed from a user perspective. That is further used to extend our approach for learning from distributed data into a theoretically sound approach to learning from semantically heterogeneous data.

More recently, [27] presented a distributed online learning scheme to classify data captured from distributed, heterogeneous and dynamic data sources. Their proposal contains a novel online ensemble learning algorithm called Perceptron Weighted Majority (PWM) to update the aggregation rule used to output the final prediction of the ensemble. This approach is able to deal with dynamic data streams due to its online nature. The authors of [28] proposed an evolutionary algorithm-based framework to generate a function for combining an ensemble in a distributed arrangement. In this framework, the models on the ensemble are trained only on a portion of the training set and later using a genetic programming evolved function they combine the classifiers composing the ensemble. A key property of their proposal is that the combine function can be recomputed in an incremental way avoiding expensive computational efforts. In [29], the authors propose a distributed learning algorithm where parts of the dataset are processed locally at every node of the distributed network, and then a consensus communication algorithm is employed to create a consolidate hypothesis. Their principal contribution is to proof the convergence of the distributed learning process in the general case where the learning algorithm is a contraction. To verify their theoretical results, they employ a binary classification problem where the update equation is based on a feed-forward neural network with backpropagation.

19.2.2 Distributed Regression Algorithms

There are only a few works dealing with this kind of problem. One of the first works was proposed by Hershberger [30, 31]. The proposal consisted in a method for distributed multivariate regression using Collective Data Mining [32] with Wavelet functions. The authors claim that this method seamesly blends machine learning and the theory of communication with the statistical methods employed in parametric multivariate regression to provide an effective data mining technique for use in a distributed data and computation environment. The authors test the proposed method with two benchmark datasets, producing results that are consistent with those obtained by applying standard parametric regression techniques to centralized datasets. [33] proposed a local distributed algorithm for multivariate regression in large peer-to-peer environments. The algorithm can be used for distributed inferencing, data compactation, data modeling and classification tasks in many peer-to-peer applications for bioinformatics, astronomy, social networking, sensor networks and web mining. The authors state that computing a global regression model from data available at the different peer-nodes using a traditional centralized algorithm can be very costly and impractical due to many reasons. The paper proposed a two-step approach to deal with these problems. Applications to Astronomy Data and Vertically Partitioned Data have been proposed in [34, 35].

The authors of [36, 37] proposed a series of algorithms based on a meta-learning approach to deal with the regression problem addressing the context heterogeneity case. The authors propose a meta-learning-based hierarchical model that is able to be successfully used in distributed scenarios with context heterogeneity. The main contribution of this proposal is that they create a scheme to estimate the context variance of the datasets, so they can add this information in the final ensemble-voting scheme. They claim that the context heterogeneity is related with the random differences in the output space, neglecting the differences that could exist in the input space.

In [38] an approach for distributed multivariate regression based on sampling and discuss its relationship with the compression method was proposed. The central idea is motivated by the observation that, although communication is limited, each individual site can still scan and process all the data it holds. Thus it is possible for the site to communicate only influential samples without seeing data in other sites. They exploit this observation and derive a method that provides trade-off between communication cost and accuracy.

In [39], an ensemble approach based on building neighborhoods of similar datasets is presented. To build the neighborhoods, it is assumed that the datasets follow a known underlying law of probability (which is possible when working with synthetic examples), and using the Hypothesis Test based on divergence measures, they form the corresponding neighborhoods. In this work it is necessary to assume a known underlying law of probability, i.e. multivariate normal distribution, in order to perform the Hypothesis tests [40]. It can be summarized as follows: At first, local models are trained with the available distributed data sets. In each distributed node, a local algorithm is trained with its corresponding data. After that, assuming that the underlying laws of probability of each distributed data sets are known (multivariate gaussian distributions), the mean and variance-covariance matrices (parameters of the underlying law of probability) are shared across all distributed nodes. With this information, in each distributed node i, an Hypothesis Test is performed with the parameters of node i and the parameters all other nodes $j = 1, \ldots, k$, where $j \neq i$ and k is the total number of distributed data sets. After performing all Hypothesis Tests, the neighborhood for node i is built if there was no evidence to reject H_0 (that the parameters of both underlying laws of probability were the same). Also all the local models are shared across the sites. Then, a second stage learner is trained, where the inputs of this learner are the outputs of *all* local models. The final output of the model is the ensemble of all second stage models, that belong to the same neighborhood, where the new data inputs are registered. In [41] an in-depth comparison between the Context-Aware Distributed Regression and other state-of-the-art models is presented, while introducing modified stacked generalization regression models and a discrete manner to represent the probability density function to generate the neighborhoods of data sources with similar contexts. The details of this model will be given in the following section.

For a more complete review on the state of the art of Distributed Machine Learning Algorithms, please refer to [14].

19.3 Distributed Machine Learning with Context-Awareness

One of the fundamental keystones of the work done in [39, 41] is to take advantage of the distributed information of similar data sources, while differencing the distributed sources that do not follow the same underlying law of probability. So a way to measure the similarity (or dissimilarity) among the underlying laws of probability of distributed data sources is needed.

19.3.1 Representation of the Probability Density Functions of the Distributed Data Sources

The probability density functions that govern the distributed data sources can be presented in two ways. In the first approach it is necessary to make an assumption about the theoretical probability density function, which as usual in the statistical literature will be abbreviated "pdf". For example, we need to assume that the pdf belongs to the multivariate Gaussian probability distributions, and the parameters are inferred from the data sets. Most of the times the pdf is unknown, but usually some assumptions are made beforehand, like normality, homoscedasticity and independence of the errors. So, as a first approach, the assumption of normality seems valid.

To work, avoiding to make this kind of assumptions (where we assume that the theoretical probability density function is unknown), we can use a discrete representation of the datasets, in order to get a rough estimation of the pdf. For this, we can use n-dimensional histograms. If the dataset is one-dimensional, the histogram is a vector. The length of the vector depends on the number of bins used to represent the histogram. A bin is a discrete interval, that reflects the frequency of the data points, that fall into it. If the dataset is 2-dimensional, the histogram is a matrix, and in n-dimensional cases the data structure used is an n-rank tensor.

19.3.2 Dissimilarity Metrics

There is an enormous number of distance/similarity measures encountered in many different fields of science. A considerable effort has been made in finding appropriate measures among such a big number of possible choices, because it is of fundamental importance to pattern classification, clustering and information retrieval [20]. From a formal point of view, the concept of distance is defined as a quantitative measurement of how apart two different objects are. Those distance measures, that satisfy the metric properties are simply called metrics, while non-metric distance measures are called divergences. Synonyms for similarity include proximity, and similarity measures are often called similarity coefficients.

If we assume that the pdf is known, we can use the well-known (h, ϕ)-divergence [40] to establish dissimilarities between the underlying laws of probability of the data sets.

The main problem when using continuous density functions is that the user has to select a specific theoretical probability density function. As explained before, we have to work under the assumption, for example, that the data follows a multivariate Gaussian distribution. This approach was used in the model presented in [39].

If we do not make any assumption about the pdf of the datasets, we may use histograms to represent the data. In order to compare several different histograms, it is necessary to count with dissimilarity metrics, that could show the differences between them. There are various definitions of distance/similarity measures and we can group them in the following families: L_p Minkowski family (Euclidean, City block, etc.), L_1 family (Sorensen, Gower, etc.), Intersection family (Intersection, Czekanowski, etc.), Inner Product family (Harmonic mean, Cosine, etc.), Fidelity family (Fidelity, Bhattacharyya, etc.), Squared L_2 family (Squared Euclidean, Pearson, etc.) and Shannon's entropy family (Kullback-Leibler, Jeffreys, etc.) In order to calculate a distance between these two data sets, it is necessary that both histograms have the same support and number of bins. These measures will aid to quantify the dissimilarity between distributed data sources, in order to build neighborhoods of data that follow similar underlying laws of probability. The type of dissimilarity measure used will depend on the representation of the data sources. For further details please refer to [20].

19.4 Building Neighborhoods According to Similarity Measures

In this section we present two ways of constructing neighborhoods according to similarity measures. If we assume that the probability density function is known, we can create the neighborhoods in the following way. We check for every data set D_i if the parameters of the known pdf is the same that the rest D_j, with $j = 1, \ldots, k$ and $i \neq j$, with an hypothesis test [40]. With this hypothesis test we build a binary vector that indicates which of the other datasets D_j belong to its neighborhood. In the case of using the histogram representation of the data, it is necessary to share the minimum/maximum per input dimension across all data sets. This is needed, in order to build histograms that are comparable with each other having the same support. To make them comparable we need bins with the same interval limits. This information shared across the system, does not contradict the restriction of sharing raw data, because it only shares the minimum and maximum global values of the examples of each distributed source. With this data, we obtain the global minimum and maximum values of all distributed sources, using this information to build histograms for all distributed sources, with the same bin limits. The idea is to use a histogram representation to build a vector of size k, that represents the dissimilarity between two datasets, using k distance measures. We then define different distance/similarity measures, using distances from different families, in order to have more diversity of distance measures.

Suppose we take and example of 5 distributed datasets. We use 3 distance measures and we build the following vectors $\text{dist}(D_i, D_j) = (d_1, d_2, d_3)$. Suppose D_1, D_3 and D_5 have the same underlying law of probability and that D_2 and D_4 have a different one. The distances from D_1 to the rest would possibly be like the following:

- $\text{dist}(D_1, D_1) = (0, 0, 0)$
- $\text{dist}(D_1, D_2) = (2.3, 4.3, 3.5)$
- $\text{dist}(D_1, D_3) = (0.2, 0.06, 0.03)$
- $\text{dist}(D_1, D_4) = (4.3, 3.5, 6.5)$
- $\text{dist}(D_1, D_5) = (0.03, 0.1, 0.19)$

If we apply a clustering algorithm to these data, and that the distances dist (D_1, D_1), dist(D_1, D_3) and dist(D_1, D_5) are in the same cluster, this gives us a similarity based neighborhood for dataset D_1. The number of clusters is fixed to 2, since we want as a result a binary decision, if the dataset D_j belongs or not to a specific neighborhood. We perform this method to dist(D_i, D_k), where $k = 1, \ldots, 5$ and i is the value to create the i-th neighborhood. The Datasets that belong to the i-th neighborhood are the D_k from dist(D_i, D_k), that belong to the same cluster dist(D_i, D_k) where $k = i$. With this method we construct 5 neighborhoods. The user should choose that one which best suits his/her needs. The representation of this neighborhood is a binary vector, that indicates which of the other datasets belong to the cluster. Using the example to illustrate this the generated binary neighborhood vector is $h_1 = [1, 0, 1, 0, 1]$.

As presented previously in both cases, the resulting neighborhoods for each of the distributed sources are crisp. The binary vector indicates if the other data sources j belong or not to the neighborhood of a data source i. Although this gives a first approach we want to know how many of the models that were trained with different data sources should weight more. So it seems natural then, to use the principle of membership functions as it is done in fuzzy sets, to define a degree of membership of the other data sources to the neighborhoods of a data source i. For this purpose we are going to use a membership function of the form $\mu_A = 1 - dist$, where $dist$ represents the dissimilarity of a pair of data sources. The dissimilarity of one data source with itself is 0, thus we define that the degree of membership of the data source i to its own neighborhood is $\mu_A = 1$. To define the degree of membership of the other data sources to data source i, we take into account only the dissimilarity measures of the data sources that belong to the neighborhood. The dissimilarity measure in each calculation is divided by the total sum of all dissimilarities. The result will give a membership function which indicates the degree of membership of each data source to the neighborhood. In [42] the authors propose a similar approach for avoiding crisp outputs using the p-value, but in a different type of application. This would work in the case that we only used the hypothesis test in order to construct the neighborhoods, but since we also proposed a solution based on histograms and a clustering algorithm, it is not directly applicable.

19.5 Distributed Regression Algorithm

The proposed algorithm consists in three phases, which are explain in detail:

1. *Phase 1—Local Learning*. Given k distributed data sets, with nodes D_i, where $i = 1, \ldots k$, we use an available learning algorithm to train a local predictive model L_i from all instances of that node (see Fig. 3.1). The choice of the learning algorithm is not restricted to any particular kind. The local data consists in an n dimensional input vector $((x_{i1}, x_{i2}, \ldots, x_{in}))$ and a continuous response variable y_i. Let $\Gamma_i = [x_{i1}, x_{i2}, \ldots, x_{in}, y_i]$, $1 \leq i \leq k$ (Fig. 19.1).

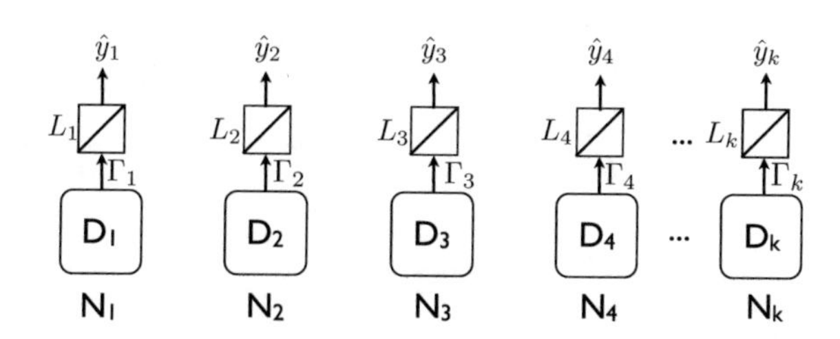

Fig. 19.1 Phase 1. Local learning

2. *Phase 2—Model and information transmission*. Each node D_i, where $i = 1, \ldots, k$ receives the model parameters (L_i learners) of all the other nodes. In case that the assumption that the real underlying law of probability is known, the mean vector, variance-covariance matrix and number of samples of the other nodes are transmitted to all distributed nodes. In case we used the discrete representation of the pdf, at first the minimum and maximum of each of the attributes $((x_{i1}, x_{i2}, \ldots, x_{in}))$ and y_i is shared to the other distributed nodes in order to establish the global values of each of the attributes and response. With this information we are able to represent the pdf as an n- dimensional histogram, that is comparable across the different sites. Then the histograms of each site are sent to each individual node. A hypothesis test is performed based on [40] for each pair of local data sets of the nodes D_i and D_j, where $i \neq j$, in the case of the continuous representation of the pdf. These hypotheses tests will check if the data of the current local node follow the same underlying law of probability of the rest of the nodes. The results are k binary variables h_{ij}, where $j = 1, \ldots, k$, which indicate the nodes j which follow the same underlying law of probability D_i. The variable h_{ii} is always 1. In the case of the histogram representation construct the neighborhoods based on Sect. 19.4.
3. *Phase 3—Algorithm combination and output*. Since every node D_i has a copy of the other local models, each local model L_l, $\ell = 1, \ldots, k$, contained in node D_i was trained with the local data from its corresponding node. Each of the local models in D_i outputs a response variable $\hat{y}_{ij}$, where $j = 1, \ldots, k$ with the available input data at that node. Each local node also trains a global learning algorithm G_i with the outputs of all the local models $(\hat{y}_{i1}, \hat{y}_{i2}, \ldots, \hat{y}_{ik})$ that were trained with the local sources that belong to the same neighborhood of site i and the real response variable y_i, obtained from the training data of the node D_i (see Fig. 19.2). The output of our model is generated in the following way: Whenever a new example arrives at a node D_i, we compute all the outputs of the local models that are stored in this node. We have an a priori information about which of the other nodes have data following the same underlying law of probability of the current node, which is reflected in the binary vector mentioned above (h_{ij}, where $j = 1, \ldots, k$). The final output of the model is the mean of the outputs of all the G_i models that received the output of all the local models in the current node that belong to the neighborhood i multiplied by a corresponding weight w_j.
E.g. in the example presented above the final output of the model should be the weighted mean of the outputs of models G_1, G_3 and G_k, because only the variables

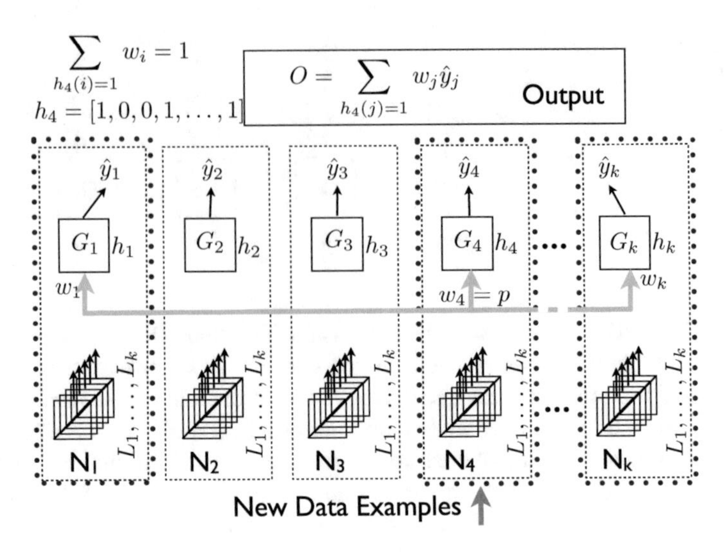

Fig. 19.2 Phase 3. Algorithm combination and output

h_{31}, h_{33} and h_{3k} are distinct from zero. Weights of the output of the same node where data arrives is user-defined as 1. The sum of weights of the other outputs that belong to the neighborhood is defined by the fuzzy membership function described in the previous section. The weights indicate the degree of membership of the nodes to the Neighborhood of node 3.

As explained before, the motivation behind constructing distributed learning models may differ depending on the use. If there is the need to construct a general model from distributed sources, stored in a Central Site, the information needed there are the global models $G_i, i = 1, \ldots, k$, trained in each individual site, the weight vector w_i (stores the degrees of membership of all the nodes to the node i) and the neighborhood vector h_i of each distributed source. When new data examples are registered in a distributed node j, the output vector $(\hat{y}_{j1}, \hat{y}_{j2}, \ldots, \hat{y}_{jk})$ of each example is transmitted to the Central Node to output the final prediction value. In the other case, if only a model valid in local context is needed, it is only necessary to transmit to the individual sites the global models G_i, the neighborhood h_i and the weight w_i vectors of the sites i that belong to that neighborhood. In this case each individual sites will have the variables to predict the outputs of the examples that are registered in site j. The output of the algorithm will be the same in both cases. The difference lies on the motivation of the user to construct the distributed regression model. In [39] the proposal that uses theoretical formulations of the pdf is presented, and the neighborhoods are constructed using Divergence Measures and Hypothesis Tests. The results presented in this work show that the proposed model outperforms other models that also work with the assumption that there exist different underlying laws of probability in different sources. The model is validated using several synthetic experiments. In [41] the proposal that uses histograms and clustering algorithms is presented. The results with synthetic and real datasets, show that the model outperforms other state-of-the-art models. Please refer to [39, 41] for details about the experiments and the results.

19.6 Concluding Remarks and Future Work

In this chapter we presented a distributed regression framework, that performs well in distributed scenarios where there are different contexts representing the distributed data sources. From a statistical learning point of view, the different contexts can be understood as different underlying laws of probability governing the data sources. By building neighborhoods of similar data sources we validate, that the use of this approach gives improved results over state of the art approaches. By merging in this work different soft computing and statistical methodologies, like degrees of membership, clustering algorithms, divergence-based test hypothesis and ensemble of artificial neural networks, we demonstrate that in solving complex problems, most of the times the solution does not come from only one area of knowledge. We validate this proposal, by running several synthetic and real world experiments, obtaining in most of the cases, very favorable results. From this we can infer that by applying our proposal methodology, we can take more advantage of the information that is naturally distributed. Also it should be added that the proposal obtained favorable results in Big Data scenarios, where the whole data set was split in different data sources. This scenario was not the core of this proposal, but nevertheless the results obtained show the potential of the proposal for this type of scenario.

There are other problems in Machine Learning that are related with changes in the underlying law of probability of the data. This method could be suitable also for Streaming scenarios, for example. The amount of data generated in streaming scenarios is so huge that it is necessary to propose methods to handle this problems. An adaptation of this proposal could be suitable to handle streaming scenarios, by creating batches of data to handle them in a distributed fashion, and taking in consideration the different underlying laws of probability that could govern the data. The changes in the underlying law of probability is referred as Concept Drift in Incremental Learning. This proposal could be extended in order to work in this type of problems and detect the changes in the statistical distribution in order to not be affected by Concept Drift. Although the proposal is a merger of Distributed Systems and Machine Learning, it is a conceptual and theoretical proposal, so in the future a real implementation on a distributed platform will be implemented. I firmly believe, that this proposal opens a huge number of possibilities to continue doing research in the field of Machine Learning and Distributed Computing.

Fig. 19.3 Cutting the tie of the Thesis Supervisor after the Thesis presentations is a tradition in the Computer Science Engineering Doctoral Program of the Santa María University (January, 2015)

Acknowledgments The core of this chapter was based in my Ph.D. Thesis, that was written and carried out under the guidance of Prof. Dr. Claudio Moraga, and presented in January 2015 at the Universidad Técnica Federico Santa María. I want to thank Profesor Moraga, for his patience and dedication. I am sure that the fulfillment of my work would not have been possible, without his help and supervision. I have known Prof. Moraga for several years, even before my PhD Thesis, and he has been always an inspiration, both academically and personally.

References

1. News, U. (2012). How much information is there in the World. http://news.usc.edu/29360/how-much-information-is-there-in-the-world/. [Online; accessed in 12/04/2016].
2. Caragea, D., Silvescu, A., and Honavar, V. (2004). A framework for learning from distributed data using sufficient statistics and its application to learning decision trees. *International Journal of Hybrid Intelligent Systems*, 1(1, 2):80–89.
3. Wirth, R., Borth, M., and Hipp, J. (2001). When distribution is part of the semantics: A new problem class for distributed knowledge discovery. In *12th European Conference on Machine Learning*, pages 3–7.
4. Lazarevic, A. and Obradovic, Z. (2002). Boosting algorithms for parallel and distributed learning. *Distrib. Parallel Databases*, 11(2):203–229.
5. Tsoumakas, G. and Vlahavas, I. (2002). Effective stacking of distributed classifiers. In *Proc. 15th European Conference on Artificial Intelligence (ECAI'02), IOS Press*, pages 340–344.

6. Kittler, J., Hatef, M., Duin, R. P. W., and Matas, J. (1998). On combining classifiers. *IEEE Transactions on Pattern Analysis and Machine Intelligence*, 20:226–239.
7. Aounallah, M. and Mineau, G. W. (2007). Distributed data mining: Why do more than aggregating models. In *IJCAI*, pages 2645–2650.
8. Brazdil, P., Carrier, C. G., Soares, C., and Vilalta, R. (2008). *Metalearning: Applications to Data Mining (Cognitive Technologies)*. Springer-Verlag.
9. Guo, Y. and Sutiwaraphun, J. (1999). Probing knowledge in distributed data mining. In *Proceedings of the Third Pacific-Asia Conference on Methodologies for Knowledge Discovery and Data Mining*, PAKDD'99, pages 443–452.
10. Hall, L. O., Bowyer, K. W., Kegelmeyer, W. P., Moore, T. E., Jr., and ming Chao, C. (2000). Distributed learning on very large data sets. In *In Proceedings of the Sixth ACM SIGKDD International Conference on Knowledge Discovery and Data Mining*, pages 79–84.
11. Hall, L. O., Chawla, N., and Bowyer, K. W. (1998). Decision tree learning on very large data sets. In *In IEEE Conference on Systems, Man and Cybernetics*, pages 2579–2584.
12. Cho, V. and Wüthrich, B. (2002). Distributed mining of classification rules. *Knowl. Inf. Syst.*, 4(1):1–30.
13. Xing, E. P., Ho, Q., Xie, P., and Dai, W. (2016). Strategies and principles of distributed machine learning on big data. *CoRR*, arXiv:1512.09295v1.
14. Peteiro-Barral, D., Guijarro-Berdiñas, B., and Pérez-Sánchez, B. (2011). On the effectiveness of distributed learning on different class-probability distributions of data. In *Proceedings of the 14th International Conference on Advances in Artificial Intelligence: Spanish Association for Artificial Intelligence*, CAEPIA'11, pages 114–123.
15. Dietterich, T. G. and Fisher, D. (2000). An experimental comparison of three methods for constructing ensembles of decision trees. In *Bagging, boosting, and randomization. Machine Learning*, pages 139–157.
16. Merz, C. J., PAZZANI, M. J., and Stolfo, S. (1998). A principal components approach to combining regression estimates. In *Machine Learning*, pages 9–32.
17. Zhang, C. and Ma, Y. (2014). *Ensemble Machine Learning: Methods and Applications*. Springer Publishing Company, Incorporated.
18. Fan, W., Stolfo, S. J., and Zhang, J. (1999). The application of adaboost for distributed, scalable and online learning. In *Pages 362-366 of: SIGKDD Conference on Knowledge and Data Mining (KDD*.
19. Breiman, L., Wolpert, D., Chan, P., and Stolfo, S. (1999). Pasting small votes for classification in large databases and on-line. In *Machine Learning*.
20. Cha, S.-H. (2007). Comprehensive survey on distance/similarity measures between probability density functions. *International Journal of Mathematical Models and Methods in Applied Sciences*, 1(4):300–307.
21. Chan, P. and Stolfo, S. J. (1993). Toward parallel and distributed learning by meta-learning. In *AAAI Workshop in Knowledge Discovery in Databases*, pages 227–240.
22. Lattner, A. D., Grimme, A., and Timm, I. J. (2010). An evaluation of meta learning and distribution strategies in distributed machine learning. In *Proceedings of the European Conference on Data Mining*.
23. Peteiro-Barral, D. and Guijarro-Berdiñas, B. (2012). An analysis of clustering approaches to distributed learning on heterogeneously distributed datasets. *Advances in Knowledge-based and Intelligent Information and Engineering Systems*, 243(1):69–78.
24. Park, B., Kargupta, H., Johnson, E., Sanseverino, E., Hershberger, D., and Silvestre, L. (2001). Distributed, collaborative data analysis from heterogeneous sites using a scalable evolutionary technique. *Applied Intelligence*, 16(1):19–42.
25. Provost, F. J. and Buchanan, B. G. (1995). Inductive policy: The pragmatics of bias selection. In *MACHINE LEARNING*, pages 35–61.
26. Tumer, K. and Ghosh, J. (2000). Robust order statistics based ensemble for distributed data mining. In *In Advances in Distributed and Parallel Knowledge Discovery*, pages 185–210. AAAI/ MIT Press.

27. Canzian, L., Zhang, Y., and van der Schaar, M. (2015). Ensemble of distributed learners for online classification of dynamic data streams. *Signal and Information Processing over Networks, IEEE Transactions on*, 1(3):180–194.
28. Folino, G. and Pisani, F. S. (2015). Combining ensemble of classifiers by using genetic programming for cyber security applications. In *Applications of Evolutionary Computation*, pages 54–66. Springer.
29. Georgopoulos, L. and Hasler, M. (2014). Distributed machine learning in networks by consensus. *Neurocomputing*, 124:2–12.
30. Hershberger, D. E. and Kargupta, H. (1999). Distributed multivariate regression using wavelet-based collective data mining. *Journal of Parallel and Distributed Computing*, 61:372–400.
31. Hershberger, D. E. and Kargupta, H. (2001). Distributed multivariate regression using wavelet-based collective data mining. *J. Parallel Distrib. Comput.*, 61(3):372–400.
32. Kargupta, H., Byung-Hoon, Hershberger, D., and Johnson, E. (1999). Collective data mining: A new perspective toward distributed data analysis. In *Advances in Distributed and Parallel Knowledge Discovery*, pages 133–184. AAAI/MIT Press.
33. Bhaduri, K. and Kargupta, H. (2008). A scalable local algorithm for distributed multivariate regression. *Statistical Analysis and Data Mining Journal*, 1(3):177–194.
34. Bhaduri, K., Das, K., Borne, K., Giannella, C., and Kargupta, H. (2011). 1 scalable, asynchronous, distributed eigen-monitoring of astronomy data streams.
35. Das, K., Bhaduri, K., and Votava, P. (2011). Distributed anomaly detection using 1-class svm for vertically partitioned data. *Statistical Analysis and Data Mining*, 4(4):393–406.
36. Xing, Y., Madden, M. G., Duggan, J., and Lyons, G. (2003). Distributed regression for heterogeneous data sets. In *Proceedings of the 5th International Symposium on Intelligent Data Analysis (IDA2003*, pages 544–553. Springer.
37. Xing, Y., Madden, M. G., Duggan, J., and Lyons, G. J. (2005). Context-sensitive regression analysis for distributed data. In *Proceedings of the First international conference on Advanced Data Mining and Applications*, ADMA'05, pages 292–299.
38. Yu, H. and Chang, E.-C. (2003). Distributed multivariate regression based on influential observations. In *Proceedings of the ninth ACM SIGKDD international conference on Knowledge discovery and data mining*, KDD'03, pages 679–684.
39. Allende-Cid, H., Moraga, C., Allende, H., and Monge, R. (2013). Context-aware regression from distributed sources. In *Proceedings of the 7th International Symposium on Intelligent Distributed Computing - IDC 2013, Prague, Czech Republic, September 2013*, pages 17–22. Springer Verlag.
40. Pardo, L. (2005). *Statistical Inference based on Divergence Measures*. Chapman and Hall/CRC.
41. Allende-Cid, H., Allende, H., Monge, R., and Moraga, C. (2015). Discrete neighborhood representations and modified stacked generalization methods for distributed regression. *Journal of Universal Computer Science*, 21(6):842–855.
42. Torres, L., Sant'Anna, S. J. S., Freitas, C. C., and Frery, A. C. (2014). Speckle reduction in polarimetric SAR imagery with stochastic and nonlocal means. *Pattern Recognition*, 47:141–157.

Chapter 20
Recent Advances in High-Dimensional Clustering for Text Data

Juan Zamora

20.1 Introduction

The clustering task has a long history in the pattern recognition literature. In a few words, this task consists in finding a structure of compact and homogeneous groups of objects according to some similarity measure adhoc to the origin of the data. In case of document collections, the Cosine and Jaccard similarity measures are often employed to compare documents, and for evaluating the quality of the clusters built in these collections, instead of using compactness within each group, other measures that employ external labels are preferred. Then, provided that external labels are available, external performance measures such as Purity and Entropy are commonly used.

The noticeable increase in the computing capacity, along with the wide use of the Internet and the ease of text content generation around the globe, has led classical data processing and information extraction techniques to their limits. Under the classical processing model, the amount of available data was rather low and the pursued aim was to extract the most of information from the data. The accelerated development of digital storage media and computational power, along with low data volume made possible the usage of techniques based on secondary memory without major concerns on efficiency. At the end of the nineties and especially during the first decade of the twenty first century, the rise of the Internet led to the need to develop new algorithms capable of processing data whose volume was much larger than the amount of available memory, and also that are capable of delivering anytime responses [9].

J. Zamora (✉)
Universidad de Valparaíso, Valparaíso, Chile
e-mail: juan.zamora@uv.cl

R. Seising and H. Allende-Cid (eds.), *Claudio Moraga: A Passion for Multi-Valued Logic and Soft Computing*, Studies in Fuzziness and Soft Computing 349, DOI 10.1007/978-3-319-48317-7_20

There exist three concepts frequently treated as equivalent, but whose meanings represent different modes of addressing data captured sequentially. The first one is *Incremental algorithm* and denotes an algorithm capable of updating its parameters as the input is captured. The second one is *Online algorithm* and represents those methods that are able to provide an answer at any time without the need of knowing the whole data. The third and last one is *Single-pass algorithm*, which denotes an algorithm that examines each input instance once, extracts a summary of it and then discards it, leaving room for the upcoming instance. Bishop [12] represents the union of these three aspects in one concept referred to as *Sequential algorithm*.

Outline

The rest of this chapter is organized as follows: In Sect. 20.2, the problems generated by the curse of dimensionality over distance based methods are described and also the existing approaches that tackle these issues are reviewed. High dimensional data clustering methods without and with special emphasis on text collections are revised in Sect. 20.3. The last two sections show a discussion of the recent contributions, the conclusion of this chapter and a reflection about the potential trends in the high dimensional clustering task over text collections.

20.2 The Curse of Dimensionality in Clustering

The Curse of Dimensionality is a concept coined by Richard Bellman in 1957 to denote the exponential growth in volume associated with the increase in the number of dimensions of the input space of the data under study. Following the example given by [31], only 100 points are needed to sample the unit interval with no more than 0.01 distance between points. An evenly spaced grid, having 0.01 inter point distances, underlying a 10-D unit hypercube would need $10^{2}0$ points in order to perform the same sampling operation than the one described in 1-D. Hence, the 10-D unit hyper-cube is 10^{18} larger than the unit interval.

The Curse of Dimensionality has several impacts on the Clustering task. First, similarity/distance measures lose their meaning (specially the ones based on the L_p norms). Also high dimensional data is generally sparse, specially text data, hence the local relevance phenomenon -in which a set of points can be grouped in several ways depending on the subset of attributes under consideration- occurs. The last impact consists in occurrence of the Hubness phenomenon, and it consists in the appearance of points that tend to be among the nearest neighbors of a large number of points. This issue is observed by measuring the positive correlation that appears between the increase in dimensionality of the space and the Skewness in the distribution of the number of times that a point appears in the neighborhood of other points in the data. Following, a more detailed description of these issues is given.

20.2.1 *Loss of Meaning of Similarity/distance Measures in High Dimensional Spaces*

Beyer et al. [11] explore the impact of the increase in dimensionality of the input space on the near neighbor search task. As the authors explain, in presence of high dimensional data, for any point, the proximities/distances to its nearest and furthest neighbors tend to be quite similar. Thus, this phenomenon, called concentration of distances, produces a loss of contrast between close and far points, especially when $L_p,\ p \in \{1, 2, \ldots, \infty\}$ based similarities/distances are used. Traditional data structures for efficient near neighbor search, such as k-d tree, fail when searching for near neighbors in high dimensional spaces, not only because of the large number of attributes, but also because the similarities loss their meaning as the dimensionality increases.

Supporting the ideas mentioned above, [6] provide theoretical and empirical results showing that for algorithms based on L_p norms, as the value of p increases, then the discriminatory power of similarity/distance measures decreases. In order to support this claim, they define a numerical coefficient called *Relative contrast*. For every point in a dataset, this coefficient allows to measure the relative difference between the distances to its nearest and furthest neighbors. Under several different distributions of points, they explore the values taken by this coefficient and find that for $L_p,\ (p > 2)$ norms, the relative distance between nearest neighbors tends to zero as p is increased. Also they notice that the tendency to degradation of contrast accelerates as p is increased. Consequently, this observed degradation of contrast between similar and dissimilar points affects also to every clustering strategy based on density and distance (e.g. DBSCAN and K-means).

20.2.2 *Sparsity and the Local Relevance Phenomenon*

Agrawal et al. [8] observed that in high dimensional spaces, points can be grouped more compactly onto different subsets of attributes or subspaces. Additionally, [7] identified the local relevance phenomenon, which consists in that for some subsets of attributes, compact groups can be found and for some other subsets of attributes, points are uniformly scattered. This local dependence behavior associated to each group of data points cannot be distinguished in the full dimensional input space. Therefore, there may exist rich information which full dimensional algorithms bypass. Finding groups hidden among the different subspaces poses two big challenges to the high dimensional data clustering task. The former consists in addressing the exponential combinatory of potential subspaces. The latter consists in undertaking this task without using traditional dimensionality reduction techniques (e.g. PCA) prior to the clustering, since these techniques work on the full dimensional space and, as they obtain a single subspace, they can obscure potentially useful groups of points spanned onto other subsets of attributes.

More recently, this problem has also been tackled in the research subject of *Sparse Topic Modeling*. As far as we could find in recent works, the computational models presented in this area correspond to generative models of text. For more detailed information refer to the works of [17, 20, 36].

20.2.3 The Appearance of Hub-Points in High Dimensional Neighborhoods

Hubness is another phenomenon that emerges in high dimensional data, and manifests as the appearance of objects, called Hubs, which tend to be among the k nearest neighbors of a large number of data points. Radovanović et al. [44] first linked the Hubness phenomenon to the concentration of distances, already described in Sect. 20.2.1. They found that accordingly to previous theoretical works that study the distance concentration phenomenon, as dimensionality increases it is expected that a considerable number of points closer to the dataset mean exist. Consequently, this would explain the appearance of Hubs, as this increasing amount of points closer to the mean has a higher probability of inclusion into the k-nearest neighbor list of other points. In order to measure the Hubness, under several distance measures, they computed the distribution of k-occurrences N_k, that is for each point, the number of times it occurs among the k-nearest neighbors of all other points in the dataset. Along with this, they showed that the increase in skewness of the distribution, i.e. S_{N_k}, highly correlates with the increase in dimensionality of the data. Recall that, given the mean μ_{N_k} and standard deviation σ_{N_k} for N_k, the skewness of the distribution of k-occurrences is calculated as

$$S_{N_k} = \frac{\mathsf{E}[(N_k - \mu_{N_k})^3]}{\sigma_{N_k}^3} \tag{20.1}$$

A positive value of S_{N_k} denotes a skew to the right, which means that the right tail is long relative to the left tail. A negative value of S_{N_k} denotes a skew to the left, which means that the left tail is long relative to the right tail.

The increasing amount of Hubs also generates an increase in the amount of points farther from the mean and in turn with much lower N_k than the rest. These points can be regarded as distance-based outliers and they are denominated by [44] as anti-Hubs. As a consequence, hub points tend to decrease the average inter-cluster distance, because they are close to many points. Furthermore, anti-Hubs lead to an increase in average intra-cluster distance. Consequently, [44] experimentally demonstrate the decrease in the clustering performance in presence of Hubs and anti-Hubs.

Consequently, a new approach used to extend traditional centroid-based methods is emerging. [46] focused on the asymmetry of the near neighbor relations. That is, given any two points $\mathbf{x}$ and $\mathbf{y}$ in a dataset, point $\mathbf{y}$ does not belong to the list of the top k nearest neighbors of $\mathbf{x}$. Nevertheless, point $\mathbf{x}$ belongs to the list of k nearest neighbors

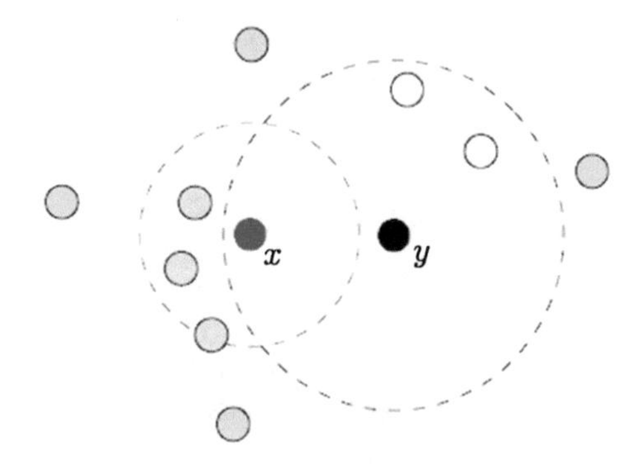

Fig. 20.1 Asymmetry in the near neighbor relation between two points. The two *dashed rings* denote neighborhoods for the top 3 nearest neighbors for each point. Point **x** belongs to the top 3 neighborhood of **y**, but not vice versa

of **y** (See Fig. 20.1). This observation impacts directly into those algorithms that use the shared-nearest-neighbors approach.

Schnitzer et al. [46] denominate theses approaches as local scaling approaches since they require knowledge of the local neighborhood of every data point in order to scale distance between data points. In contrast, they propose a global scaling approach that uses a novel distance measure that enforces symmetric nearest neighbor relations. This measure is called *Mutual Proximity* and it reinterprets the distance between two objects as a mutual proximity between both distribution of distances of all points to each one of the two points.

In a more recent contribution, [50] studied the relation between centroids and Hubs as dimensionality is increased. Moreover, they propose iterative methods that exploit hubness to detect hyper-spherical clusters by selecting Hubs as cluster prototypes. The attained results over synthetic and real datasets show that using Hubs to approximate data centers leads to better performance in contrast to several centroid-based approaches.

20.3 Related Work

The document clustering task involves contributions from two fields, namely high-dimensional data clustering and text processing. The former is more concerned with the design of methods capable of dealing with the problems attached to high-dimensional data as mentioned in Sect. 20.2. The latter includes a vast amount of techniques designed especially to exploit the nature of text data and to cope with the efficiency issues inherent to its processing. Hence, the contributions in this field are often concerned with enabling the construction of vector representations suitable to the application of machine learning methods.

The overall topic of this section is about the previous works in text data clustering. Nevertheless, in order to enrich the further discussion about the contributions made, related works in the task of high dimensional data clustering are also included. Consequently, the Sect. 20.3.1 is about the most outstanding works -in the opinion of this author- in the general task of high dimensional data clustering, and the Sect. 20.3.2 deals with more specific contributions made in the task of text data clustering.

20.3.1 Previous Advances in High-Dimensional Data Clustering

Previous progress in the high dimensional data clustering task has been developed in three lines of research: *Subspace clustering*, *Random sampling* of data points or features, and finally, by using similarity measures based on shared nearest neighbors (*SNN*).

20.3.1.1 Subspace Clustering Approach

The *Subspace clustering* task addresses the curse of dimensionality by integrating dimensionality reduction into the process of clustering. As [34] explain, in this approach the combinatorial explosion involved in the selection of a subspace or subset of features made in order to identify groups, is mainly tackled in two ways: either searching in *parallel axes*—approach known as *Projected* clustering—or searching for subspaces in *arbitrarily oriented axes*—approach known as *Oriented* clustering. *Projected* clustering directs its search to subsets of the original attributes, which involves an exponential number of potential feature spaces. Whereas that *Oriented* clustering attempts to explore feature spaces built from the combination of original features, thus projecting the data points in an optimal way. This last approach may involve a potentially infinite number of subspaces.

In the static setting, where it is assumed that data fits in main memory, several methods have been proposed since the seminal work of [8] (CLIQUE algorithm). Among these contributions, the most renowned are PROCLUS [7], ORCLUS [5] and COSA [22]. These methods make several scans over the data collection and thus they are not able to cope with big data domains or with the streaming setting. For instance the CLIQUE algorithm generates an initial set of medoids by using the classical K-medoids methods, and then, by following a *Hill Climbing* strategy, it starts to improve this set of medoids iteratively. A very detailed description of the methods proposed for the static setting and without massive data considerations is developed by [10, 34].

In dynamic environments where the data collection does not fit in main memory because of its large volume, which can even be undetermined, the amount of contributions is far more reduced in comparison with the above mentioned static or offline setting. Nonetheless, those works whose proposals could be considered as distinctive in terms of their contribution and more prominent in terms of their impact on the research community are HPStream [4] and HDDStream [42]. These two methods employ summarization structures or synopses, derived from the BIRCH system [55], in order to fulfill the single-pass constraint. Then the clustering process is performed on demand using solely the information already stored in the synopsis. Specifically, HPStream [4] requires the number of clusters to find as a parameter and then uses a *Fading-cluster-structure* to represent each one. This structure allows to maintain updated the first two statistical moments along each feature, together with a bit vector that indicates the features or dimensions identified as relevant to the current cluster. On the other hand, HDDStream [42] adapts the number of clusters depending on the stream, and employs extensions of the *micro-clusters* proposed by [3]. These extensions allow the maintenance of information about the relevant dimensions to each cluster (those features presenting the lowest variance), which is found by an offline density clustering algorithm (PreDeCon [13]). As mentioned by [53], besides that these algorithms are designed to process high dimensional data streams, the synopses that they employ are not scalable to high dimensional data, such as text, because of their feature based structures.

There are few studies focused in processing text streams by using the *Subspace clustering* approach [35]. Besides the interesting similarities between the *Subspace clustering* and the *Text clustering* problems, the subspace techniques do not seem to be directly applicable to the latter problem as [35] also mention. Lastly, it is important to emphasize that, as far as we know, in the literature corresponding to offline and streaming *Subspace-clustering* algorithms, the dimensionality of the datasets employed in the experimental setups is not beyond the order of tens of features. In contrast with this, in text processing scenarios the number of features involved can easily reach the order of hundreds of thousands.

20.3.1.2 Row and Column Sampling Approach

In the classic data processing scenario, where a data collection fits in main memory, the input to any algorithm consists in a matrix having one row per each instance and one column per each feature with which data are represented. An initial approach to extend classical clustering algorithms under the single-pass constraint, was to use a random sample of instances or features by means of the *Random-projection* method [47]. Both schemes have enabled to address problems with massive and high dimensional data without making important accuracy losses in performance.

Guha et al. [23], O'Callaghan et al. [43] propose single-pass algorithms that approximately solve the k-median problem under the streaming model of computation. This problem is to find k centers in n points minimizing the sum of distances from each *datum* to its nearest center. The aforementioned algorithms process data in a batch manner, keeping also a fixed number of medians meanwhile the stream is processed. In order to build the final set of medians, these techniques collapse the list of medians. Scalability is attained by always keeping in memory a fixed amount of data.

Ackermann et al. [2] propose another interesting method based on sampling named *StreamK++*. *StreamK++* uses small samples of the data, then at any time it builds a *coreset* for the sample and solves the k-median problem on the *coreset*. A *coreset* is a small weighted point set that approximates the original input point set with respect to a given optimization problem, in this case the k-median. The main contribution of this work consists in the mechanism employed to build *coresets* at any time, using a reduced fraction of the whole data.

20.3.1.3 Shared Nearest Neighbors (*SNN*) approach

In [49], the authors compare several Web page clustering methods using various similarity measures (Jaccard, cosine, correlation and Minkowski). Using labeled datasets in the evaluation, they observe that when the Jaccard and cosine measures are used, the obtained groups better resemble the manual categorizations. Furthermore, they empirically show that a graph partitioning method obtains better performance than other techniques such as k-means when documents are represented by vectors of term frequencies.

Moreover, the curse of dimensionality impacts distances, generating a poor level of discrimination between near and far neighbors, especially when the similarity value is low [21]. Ertöz et al. [21] also show for the TREC collection, that approximately 15 % of labeled documents have distinct labels than their near neighbors under the cosine similarity.

Long time ago and without regard to the high dimensional data problem, [30] addressed the clustering problem with a neighborhood based approach. To this end, they proposed a method that uses a similarity measure between two points that depends on the number of k-nearest-neighbors that they have in common. This is, the degree in which two points resemble each other is given by the overlap level between their neighborhoods and eventually by the fact that both points have each other in their corresponding near neighbor list. The main advantage of this approach is that the similarity measure takes into consideration the density of points in the space, thus it is not necessary to use any distance threshold to define the neighborhoods. The only parameter used is k and its use allows to find groups either in less dense and in more compact areas in the feature space.

More recently, [21] use the same abovementioned approach to extend an existing density-based clustering algorithm for high dimensional data. Regarding the attained results, they suggest that *shared-nearest-neighbors* based similarity measures reflect better the local spatial configuration of points and also they tend to be less sensitive to high dimensionality and density variations along the feature space. Some time later, [28] study the performance of classical distance measures over high dimensional data. Moreover, they propose different secondary (built over underlying classical measures) similarity measures based on SNN, which show an interesting performance even when high dimensional data appear to obscure the discrimination potential of underlying measures. The results obtained in this work suggest that SNN based similarity built over classical metrics (cosine or euclidean) appears as a more robust option for measuring closeness between high dimensional points.

Other clustering methods employing this family of measures are ROCK [24], *Path Model* [26], *SFN* [45], *Relevant-set correlation* [27], [40] and *CMUNE* [1].

20.3.2 *Scalable Clustering over High Dimensional Text Data Using Hashing*

Considering the nearest neighbors task for high dimensional data, part of the data mining community has focused its attention on developing techniques based on hash functions [14, 15, 29] (refer to [51] for more details). These functions have a solid mathematical formalism provided by previous advances in randomization methods, and also they have very efficient computational implementations due to their long history in cryptography and data structures. The advances in this area have continued introducing interesting improvements to the previously mentioned works especially highlighting [16, 38, 39, 48]. Consequently, some other massive and high dimensional tasks have been nourished with this progress, such as automatic classification with kernel methods [52], dimensionality reduction techniques [19], news recommendation systems [18, 37] and duplicate detection algorithms [41].

In contrast with the aforementioned tasks and as far as we know, the application of these methods to enhance high dimensional data clustering techniques has been scarce. Among the specific literature referring to this subject it is possible to identify the combined use of Minwise functions for set similarity with graph partitioning methods [15, 25] and also the use of the functions defined by [29] together with a hierarchical agglomerative clustering algorithm proposed by [32, 33].

Broder et al. [15] use the Resemblance and Containment measures, previously defined by [14] in the context of duplicate detection on large document databases, into the task of syntactic clustering of Web documents. With this aim, and from a feature set representation for documents, they build a summarized representation for each document by performing several random permutations over the set of features. Then, they compare every pair of documents and, by means of secondary memory, they sort the resulting list of pairs by the matching degree between their signatures. Next, this list is employed in order to build a document net, into which two nodes are linked if the estimated Resemblance score from the match between their signatures exceeded a certain threshold. Finally, the document clusters are identified by detecting connected components in the net.

Haveliwala et al. [25] propose an efficient clustering method capable of dealing with large Web collections by using Min-hash signatures. Each document is represented using the Bag-of-Words model over the terms occurring within it and within the documents that reference it (anchor-text in links). Over this enriched text representation several signature values are computed iteratively. In every iteration, the whole collection is sorted by the current signature value and then, using this sorted list, each document is linked to the other documents that have the same signature value in the iteration. Finally, a graph partitioning algorithm is performed over the document network graph or net to obtain the desired clusters.

Koga et al. [32, 33] followed a different approach than previous contributions. In both works, a locality sensitive hashing scheme is applied in order to avoid performing distance computation between every document pair in a single-linkage hierarchical clustering algorithm. The proposed algorithm operates in several stages, each one corresponding to a single level in the final hierarchy. In each stage, a new locality sensitive structure (array of hash tables) is generated and thus new signatures are built for the whole collection. Then, at the end of each stage, the parameters of the locality sensitive structure are modified in order to increase the neighborhood radius and then to provide the algorithm the capability of detecting varying neighborhoods for each point in every hierarchy. Finally, a complete hierarchy of point distances is obtained. The proposed method, called LSH-Link, is favorably compared in terms of speed to the exact Single Link Agglomerative Hierarchical Clustering, and also obtains close performance results.

Recently, [54] propose a clustering method for text data in which a hashing-based sketch is built in one pass and then, when clustering is required, pairwise similarities are estimated in order to build a near neighbor net. Finally, this net is partitioned and the clusters are identified by performing a single scan over the sketched collection. The obtained results show that it is possible to obtain an approximate clustering solution which is very close to the exact one over different text collections by maintaining a low-dimensional embedding of the overall collection.

20.4 Discussion

The advances made in the Subspaces clustering task that were presented in Sect. 20.3.1 are of great interest in presence of data having dimensionality $\sim$50 attributes, especially due to its gain in interpretability of the clusters. Nonetheless, despite that this benefit could be valuable on text clustering, the cardinality of the feature set (vocabulary) in document collections is several orders of magnitude over the datasets currently used with these techniques. Thus, there exist scalability issues that still have to be addressed in order to successfully apply these methods for this task and especially considering the single-pass constraint.

K-means based algorithms have been successfully used on document collections of moderate size besides the scalability issues and loss of meaning of distances due to high dimensionality. The methods described in Sect. 20.3.1 have enabled, by using sampling, the extension of classical algorithms in order to deal with the processing of large collections. For instance, K-means extensions based on corsets identify representative points in an efficient manner and have been successfully evaluated on datasets containing even million of objects. Nevertheless, besides the scalability improvements, given that the base technique suffers from the curse of dimensionality, these techniques inherit the same issue.

As far as we know, the *Single-pass Shared-Nearest-Neighbor clustering* has been less explored than the previous two approaches presented in Sect. 20.3.1. Notwithstanding, it offers interesting advantages to process high dimensional data as it is less prone to suffer from the loss of meaning of distances. The main obstacle for the methods presented in Sect. 20.3.1 lies in the expensive cost involved in the number of distance calculations that must be performed to build the neighborhoods, which explodes exponentially as the number of documents grows. This scenario worsens when the single-pass constraint is applied and a light procedure for building the neighborhoods is needed.

Despite the progress already made, high dimensionality and the sparsity found in text representations still pose significant challenges to an efficient performance of the clustering algorithms, such as those presented in Sect. 20.3.1. Additionally, as text data is generated in increasingly large volumes in several domains, e.g. the WWW and social networks such as Facebook, there is the need for algorithms capable to process massive collections, i.e. those whose sizes exceed the available main memory, and that are also able to perform in an online fashion, attaining a performance comparable to exact multi-pass algorithms.

The processing of massive text data implies that each collected document must be processed by making a single scan, meanwhile the online operation restricts the usage of secondary memory. This also denotes that the algorithm must be able to return an answer at any time considering the collection of documents received until then.

The contributions shown in Sect. 20.3.2 highlight the benefits of using approximate near neighbor strategies for text clustering. Among all these works except the last, the pending task is to address the single pass and the anytime response constraints. Zamora et al. [54] poses that the approximate construction of neighborhoods, using a locality sensitive scheme, allows the design and implementation of algorithms capable of clustering high dimensional and massive data by performing a single-pass over the collection. The results attained by all these contributions suggest that the approximate neighborhoods approach constitute a promising strategy to address the task of high-dimensional and massive data clustering.

20.5 Conclusions and Future Directions

This article aims to present a detailed survey about the clustering task for high dimensional and massive text data. For this purpose, it first details the computational issues produced by the curse of dimensionality concerning the processing of text data when it is represented as multidimensional vectors, which is a common scenario. Additionally, the contributions for this task were categorized in two groups: High-dimensional data clustering, in which general works tackling the computational issues aforementioned are listed and High-dimensional text data clustering, in which the emphasis is on the scalability issues that arise from the sparse and (really) high dimensional nature of the computational representations.

In spite of the fact that the definition of the clustering task applies equally to any kind of data source, the subtleties of the computational representation for text data poses interesting challenges not yet solved, specially when larger and larger document collections are available and more computational constraints are included. Designing online methods that only perform a single scan over the collection is a promising approach to enable the processing of very large collections. Additionally, the distributed computing approach seems an appealing strategy to process large

Fig. 20.2 Professor Moraga is standing in the second position from *left* to *right*

collections but also to work in distributed data scenarios where maybe the document collection is generated in different geographical locations and there exist constraints on the size of the transmitted data.

Acknowledgments It is a great honor to me to become part of this homage book for a person who I sincerely admire. I met Professor Moraga in 2010 because of a seminary course that he gave at the Universidad Técnica Federico Santa María for doctoral students. After that I could appreciate his human warmth and constant eagerness to help others (me among them). I am quite sure he will not be comfortable if I write a long acknowledgement text, hence I just want to express my gratitude for his advices, discussions and great willingness. Finally, I just wanted to put a very important photo to me that was taken in my doctoral defense. Professor Moraga participated in the Commission, his observations were invaluable and made a much better work of my thesis.

References

1. Abbas, M. A. and Shoukry, A. A.: Cmune: A clustering using mutual nearest neighbors algorithm, In *Information Science, Signal Processing and their Applications (ISSPA), 2012 11th International Conference on*, pp. 1192–1197.
2. Ackermann, M. R., Märtens, M., Raupach, C., Swierkot, K., Lammersen, C., and Sohler, C.: Streamkm++: A clustering algorithm for data streams. *J. Exp. Algorithmics*, 17, 2012.

3. Aggarwal, C., Han, J., Wang, J., and Yu, P.: A framework for clustering evolving data streams, Proceedings of the 29th international conference on Very large databases (VLDB '03), Morgan Kaufmann, 2003, pp. 81–92.
4. Aggarwal, C., Han, J., Wang, J., and Yu, P.: A framework for projected clustering of high dimensional data streams, Proceedings of the 30th international conference on Very large data bases (VLDB '04), 2004, pp. 852–863.
5. Aggarwal, C. and Yu, P.: Finding generalized projected clusters in high dimensional spaces. *SIGMOD Rec.*, Vol. 29 (2), 2000, pp. 70–81.
6. Aggarwal, C. C., Hinneburg, A., and Keim, D. A.: *On the surprising behavior of distance metrics in high dimensional space*, Springer, 2001.
7. Aggarwal, C. C., Wolf, J. L., Yu, P. S., Procopiuc, C., and Park, J. S.: Fast algorithms for projected clustering, Proceedings of the 1999 ACM SIGMOD International Conference on Management of Data (SIGMOD '99), ACM, 1999, pp. 61–72, New York.
8. Agrawal, R., Gehrke, J., Gunopulos, D., and Raghavan, P.: Automatic subspace clustering of high dimensional data for data mining applications, In *Proceedings of the 1998 ACM SIGMOD International Conference on Management of Data*, SIGMOD '98, ACM, 1998, pp. 94–105, New York.
9. Albers, S. and Leonardi, S.: On-line algorithms, *ACM Computing Surveys*, Vol. 31 (3), 1999.
10. Assent, I.: Clustering high dimensional data, *Wiley Interdisciplinary Reviews: Data Mining and Knowledge Discovery*, Vol. 2 (4), 2012, pp. 340–350.
11. Beyer, K., Goldstein, J., Ramakrishnan, R., and Shaft, U.: When is "nearest neighbor" meaningful? In *Database Theory—ICDT'99*, Springer, pp. 217–235, 1999.
12. Bishop, C.: *Pattern recognition and machine learning*, Vol. 4., Springer New York, 2006.
13. Bohm, C., Railing, K., Kriegel, H., and Kroger, P.: Density connected clustering with local subspace preferences. In *Data Mining, 2004. ICDM'04. 4th IEEE International Conference on*, pp. 27–34.
14. Broder, A. Z.: On the resemblance and containment of documents, In *Compression and Complexity of Sequences 1997. Proceedings*, pp. 21–29.
15. Broder, A. Z., Glassman, S. C., Manasse, M. S., and Zweig, G.: Syntactic clustering of the web, *Computer Networks and ISDN Systems*, Vol. 29 (8), 1997, pp. 1157–1166.
16. Charikar, M. S.: Similarity estimation techniques from rounding algorithms, Proceedings of the 34th annual ACM symposium on Theory of computing, ACM, 2002, pp. 380–388.
17. Chien, J.-T. and Chang, Y.-L.: Bayesian sparse topic model, *Journal of Signal Processing Systems*, Vol. 74 (3), 2014, pp. 375–389.
18. Das, A., Datar, M., Garg, A., and Rajaram, S.: Google news personalization: scalable online collaborative filtering, Proceedings of the 16th international conference on World Wide Web, ACM, 2007, pp. 271–280.
19. Dasgupta, A., Kumar, R., and Sarlós, T.: A sparse Johnson–Lindenstrauss transform, Proceedings of the 42nd ACM symposium on Theory of computing, 2010, pp. 341–350.
20. Eisenstein, J., Ahmed, A., and Xing, E.: Sparse additive generative models of text, Proceedings of the 28th International Conference on Machine Learning (ICML-11), New York, ACM, 2011, pp. 1041–1048
21. Ertöz, L., Steinbach, M., and Kumar, V.: Finding clusters of different sizes, shapes, and densities in noisy, high dimensional data, *SDM*, SIAM, 2003, pp. 47–58.
22. Friedman, J. and Meulman, J.: Clustering objects on subsets of attributes (with discussion). *Journal of the Royal Statistical Society: Series B (Statistical Methodology)*, Vol. 66 (4), 2004, pp. 815–849.
23. Guha, S., Meyerson, A., Mishra, N., Motwani, R., and O'Callaghan, L.: Clustering data streams: Theory and practice, *IEEE Trans. on Knowl. and Data Eng.*, Vol 15 (3), 2003, pp. 515–528.
24. Guha, S., Rastogi, R., and Shim, K.: Rock: a robust clustering algorithm for categorical attributes. Data Engineering, 1999. Proceedings 15th International Conference on, pp. 512–521.
25. Haveliwala, T., Gionis, A., and Indyk, P.: Scalable techniques for clustering the web, Proceedings of the 3rd International Workshop on the Web and Databases, 2000, pp. 129–134.

26. Houle, M. E.: Navigating massive data sets via local clustering, Proceedings of the 9th ACM SIGKDD International Conference on Knowledge Discovery and Data Mining (KDD '03), New York: ACM, 2003, pp. 547–552.
27. Houle, M. E.: The relevant-set correlation model for data clustering, *Statistical Analysis and Data Mining*, Vol. 1(3), 2008, pp. 157–176.
28. Houle, M. E., Kriegel, H.-P., Kröger, P., Schubert, E., and Zimek, A.: Can shared-neighbor distances defeat the curse of dimensionality? Proceedings of the 22nd International Conference on Scientific and Statistical Database Management (SSDBM' 10), Berlin, Heidelberg: Springer-Verlag, 2010, pp. 482–500.
29. Indyk, P. and Motwani, R.: Approximate nearest neighbors: towards removing the curse of dimensionality, Proceedings of the 30th annual ACM symposium on Theory of computing, 1998, pp. 604–613.
30. Jarvis, R. and Patrick, E. A.: Clustering using a similarity measure based on shared near neighbors, *Computers, IEEE Transactions on*, Vol. C-22 (11), 1973, pp. 1025–1034.
31. Keogh, E. and Mueen, A.: *Curse of Dimensionality*, Springer US, Boston, MA., 2010, pp. 257–258.
32. Koga, H., Ishibashi, T., and Watanabe, T.: Fast hierarchical clustering algorithm using locality-sensitive hashing, In *Discovery Science*, 2004, pp. 114–128.
33. Koga, H., Ishibashi, T., and Watanabe, T.: Fast agglomerative hierarchical clustering algorithm using locality-sensitive hashing, *Knowledge and Information Systems*, Vol. 12 (1), 2007, pp. 25–53.
34. Kriegel, H., Kröger, P., and Zimek, A.: Clustering high-dimensional data: A survey on subspace clustering, pattern-based clustering, and correlation clustering, *ACM Trans. Knowl. Discov. Data*, Vol. 3 (1), 2009, pp. 1:1–1:58.
35. Kriegel, H.-P. and Ntoutsi, E.: Clustering high dimensional data: Examining differences and commonalities between subspace clustering and text clustering—a position paper, *SIGKDD Explor. Newsl.*, Vol. 15(2), 2014, pp. 1–8.
36. Larsson, M. O. and Ugander, J.: A concave regularization technique for sparse mixture models, In *Advances in Neural Information Processing Systems*, 2011, pp. 1890–1898.
37. Li, L., Wang, D., Li, T., Knox, D., and Padmanabhan, B.: Scene: a scalable two-stage personalized news recommendation system. Proceedings of the 34th international ACM SIGIR conference on Research and development in Information Retrieval, ACM, 2011, pp. 125–134.
38. Li, P. and König, C.: Theory and applications of b-bit minwise hashing, *Communications of the ACM*, Vol. 54 (8), 2011, pp. 101–109.
39. Li, P., Owen, A., and Zhang, C.-H.: One permutation hashing. *Advances in Neural Information Processing Systems*, 2012, pp. 3113–3121.
40. Luo, C., Li, Y., and Chung, S. M.: Text document clustering based on neighbors, *Data & Knowledge Engineering*, Vol. 68 (11), 2009, pp. 1271–1288.
41. Manku, G. S., Jain, A., and Das Sarma, A.: Detecting near-duplicates for web crawling, Proceedings of the 16th international conference on World Wide Web, ACM, 2007, pp. 141–150.
42. Ntoutsi, I., Zimek, A., Palpanas, T., Kröger, P., and Kriegel, H.: Density-based projected clustering over high dimensional data streams, Proceedings of SDM '12, SIAM, 2012, pp. 987–998.
43. O'Callaghan, L., Mishra, N., Meyerson, A., Guha, S., and Motwani, R.: Streaming-data algorithms for high-quality clustering. Proceedings 18th International Conference on Data Engineering (ICDE '02), IEEE Computer Society. 2002, pp. 685–694.
44. Radovanović, M., Nanopoulos, A., and Ivanović, M.: Hubs in space: Popular nearest neighbors in high-dimensional data. *The Journal of Machine Learning Research*, Vol. 11, 2010, pp. 2487–2531.
45. Rovetta, S. and Masulli, F.: Shared farthest neighbor approach to clustering of high dimensionality, low cardinality data, *Pattern Recognition*, Vol. 39 (12), 2006, pp. 2415–2425.
46. Schnitzer, D., Flexer, A., Schedl, M., and Widmer, G.: Local and global scaling reduce hubs in space. *The Journal of Machine Learning Research*, Vol. 13 (1), 2012, pp. 2871–2902.
47. Schulman, L. J.: Clustering for edge-cost minimization (extended abstract), Proceedings of the 32nd Annual ACM Symposium on Theory of Computing (STOC '00), New York, ACM, 2000. pp. 547–555,

48. Shrivastava, A. and Li, P.: Fast near neighbor search in high-dimensional binary data, In *Machine Learning and Knowledge Discovery in Databases*, 2012, pp. 474–489.
49. Strehl, A., Ghosh, J., and Mooney, R. (2000). Impact of similarity measures on web-page clustering, Workshop on Artificial Intelligence for Web Search (AAAI 2000), pp. 58–64.
50. Tomasev, N., Radovanović, M., Mladenic, D., and Ivanović, M. (2014). The role of hubness in clustering high-dimensional data. *Knowledge and Data Engineering, IEEE Transactions on*, Vol. 26 (3), 2014, pp. 739–751.
51. Wang, J., Shen, H., Song, J., and Ji, J.: Hashing for similarity search: A survey, arXiv preprint arXiv:1408.2927, 2014.
52. Weinberger, K., Dasgupta, A., Langford, J., Smola, A., and Attenberg, J.: Feature hashing for large scale multitask learning. Proceedings of the 26th Annual International Conference on Machine Learning, 2009, pp. 1113–1120. ACM.
53. Xu, R. and Wunsch, D. I.: Survey of clustering algorithms. *Neural Networks, IEEE Transactions on*, Vol. 16 (3), 2005, pp. 645–678.
54. Zamora, J., Mendoza, M., and Allende, H.: Hashing-based clustering in high dimensional data. *Expert Systems with Applications*, 2016.
55. Zhang, T., Ramakrishnan, R., and Livny, M. (1996). Birch: an efficient data clustering method for very large databases. Proceedings of the 1996 ACM SIGMOD International Conference on Management of Data (SIGMOD '96), pp. 103–114. ACM Press.

Chapter 21
From Lisp to FuzzyLisp

Luis Argüelles Méndez

21.1 Introduction

I met Claudio Moraga at the *European Centre for Soft Computing* back in 2008. My friend and colleague Enric Trillas introduced him to me and soon I understood Claudio was not only a scientist in the truest and highest meaning of the word but also a very special human being: friendly, empathic, cordial and friend of his friends. Among his many educational achievements, he got a master degree in engineering in his youth from the *Massachusetts Institute of Technology*, MIT. This is important because computing is a helping technique in engineering disciplines and at MIT, Lisp has always been the "in-house" computing language. Even more interesting: Lisp can be an excellent tool for teaching Fuzzy Logic theories from a practical viewpoint.

21.2 Why Lisp Today?

Java, Python or PHP are hot these days as programming languages. Other languages such as C or even actual improved versions of it such as C# or C++ seem to have lost some momentum when in comparison to the former ones. Java is highly portable. This means that a program written in Java under Windows can be executed under Mac OS, or even in your smartphone with some slight additions of XML and some tricky organization. Java, Python and PHP work extremely well in today's world of Internet. So, the question seems obvious: why to use Lisp today?

Lisp is the second older high-level language in the history of computer languages, so it can seem a strange decision at first to use Lisp as a vehicle for teaching Fuzzy Sets theory and Fuzzy Logic. In some way it could be seen almost as a contradiction:

L.A. Méndez (✉)
Oviedo, Spain
e-mail: arguelles314@gmail.com

R. Seising and H. Allende-Cid (eds.), *Claudio Moraga: A Passion for Multi-Valued Logic and Soft Computing*, Studies in Fuzziness and Soft Computing 349, DOI 10.1007/978-3-319-48317-7_21

Using a very old language for teaching and spreading forefront computing concepts that are applied today in robotics, machine learning, adaptive systems and other fields of Artificial Intelligence (AI).

There are some strong technical reasons for selecting Lisp as the ideal programming tool for learning and understanding Fuzzy Logic in a practical way. First of all, Lisp offers the user or programmer an automatic system of memory management. Other computer languages demand from programmers a special and constant attention for seizing and liberating memory while writing programs. Lisp manages memory allocations automatically for the user. This means more confidence in the code and more speed in writing programs. It means superior productivity. Lisp is relatively simple. As with chess, the mechanics of the language are simple and powerful, yet the possibilities of play are practically infinite. Needless to say, as it happens with chess, it takes time to be a sophisticated Lisp programmer, but the essential movements are easy to grasp.

Lisp is elegant. Take recursion as an example: Recursion is a programming paradigm (speaking properly it is a mathematical paradigm) where a function written by the user in Lisp is called inside the same function. Using recursion it is possible to write short, yet powerful programs. Lisp builds Lisp-thinking on the programmer. In some way, it happens the same after learning a human language: Learning the language allows a person to understand the culture of the country that speaks the language. This feature helps him or her to think on creating programs with Lisp style, with Lisp organization, with Lisp freedom.

Lisp is highly interactive. While the user writes code he or she can take apart one function or a set of them and try how they do behave. A Lisp programmer can test functions or fragments of a Lisp program without effort. Even better, the user can interactively modify functions in a Lisp session, make some improvements and then incorporate the transformed functions into the main program. This is described as the "read-evalprint loop" paradigm. The programmer types Lisp expressions at the keyboard, Lisp interprets and evaluates what the user has entered and after evaluation, it prints the results. Not only it is possible to run complete programs, but parts of it. The Lisp programmer feels that he or she is talking with the system, and that talk creates a special relationship with the system and the Lisp language itself. Finally, Lisp is probably one of the best general-purpose computer languages for representing fuzzy-sets, as we shall see later.

21.3 A Short History on Lisp

As previously stated, LISP is one of the oldest high-level languages in the history of computer science and only FORTRAN, a language whose main goal is to write technical programs where numbers are the main data type, is older. LISP was born from the works of John McCarthy and his colleagues at MIT back in the last fifties of century XX in an attempt to develop a language for symbolic data processing, deriving in an extraordinary tool for **LIS**t **P**rocessing.

The seminal paper of LISP, "Recursive Functions of Symbolic Expressions and Their Computation by Machine" was written by McCarthy himself as a programming system for the IBM 704 (the first mass-produced computer with floating point hardware), in order to facilitate experiments with a proposed system called "Advice Taker", that could exhibit some class of "common sense" in carrying out its instructions [6]. Soon after this theoretical paper, the first real implementation of Lisp was created by Steve Russell on the 704 using its own machine language, and not much later, in 1962, Tim Hart and Mike Levin wrote the first Lisp compiler, written in Lisp, for the IBM 7090, a more advanced (and really expensive) computer at that time.

Two important documents were published by MIT Press afterwards: "LISP 1.5 Programmer's Manual" [7] and "The Programming Language LISP: Its Operations and Applications" [1]. These documents, together with the seminal paper by McCarthy are easily found today on the Internet and constitute the real cradle of Lisp.

In the late sixties, people at MIT developed an enhanced version of Lisp named MACLISP for the PDP-6/10, a family of 32 bits computers manufactured by Digital Equipment Corporation. MACLISP included new data types such as arrays and strings of characters. It helped to enhance the development of Artificial Intelligence (AI) until the early 80s. The Reference manual for MACLISP [8] has been preserved and can be consulted freely in the Internet.

While MIT was the cradle of LISP and even today is considered an in-house development for completing a curriculum in computer science, it soon spread to other regions, both academic and commercial. Interlisp [10] was, seen with the perspective of an historian, a transfer of computer language from MIT to the *Palo Alto Research Center* (PARC) in California, a division of *Xerox Corporation* at mid seventies last century. Also from coast to coast in the United States, another dialect of Lisp based on MACLISP and named Franz Lisp, appeared in Berkeley in the seventies and eighties, becoming one of the most commonly available Lisp dialects on machines running under the Unix operating system [5].

The language didn't take too much time to reach Europe. One of the first dialects flourished in France under the name of VLISP in 1971. It was developed at the *University of Paris VIII* at Vincennes, giving name to the dialect, V(incennes)LISP. After VLISP soon came Le-LISP a close variation from VLISP, also in France. Under an historic point of view Le-LISP is a milestone that deserves some attention because it was one of the first implementations of the Lisp language on the IBM PC, thus, inaugurating an epoch of, let's say, personal programming in Lisp on microcomputers [4].

Since Lisp can be written using Lisp itself, it is easy to understand the easy and quick blooming of versions and dialects of the language, so soon it became evident that some type of standardization was needed in the community of Lisp programmers and users. Due to this, the *American National Standard Institute* published in 1994 a language specification document that gave birth to Common Lisp, also known as ANSI Common Lisp.

21.4 FuzzyLisp

In the nineties last century I was asked to prepare an introductory course on Fuzzy Logic to engineering students that were eager to take the fundamentals of the theory, especially from a practical approach. As a teaching vehicle specifically suited to the task I wrote a small set of Lisp functions that I designed from scratch in such a way that students could easily understand the theory and at the same time build simple fuzzy models. Almost twenty years later, and as a complementary tool for writing a book titled "A Practical Introduction to Fuzzy Logic using Lisp" [2] I took the same approach, completely rewriting the code and enhancing it until developing a complete computing tool for helping readers to understand fuzzy sets and fuzzy logic theories, providing them with a tool for developing from small to complex fuzzy logic based applications. In other words, FuzzyLisp is, simply put, a collection of Lisp functions that allows to explore the world of Fuzzy Logic theories and to develop Fuzzy Logic applications with relative little effort. It is in fact a small and compact metalanguage that permits the user to concentrate in the construction of fuzzy models while still retaining full control of all the Lisp features. FuzzyLisp is freely available on the Internet [3] and runs in any ANSI Common Lisp compiler or in the NewLisp environment [9].

21.4.1 Fuzzy Sets Representation in FuzzyLisp

For representing fuzzy sets within FuzzyLisp, both triangular and trapezoidal shaped membership functions can be defined with only four singular points on the real axis, $x1$, $x2$, $x3$, and $x4$:

$$(fuzzy - set - name\ x1\ x2\ x3\ x4)$$

The support of a fuzzy set is thus represented by the interval $[x1, x4]$, while the nucleus is represented by the interval $[x2, x3]$. By adequately combining these values we can obtain any triangular and trapezoidal characteristic function as follows:

$x1 \neq x2 = x3 \neq x4 \rightarrow$ Standard triangle
$x1 = x2 = x3 \neq x4 \rightarrow$ Right triangle by-left
$x1 \neq x2 = x3 = x4 \rightarrow$ Right triangle by-right
$x1 \neq x2 \neq x3 \neq x4 \rightarrow$ Normal trapezium
$x1 = x2 \neq x3 \neq x4 \rightarrow$ Right trapezium by-left
$x1 \neq x2 \neq x3 = x4 \rightarrow$ Right trapezium by-right

Aside triangular and trapezoidal shaped membership functions, discrete fuzzy sets can be also represented within FuzzyLisp using the following structure:

$$(fuzzy - set - name(x1\ \mu(x1))\ (x2\ \mu(x2)) \ldots (xn\ \mu(xn)))$$

21.4.2 FuzzyLisp Functions

The complete actual version of FuzzyLisp contains more than forty functions that can be categorized in the following sections: Membership functions, Alpha-cuts management, Truth values, Interval arithmetic, Fuzzy arithmetic, Fuzzy sets management, Fuzzy Rule Based System functions and then other miscellaneous functions. The following paragraphs give a concise description of them:

Membership functions:

$(fl - belongs?\ f\ x)$ returns true if a crisp value x defined on the real axis belongs to the fuzzy set f, else returns nil.
$(fl - set - membership?\ f\ x)$ returns the membership degree of a crisp value x defined on the real axis to the fuzzy set f.
$(fl - belongs2?\ f\ x)$ is a sort of mix of the functions $(fl - belongs?)$ and $(fl - set - membership?)$. If the crisp value x is contained in the support of f it returns the membership degree of x to f, otherwise, it returns nil.
$(fl - dset - membership?dfsetx)$ returns the interpolated membership degree of a crisp value x defined on the real axis to the discrete fuzzy set *fset*.
$(fl - lv - membership?\ lv\ x)$ prints all the membership degrees of a crisp value x to every fuzzy set contained in a linguistic variable lv in the Lisp console.
$(fl - lv - membership2?\ lv\ x)$ returns as a list all the membership degrees of a crisp value x to every fuzzy set contained in a linguistic variable lv.
$(fl - dlv - membership2?\ lv\ x)$ returns as a list all the membership degrees of a crisp value x to every fuzzy set contained in a linguistic variable lv. All the fuzzy sets from the linguistic variable lv must have a discrete characteristic function.
$(fl - set - complement - membership?\ f\ x)$ returns the membership degree of a crisp value x to the complementary set of f.
$(fl - set - intersect - membership?namef1f2x)$ returns the membership degree of the crisp value x to the intersection of fuzzy sets $f1$ and $f2$. Parameter name is a symbol for associating a name to the function's resulting list.
$(fl - set - union - membership?\ name\ f1\ f2\ x)$ returns the membership degree of the crisp value x to the union of fuzzy sets $f1$ and $f2$. Parameter name is a symbol for associating a name to the function's resulting list.

Alpha-cuts management:

$(fl - alpha - cut\ f\ alpha)$ scans a fuzzy set f represented by a trapezoidal or triangular membership function from left to right and returns the obtained alpha-cut alpha as a list, including the name of the original fuzzy set.
$(fl - def - set\ name\ a - cut1\ a - cut2)$ defines and creates a fuzzy set by means of two alpha-cuts $a - cut1$, $a - cut2$. The returned fuzzy set has either a triangular or trapezoidal membership function. The parameter name is a symbol for associating a name to the resulting fuzzy set.

Truth values:

$(fl - truth - value - negation - p?\ P\ x)$ returns the truth-value of the negation of a fuzzy proposition P where x is a real number expressing the subject of P.
$(fl - truth - value - p - and - q?\ P\ Q\ x\ y)$ returns the truth-value of a compound fuzzy proposition containing the logical connective "and". P and Q are represented by fuzzy sets with triangular or trapezoidal membership functions.
$(fl - truth - value - p - or - q?PQxy)$ returns the truth-value of a compound fuzzy proposition containing the logical connective "or". P and Q are represented by fuzzy sets with triangular or trapezoidal membership functions.
$(fl - truth - value - fuzzy - implication - p - q?\ P\ Q\ x\ y)$ returns the truth-value of a compound fuzzy implication $p \Rightarrow q$. P and Q are represented by fuzzy sets with triangular or trapezoidal membership functions.
$(fl - dtruth - value - negation - p?\ P\ x)$ returns the truth-value of the negation of a fuzzy proposition P where x is a real number expressing the subject of P. In this case P is represented by a discrete fuzzy set.
$(fl - dtruth - value - p - and - q?\ P\ Q\ x\ y)$ returns the truth-value of a compound fuzzy proposition containing the logical connective "and". In this case P and Q are represented by discrete fuzzy sets.
$(fl - dtruth - value - p - or - q?\ P\ Q\ x\ y)$ returns the truth-value of a compound fuzzy proposition containing the logical connective "or". In this case P and Q are represented by discrete fuzzy sets.
$(fl - dtruth - value - fuzzy - implication - p - q?\ P\ Q\ x\ y)$ returns the truth-value of a compound fuzzy implication $p \Rightarrow q$. P and Q are represented by discrete fuzzy sets.

Interval arithmetic:

$(flu - intv - add\ x1\ x2\ x3\ x4)$ returns a list representing the addition of two intervals $[x1, x2]$, $[x3, x4]$.
$(fl - int - sub\ x1\ x2\ x3\ x4)$ returns a list representing the subtraction of two intervals $[x1, x2]$, $[x3, x4]$.
$(fl - intv - mult\ x1\ x2\ x3\ x4)$ returns a list representing the multiplication of two intervals $[x1, x2]$, $[x3, x4]$.
$(fl - intv - div\ x1\ x2\ x3\ x4)$ returns a list representing the division of two intervals $[x1, x2]$, $[x3, x4]$.

Fuzzy arithmetic:

$(fl - fuzzy - add\ name\ A\ B)$ returns a fuzzy number as the result of adding two fuzzy numbers A, B.
$(fl - fuzzy - sub\ name\ A\ B)$ returns a fuzzy number as the result of subtracting two fuzzy numbers A, B.

$(fl - fuzzy - mult\ name\ A\ B\ n)$ returns a fuzzy number as the result of multiplying two fuzzy numbers A, B. A and B are represented by triangular or trapezoidal shaped membership functions. The resulting fuzzy number $A.B$ is represented by means of a discrete characteristic function. The parameter n expresses the desired resolution.
$(fl - fuzzy - div\ name\ A\ B\ n)$ returns a fuzzy number as the result of dividing two fuzzy numbers A, B. A and B are represented by triangular or trapezoidal shaped membership functions. The resulting fuzzy number A/B is represented by means of a discrete characteristic function. The parameter n expresses the desired resolution.
$(fl - fuzzy - add - sets\ fsets\ name)$ returns a fuzzy number as the result of adding all the fuzzy numbers contained in *fsets*.
$(fl - fuzzy - factor\ fset\ k)$ takes a fuzzy number A and then multiplies it by a crisp number k, returning the fuzzy number kA. In practical terms when $k > 1$ it per-forms a multiplication by k and when $k < 1$ it performs a division by k.

$(fl - fuzzy - average\ fsets\ name)$ returns a fuzzy number as the average of n fuzzy numbers represented by *fsets*.

Fuzzy sets management:

$(fl - dset - hedge\ dset\ hedge)$ applies a fuzzy hedge (linguistic modifier) to a fuzzy set, where *dset* is a list representing a discrete fuzzy set and *hedge* is a Lisp symbol, either VERY or FAIRLY.
$(fl - expand - contract - set\ f\ k)$ expands or contracts a fuzzy set f. The returned fuzzy set is still placed over its original position, but its support and nucleus are expanded or contracted accordingly.
$(fl - fuzzy - shift\ f\ x)$ shifts (moves horizontally) a fuzzy set f towards left or right over the real axis X by an amount given by a real value x, returning the shifted fuzzy set.
$(fl - discretizefsteps)$ takes a fuzzy set f with triangular or trapezoidal characteristic function and discretizes it with a resolution given by steps. In other words, it transforms a FuzzyLisp Standard Set Representation into a FuzzyLisp Discrete Set Representation.
$(fl - discretize - fx\ name\ fx\ steps\ a\ b)$ discretizes any continuous function $y = f(x)$ in n steps between $x = a$ and $x = b$, producing a discrete fuzzy set.

Fuzzy Rule Based System functions:

$(fl - translate - ruleheaderrulexy)$ takes an expert rule at a time from a Fuzzy Rule Based System, performs the adequate inferences and translates the rule into membership degrees, that is, into numerical values.
$(fl - translate - all - rules\ set - of - rules\ x\ y)$ evaluates all the fuzzy rules contained in the knowledge database of a Fuzzy Rule Based System, calling iteratively to the function $(fl - translate - rule)$.

$(fl - defuzzify - rules\ translated - rules)$ takes as input the list obtained from either $(fl - dtranslate - all - rules)$ or $(fl - dtranslate - all - rules)$ and then converts that fuzzy information into a crisp numerical value.
$(fl - inferenceset - of - rules\ x\ y)$ is an automatic call to the functions $(fl - translate - all - rules)$ and $(fl - defuzzify - rules)$ in a sort of black box that directly transforms two input crisp $x\ y$ values entering a Fuzzy Rule Based System into a resulting crisp value. Conversely, the FuzzyLisp function $(fl - dinferenceset - of - rulesxy)$ is suited to deal with FRBS where input linguistic variables are composed by fuzzy sets with discrete membership functions.
$(fl - 3d - meshnamefile\ set - of - rules\ nxny)$ creates an ASCII output file in comma-separated values format (CSV) where every line adopts the following structure: xi, yi, zi. Both xi, yi are input crisp values from the universes of discourse of their respective linguistic variables from a Fuzzy Rule Based System. On the other hand, zi is the inferred value from every possible pair (xi, yi). The output file is in fact a discretized geometrical 3D mesh. Conversely, the FuzzyLisp function $(fl - 3d - dmesh)$ is suited to deal with FRBS where input linguistic variables are composed by fuzzy sets with discrete membership functions.

Miscellaneous functions:

$(fl - list - sets\ lv)$ prints all the fuzzy sets belonging to a linguistic variable lv at the Lisp console.
$(fl - simple - defuzzification\ f\ mode)$ takes a fuzzy number f and produces a crisp number for it with a simple algorithm.

21.5 A Simple Example of Fuzzy Control with FuzzyLisp

As it is well known, an air-conditioner system's goal is to maintain the temperature of a given enclosure at a constant value. The basic information needed to accomplish the goal is usually based on two values: actual temperature t and temperature variation *delta-t* obtained from two readings from a thermometer obtained every s seconds. After the readings are made, the proposed fuzzy controller calculates the needed output airflow, AFT, to stabilize the enclosure's temperature at a temperature goal, let us say, $T = 21\,°C$.

In this example, the universe of discourse representing temperature has a range from 0 to 50C, while the universe of discourse representing temperature-variation has a range from −2.0 to 2.0 °C/min. For output, five single-tons, representing airflow percentage and cold (negative) and hot (positive) temperatures are used. Figures 21.2, 21.3 and 21.4 show a graphical representation of the input and output variables of the air-conditioner controller:

Fig. 21.1 Luis Argüelles (first from *Left*) and Claudio Moraga (third from *left*) during the First "Alfredo Dean" Seminar on Ordinary Reasoning, to be held at the Cultural Center "Muralla Romana" of Gijon (Asturias), Spain, April 29, 2011.; photograph by Rudolf Seising

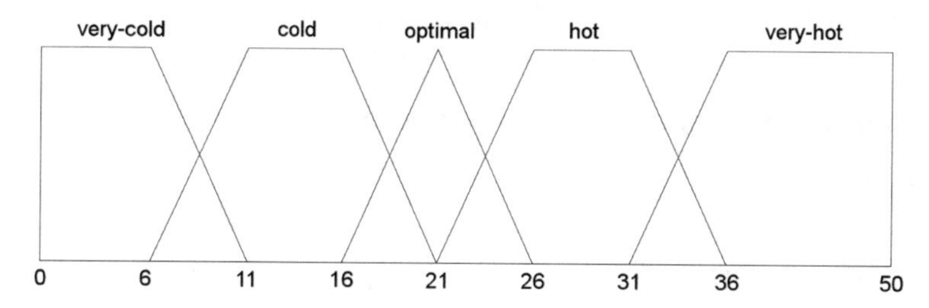

Fig. 21.2 Input linguistic variable lv-temperature

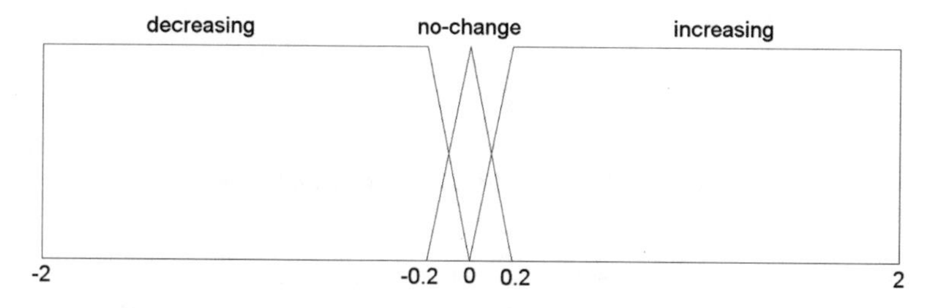

Fig. 21.3 Input linguistic variable lv-delta-t

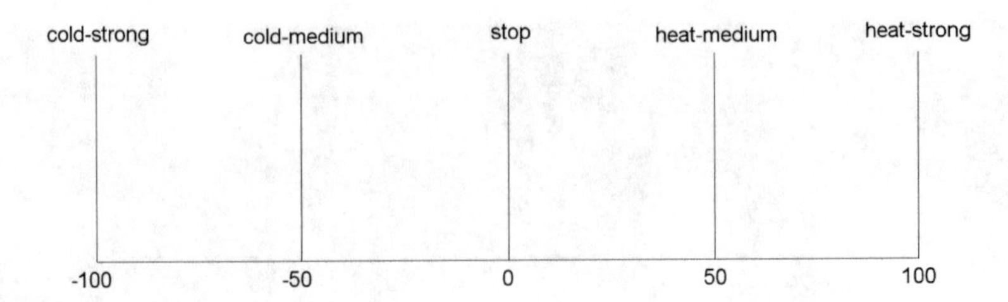

Fig. 21.4 Output linguistic variable AFT

The following FuzzyLisp code entirely models the air-conditioner fuzzy controller:

```
;fuzzy sets and linguistic variables definitions:

(setq T1 '(very-cold 0 0 6 11))
(setq T2 '(cold 6 11 16 21))
(setq T3 '(optimal 16 21 21 26))
(setq T4 '(hot 21 26 31 36))
(setq T5 '(very-hot 31 36 50 50))
(setq lv-temperature '(T1 T2 T3 T4 T5))

(setq dT1 '(decreasing -2 -2 -0.2 0))
(setq dT2 '(no-change -0.2 0 0 0.2))
(setq dT3 '(increasing 0 0.2 2 2))
(setq lv-delta-t '(dT1 dT2 dT3))

;singletons representing output:

(setq AFT '(

  (cold-strong -100)
  (cold-medium -50)
  (stop 0)
  (heat-medium 50)
  (heat-strong 100))

)

;fuzzy rules section:

(setq rules-controller '((lv-temperature lv-delta-t
AFT)

  (very-cold decreasing heat-strong AND-product)
  (very-cold no-change heat-strong AND-product)
  (very-cold increasing heat-strong AND-product)
  (cold decreasing heat-strong AND-product)
  (cold no-change heat-medium AND-product)
  (cold increasing heat-medium AND-product)
```

```
    (optimal decreasing heat-medium AND-product)
    (optimal no-change stop AND-product)
    (optimal increasing cold-medium AND-product)
    (hot decreasing cold-medium AND-product)
    (hot no-change cold-medium AND-product)
    (hot increasing cold-strong AND-product)
    (very-hot decreasing cold-strong AND-product)
    (very-hot no-change cold-strong AND-product)
    (very-hot increasing cold-strong AND-product))
)
```

For testing the controller for actual temperature $t = 22\,\mathrm{C}$ and temperature variation *delta-t* $= 0.25\,\mathrm{C/min}$, the user only needs to type the following at the Lisp con-sole: $(fl - inferencerules - controller\ 22\ 0.25)$ producing a result AFT = -60. This means the controller supplies cold air at %60 of the available output airflow.

21.6 Summary

Lisp is far from being dead. Although it is the second older high-level programming language it still holds all the great features designed by John McCarthy and his colleagues at MIT back in the last fifties of century XX: Lisp offers the user or programmer an automatic system of memory management, it is relatively simple and easy to learn, it is elegant and even short programs can be extremely powerful. Moreover Lisp is highly interactive and helps to build Lisp-thinking on the programmer. Even more interesting for the Fuzzy Logic community, Lisp is probably one of the best general-purpose computer languages for representing and managing fuzzy-sets.

FuzzyLisp is a small and compact metalanguage composed by more than forty functions that permits the user to experiment with fuzzy sets and concentrate in the construction of fuzzy models while still retaining full control of all the Lisp features. It is extremely well suited to teach fuzzy theories from a practical approach and experience has shown it is an excellent complementary tool in engineering and computer sciences courses. FuzzyLisp is freely available on the Internet and runs in any ANSI Common Lisp compiler or in the NewLisp environment. I'm sure Claudio will have sweet memories from MIT when observing the structure of FuzzyLisp.

References

1. Abrahams, P. et al.: *The Programming Language LISP: Its Operations and Applications*, The M.I.T. Press, Cambridge, Mass., 1964 http://www.softwarepreservation.org/projects/LISP/book/III_LispBook_Apr66.pdf.

2. Argüelles, L.: *A Practical Introduction to Fuzzy Logic using Lisp*, Berlin [et al.]: Springer, 2015.
3. Argüelles, L.: FuzzyLisp, www.fuzzylisp.com, acceded August, 2016.
4. Chailloux, J.; Devin, M; and J. M. Hullot, J. M.: *Le Lisp, a portable and efficient Lisp system*, INRIA, 1984.
5. Gabriel, Richard P.: *Performance and evaluation of Lisp systems*. MIT Press (Computer Systems Series, LCCN 85-15161), May 1985.
6. McCarthy, J.: Recursive Functions of Symbolic Expressions and Their Computation by Machine, *Communications of the ACM*, April, 1960 http://www.cs.berkeley.edu/~christos/classics/lisp.ps.
7. McCarthy, J. et al.: LISP 1.5 *Programmer's Manual*, The M.I.T. Press, Cambridge, Mass., 1962. http://www.softwarepreservation.org/projects/LISP/book/LISP%201.5%20Programmers%20Manual.pdf
8. Moon, David: MACLISP *Reference Manual*, The M.I.T. Press, Cambridge, Mass., 1974. http://www.softwarepreservation.org/projects/LISP/MIT/Moon-MACLISP_Reference_Manual-Apr_08_1974.pdf.
9. Müller, L.: NewLisp, www.newlisp.org, acceded August, 2016.
10. Teitelman, W.; Masinter, L.: The Interlisp Programming Environment, *IEEE Computer*, April 1981. http://larry.masinter.net/interlisp-ieee.pdf.

Claudio Moraga had an hour's rest during Saturday's Scientific Conversations "Thinking and Fuzzy Logic" in Palermo (Sicily, Italy), 14th of May, 2011, photograph by Rudolf Seising.

Appendix A
You, Claudio

Abstract This speech was delivered to mark the 65th birthday of Claudio Moraga. The speech tried to be of a festive and joking character, and came not from the head but from the heart.

0.

Before to begin, let me thank all friends at Dortmund University, and specially Professor Bernd Reusch, for giving me this opportunity of publicly addressing to Professor Claudio Moraga. This speech is like a letter to you, Claudio, but it is not exactly a letter. It will try to be an average between a letter, a personal reflection endowed with some touches of humor, and some manifestation of sentiments. Perhaps, better than "You, Claudio", the title should be "To you, Claudio", in the manner a bullfighter offers to someone his next fight with the bull. Of course, there is no any similarity between this speech and the 1934 Robert Graves' famous novel "I Claudio", a novel that, 'unlike' any university, is lull of gossipery and friend's treachery.

1.

If I remember it well, I met you by the first time, Claudio, in Chicago during an International Symposium on Multiple-Valued Logic, in 1981. Hence, our acquaintance is not too young: more than 21 years! Furthermore, after 1981 our contacts have been not only continuous but transformed in what for me is most important, in a true friendship. A friendship that, happily, was extended to many other Spanish researchers. In fact, I have completely lost the count of your stays in Spain: Barcelona, Girona, Palma de Mallorca, Madrid, Granada, Santiago de Compostela, etc., are cities in which you have a lot of friends that have a big respect for you. A respect that is shared by many other people around the world.

2.

When I began to think on what I could say with this speech, three phrases came to my mind. The first, by Albert Einstein, "Most people say that it is the intellect which makes a great scientist. They are wrong: it is the character". The second, by Samuel Taylor Coleridge, "Men, I think, have to be weighted,pg not counted". And

R. Seising and H. Allende-Cid (eds.), *Claudio Moraga: A Passion for Multi-Valued Logic and Soft Computing*, Studies in Fuzziness and Soft Computing 349, DOI 10.1007/978-3-319-48317-7

Fig. A.1 Title page of the document concerning the commemorative event on the 65th birthday of Claudio Moraga

FESTIVE COLLOQUIUM

FOR

PROF. DR. CLAUDIO MORAGA

University of Dortmund
February, 8th, 2002

the third, by Isabel Allende, "On a wine's taste you can say three or four adjectives and nothing else. On the ideas, you dispose of all the words in the Dictionary". These phrases directed me not to your intellectual work but to find adjectives to weight your character.

Hence, I asked some of our common colleagues to select these adjectives and, I don't know if it is surprising or not, but all of them were coincidental with me in the following seven (more than twice the number Allende would apply to wine): Honest, Heavy worker, Rigorous, Discrete, Marvelous Lecturer, Always ready to help anyone, Good scientist. To this list, I will add a characteristic that I don't know how to compact in an English adjective: your ability to grasp the personality of people. An ability that some of your characteristics compacted in the former adjectives don't allow you to show it too much frequently.

3.
Let me stop for a while to ask myself "What are we doing here today?". I believe that we are just taking benefit of the opportunity given by the end of the contract between you, Claudio, and the University of Dortmund, to pay homage to a distinguished scholar that now will pass to a new time in his life, a time in which he will have freedom enough for doing new activities, either of research or of other type, in any part of the world.

Today people can expect to live more than eighty years. Why retirement at 65? Which marvelous property has number 65, or 70 it does not matter at all, to legally cut a contract of professorship and not, for example, one of marriage? Why to punish Social Security with a new pension? The only reason I can foresee is that of making room for younger people without putting more money in the system. Well, as I am approaching 65 I think that this is not a good solution to the problem, it is just a trivial solution.

As I think that trivial solutions are for nothing, I tried to look at some arithmetical properties of 65, and the only I found (perhaps someone here knows another) is that 65 is, when written in the bases 2, 4 and 12, a symmetrical number:

$$65 = 1000001_2 = 1001_4 = 55_{12}! \qquad \text{(A.1)}$$

As in Spain we retire at 70, I also looked at number 70 and I found that in no base until 12 it is symmetrical. Hence, it seems I found the deep reason of the retirement at either 65 or 70: German Law is for palindromy and Spanish Law is against palindromy! Certainly, this is not politically correct in today's European Union, and perhaps it can be a good subject for the European Parliament.

In my opinion, 81 is a better age to retire professors: 81 is the square of 9, 9 in base 2 is 1001, palindrome, and the probability of charging the Social Security with a long-term pension is very low at 81.

4.

Well, after this "serious" argumentation, let me come back again to you, Claudio, to report a story that is not exactly true but that could be true if we put the imagination in gear.

All of you know both the Spanish Professor Justo de Montealto and the German Professor Justus von Hoheberg, as well as their fame as important scientists and their, let me say, haughty personality, I would say, how proud of themselves they are. The story can be summarized in the following transparencies. In the first and second, you can see de Montealto and his team and von Hoheberg and his team, when they are going to send a paper to a conference; an excellent one, of course, in both cases. In the third, you can see what happens with the respective teams when it is spread out the rumor that you, Claudio, are reviewing the papers by de Montealto and von Hoheberg: It is Panic! In the fourth, you can see anyone of the two important professors after the rumor's confirmation: Moraga is reviewing my paper!, a confirmation that guarantees that no word, no formula, no computation, no reference, no definition, will remain without a rigorous checking. What the story does not tell us, is the final report by Moraga; but I can imagine it as a large list of corrections to be made, and the final acceptance of the paper. A paper that, after the corrections will be, actually, a new one.

This is, of course, a "kind" of joke. Anyway, as you Claudio are now a free citizen, God will not give to any EU bureaucrat the idea of appointing you as the Boss of Security for European Airports. It will mean to be at the Checking Desk more than four hours before the flight departure. This, of course, only in domestic flights!

5.

You, Claudio, you are a person who likes precision, but not any kind of precision. I think that there is a story that reflects this statement of mine. When we met by the first time in 1981, you were the Chairman of a Session at which one of the speakers referred several times to the famous logician "Łukasiewicz" with this usual pronunciation. And I perfectly remember your face when, at the end of the paper and after yours "Is there any question?", one person in the room said "Well, it is not properly a question, it is only to advice that it is not 'Lukasiewicz' but 'Fukasiewicz', as the first letter is not an English *L* but a Polish *W*". Your face completely betrayed what you were thinking, "What a fretful person appeared here this year!". I was that fretful person, and sure that, when at the end of the Session I approached you, you accepted my conversation with the aim of helping me to non repeat such kind of, let me say, scrupulous comments. You can believe me, Claudio: Since then I always tried to avoid such kind of remarks!

6.

After studying Engineering at the *Federico Santamaria University*, in Valparaiso (that perhaps we can translate as Paradise Valley), the nice city in the Pacific Ocean Coast in which you were born 65 years ago, you moved to the MIT for doing the graduate studies and the Ph.D. Dissertation. After this, you Claudio came back to Valparaiso where, to properly exploit your research's capabilities, you were appointed as Vice-Rector, a position that probably would be the end of Moraga as a researcher if there were not a General that "firmly supported" a very different opinion. And, as forced by the General's insistence, you should immediately flight to Germany, where you arrived with a von Humboldt grant. And you quickly realized that this is a country with a deep love for university professors. In fact, in the passports-line a civil servant looked at your passport and, after confirming that you were a university professor, introduced you (alone) in an office where someone did for you all the bureaucratical process for your entrance in Germany. This is actually marvelous, and unbelievable in Spain. When the von Humboldt grant finished and you had to remain in Germany, you Claudio realized that Germans have also a deep love for doctorates as, until the day the grant expired you were called "Professor Moraga" and the day after you were called "Doctor Moraga". Curiously, this is perfectly believable in Spain and, hence, we can establish as 0.5 the degree of similitude between our respective countries in what concerns the love for titles. Who knows if this can be a subject for a future European Program in Social Sciences.

7.

Well, as I can suppose that your life after this time is well known by everybody here, let me jump to this Festive Colloquium. A Festive Colloquium that clearly shows the affection this University and your colleagues feel for you, Claudio. Today you can appreciate the fondness of your German colleagues. Sure than your more than 27 years in Germany have shown you the same I learned when, being a child, I was educated by a grandmother of mine who was a German lady born in Hamburg: That at least at the beginning and at the end, this is a country with a people for which sentiments are very important. May I say that this is very nice, Claudio? Or, even more, that it is touching? This fondness for you is not closed to Germany, your country, and to exhibit it let me show a few transparencies containing the messages of some of your Spanish friends.

Generalitat de Catalunya
Departament d'Universitats, Recerca
i Societat de la Informació
Secretaria del Consell Interuniversitari
de Catalunya
Oficina de Coordinació i d'Organització de les PAU de Catalunya

Dear Claudio Moraga,

Using our friend Enric I want to send you our best wishes for the present festivities and for your near future. As Woody Allen says "the future is a very interesting place because we are planning to locate ourselves in it the rest of our life's"

Since both of us are thinking to write our autobiographies under the title: "I, Claudio" we will need to introduce some additional item.

Thanks for your friendship!

Sincerely

Claudi Alsina

Barcelona, 4 de febrer de 2002

Via Laietana, 33, 2n 1a
08003 Barcelona
Tel. 93 268 92 20
Fax 93 268 92 25
coordinaciopau@correu.gencat.es
http://durei.gencat.es/cat/un/pau.htm

It is thanks to people like Professor Moraga that our world still has a chance of "surviving"; not only for what they themselves do, which is a great deal, but also, and above all, because they transmit to the rest of us their desire to know more and more, and their determination in carrying out constant, meticulous and honest work.

Senén Barro, Alberto Bugarín

Dear Claudio:

In you we have always seen dedication to work well-done, intellectual diligence in the search for solutions, honesty in commitments undertaken and generous support for those who were in need of it from you. But, further on this, you have always shown us that you are a true

FRIEND

My friend, take what you want,
Your penetrating gaze reaches into all corners,
And if it is your wish, I give to you my entire soul,
With its white avenues and its songs.
...
Everything, my friend, I have done for you, all this,
that without looking, you will see in my naked room,
all this which rises up the straight walls -
-like my heart- always seeking the high places.

Pablo Neruda, Crepusculario

[Amigo, llévate lo que tu quieras,
penetra tu mirada en los rincones,
y si así lo deseas, yo te doy mi alma entera,
con sus blancas avenidas y canciones.
...
Todo, amigo, lo he hecho para ti. Todo esto
que sin mirar verás en mi estancia desnuda:
todo esto que se eleva por los muros derechos
-como mi corazón- siempre buscando altura.]

Congratulations,

Alejandro Sobrino

Granada January the 30th, 2002-01-15

Prof. Dr. Claudio Moraga
Univ. of Dortmund
Dortmund
Germany

Dear Claudio,

It is for us a pleasure and a honour to join (at least by mean of a letter) to your retirement celebration.

All of us have received your friendship as well as your advice and scientific direction, and at this moment is very difficult to distinguish the friend from the master. We are happy because of your retirement as you will have more time to spend with us.

Our heart and mind is now and always with you and you can be sure that here, at Granada, is you home for ever.

J.L. Castro M. Delgado

F.Herrera M. Lozano

C.J. Mantas J.L. Verdegay

Bellaterra, January 21, 2002

Dear Claudio,

We are happy about your retirement because at the IIIA we expect that now you will have plenty of time to work even more on science (avoiding burocracy but not musical activities and dancing!!!!) and this will be beneficial for all of us!

On behalf of the IIIA members,

Francesc, Ramon and Lluis (or any permutation of them)

Madrid, den 30.Januar 2002

Lieber Claudio:

Ich vollte auch Dir ein paar Worte zuschicken, indem ich meine Zuneigung zu Dir ausdrücke.

Und was soll ich Dir sagen? Sicher ist es etwas, was nich nötig ist: dass Du Gott für Dein ganzes Leben danken sollst, und, in diesen Momenten besonders für Dein Berufsleben. Wenn es wahr ist, dass wir jederzeit dem Herrn für so vielen guten Sachen danken sollten, wie wirvon Ihm empfangen haben, ist es doch noch wichtiger, am Ende einer Ettape dies zu tun. Es ist gut nach hinten zu schauen, aber ohne "Heimweh", nur soll man in jeden Moment den vorsorglichen Han deseen entdecken, der unsere Schritte lenkt und alles zu unserer Gute tut. Wie gross ist das Geschenk des Glaubens, Claudio! Es ist dieser Glaube, den Du und ich zusammenteilen. Niemals werden wir Ihn genug für diese grossartige Gnade danken, weil wir häufig an viel zu menschliche Gedanken festhalten.

Deine,

Susana.

My dear friend Claudio, over all a friend and, always, a scientist. Always loyal to tour principles, to your profession, to science …

You have covered an important and successful step in your life, but your young spirit is plenty of space for much more fruits. Your friends in Spain, as myself, are waiting for your invaluable partnership.

We will continue together, because, as you know, what make people to be united is all what they are able to build together.

With my best congratulations,

Julio Gutiérrez Ríos
Facultad de Informática
Universidad Politécnica de Madrid

Departamento de Inteligencia Artificial
Facultad de Informática
Universidad Politécnica de Madrid

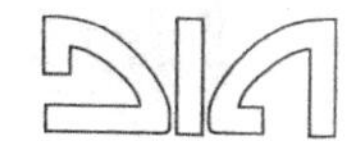

Madrid, 5 de febrero del 2002

Estimado Claudio,

Tienes nuestro respeto como profesor y somos dos de los muchos que reconocen tu prestigio como investigador. Estamos seguras de que tus tareas futuras los incrementarán.

Un fuerte abrazo,

Josefa Z. Hernández y Ana García-Serrano

English version:

You have won our respect as a teacher and you are widely recognized as a researcher. We are sure that your further tasks will increase these.
With our best wishes

Campus de Montegancedo, s/n :-: Boadilla del Monte :-: 28660 MADRID

UPC
UNIVERSITAT POLITÈCNICA DE CATALUNYA

Secció de
Matemàtiques
i Informàtica

Dept. Estructures a
l'Arquitectura
E.T.S. d'Arquitectura
Av. Diagonal, 649
Tel. (93) 401 63 72
08028 Barcelona

Dear Claudio,

On behalf of all members of our department and especially of mine personally, I wish you that your next steps in life be (at least) so fruitful and exciting as your previous ones.

Personally I am convinced that liberation of certain academic activities will allow you to much better devote and concentrate on all those matters we all preserve for this new way in life

Yours always, a friend

Joan [illegible]

Desperta, és un nou dia,
la llum
del sol llevant, vell guia
pels quiets camins del fum.
No deixis res
per caminar i mirar fins al ponent.
Car tot en un moment,
et serà pres.

Recordant Goethe
Salvador Espriu

Wake up, it is a new day,
the rising sunshine, old guide
through the quiet misty ways.
Don't leave anything
to walk and look until the sunset.
Because in a moment,
everything will be gone.

Gaspar Mayor
University of the Balearic Islands
Mallorca (Spain)

A LA ATENCIÓN DE: ENRIC TRILLAS
FAX: 91 352.48.19

DE: ANA PRADERA

Dear Claudio,

This was supposed to be a letter of congratulation because of your retirement... I am sorry for my self-interest, but, actually, I will mainly congratulate myself: if your contribution to the scientific formation of many of us has been crucial until now (personally, I specially remember the time you devoted to my Ph.D. thesis), I believe that your retirement will allow us to enjoy your company and to take profit of your widespread knowledge much more than before!

Looking forward to seeing you very soon,

Ana

Ana Pradera.

Universitat de Girona
Càtedra Ferrater Mora
de Pensament Contemporani

Girona, den 28. Januar 2002

Professor Dr. Claudio Moraga
Universität Dortmund

Lieber Claudio,

Anlässlich Deiner Emeritierung wünsche ich Dir das Beste. Nach Berichten anderer Kolleguen, die Emeritierung bedeutet nicht automatisch, dass man mehr freie Zeit zu Verfügung hat. Dieses Scheinparadoxon lässt sich aber schnell auflösen, da nur Eines wichtig ist: die Zeit, die man jeweils hat, zu geniessen. Das hast Du immer gut gekonnt. Mach so weiter!

Mit besten, besten Wünschen für eine lange und fruchtbare Lebenszeit,

Dein,

Josep-Maria Terricabras

Pl. Ferrater Mora, 1
17071 Girona
Tel. 34-72-41 80 19
Fax 34-72-41 82 30

8.

To finish, Claudio, let me offer you a poem by the late Catalan poet Joan Maragall that, again in my poor translation into English reflects, I think, the main traits of your character.

Do the best job
as if, of each detail you think,
of each word you say,
of each piece you place,
of each hammer's blow you give,
will depend mankind's salvation,
because on it depends. Believe it.

Thank you very much!

Abstracts

Multiple-valued Logic and Complex-Valued Neural Networks
Igor Aizenberg

In classical multiple-valued logic its values are encoded by integers. This complicates the use of multiple-valued logic as a basic model, which can be utilized in an artificial neuron, because the values of k-valued logic encoded by integers 0, 1, 2, …, k are not normalized. To overcome this obstacle, it was suggested to encode the values of k-valued logic by complex numbers located on the unit circle, namely by the kth roots of unity. It is described in the paper how this model of multiple-valued logic over the field of complex numbers was suggested and how it was used to develop a multi-valued neuron (MVN). Then it is considered how a feedforward neural network based on MVN—a multilayer neural network with multi-valued neurons (MLMVN) was designed and its derivative-free learning algorithm based on the error-correction learning rule was presented. Different applications of MLMVN, which outperforms many other machine learning tools in terms of learning speed and generalization capability are also observed.

Ensemble Methods for Time Series Forecasting
Héctor Allende and Carlos Valle

Improvement of time series forecasting accuracy is an active research area that has significant importance in many practical domains. Ensemble methods have gained considerable attention from machine learning and soft computing communities in recent years. There are several practical and theoretical reasons, mainly statistical reasons, why an ensemble may be preferred. Ensembles are recognized as one of the most successful approaches to prediction tasks. Previous theoretical studies of ensembles have shown that one of the key reasons for this performance is diversity among ensemble members. Several methods exist to generate diversity. Extensive works in literature suggest that substantial improvements in accuracy can be achieved by combining forecasts from different models. The focus of this chapter will be on ensemble for time series prediction. We describe the use of ensemble methods to

R. Seising and H. Allende-Cid (eds.), *Claudio Moraga: A Passion for Multi-Valued Logic and Soft Computing*, Studies in Fuzziness and Soft Computing 349, DOI 10.1007/978-3-319-48317-7

compare different models for time series prediction and extensions to the classical ensemble methods for neural networks for classification and regression prediction by using different model architectures. Design, implementation and application will be the main topics of the chapter, and more specifically: conditions under which ensemble based systems may be more beneficial than their single machine; algorithms for generating individual components of ensemble systems; and various procedures through which they can be combined. Various ensemble based algorithms will be analyzed: Bagging, Adaboost and Negative Correlation; as well as combination rules and decision templates. Finally, future directions will be time series forecasting, machine fusion and others areas in which ensemble of machines have shown great promise.

Distributed Machine Learning with Context-Awareness for the Regression Task
Héctor Allende-Cid

The amount of information available nowadays is almost incalculable, presenting new opportunities to gain insight from this data. In this chapter we present some of the work done in field of Distributed Machine Learning and discuss a problem not often mentioned in the literature. The problem is related when the distributed information comes from different contexts. Different contexts can be defined as the different underlying laws of probability governing the data. This is a problem not always addressed, where the majority of the contributions assume that between distributed sources, there is no difference in the underlying law of probability. In this chapter a distributed regression model is presented that addresses this problem.

From Lisp to FuzzyLisp
Luis Argüelles Méndez

Researchers in Fuzzy Logic have in theory many computer languages at his or her disposal for developing the work they have initially in mind. However, and maybe surprisingly, many times they use standard commercial packages that, although powerful, are not as flexible as a pure computer language. The Lisp programming language is the second oldest high-level computer language in the history of computing, but its elegance, powerful memory management model and interactivity converts it in one of the most interesting programming languages for exploring Fuzzy Logic theories. This chapter shows the basic description of FuzzyLisp, a small and compact metalanguage that allows the researcher to experiment with fuzzy sets and concentrate in the construction of fuzzy models while still retaining full control of all the Lisp features.

Claudio Moraga and the University of Santiago de Compostela: Many Years of Collaboration
Senén Barro, Alberto Bugarín and Alejandro Sobrino

In this chapter we summarize the fruitful relationship between Prof. Claudio Moraga and the University of Santiago de Compostela. For almost two decades Prof. Moraga was a regular visitor at our departments as well as a lecturer at our doctoral and summer courses. A milestone in this long term relationship was his key role in the hosting and organization of one of the most important international scientific meetings the Intelligent Systems Group at the University of Santiago de Compostela ever organized.

Class-Memory Automata Revisited
Henrik Björklund and Thomas Schwentick

Data words are an extension of traditional strings that have at each position, besides a symbol from some finite alphabet, a data value from some infinite domain. Class-memory automata constitute an automata model for data words with a decidable emptiness problem. The paper improves the previous result that class-memory automata are strictly more expressive than register automata, another automata model for data words. More specifically, it shows that weak class-memory automata, a restriction of class-memory automata introduced by Cotton-Barratt, Murawski, and Ong are strictly more powerful than the extension of register automata by a data-guessing facility. While weak deterministic class-memory automata yield a restriction of class-memory automata which is closed under Boolean operations, the paper also proposes an extension of deterministic class-memory automata with this property.

Associative Globally Monotone Extended Aggregation Functions
Tomasa Calvo, Gaspar Mayor and Jaume Suñer

In this contribution we deal with a global monotonicity condition for the class of associative extended aggregation functions. We insist on the idea that global monotonicity can be taken as a minimum requirement for an extended aggregation function to be considered consistent.

Using Background Knowledge for AGM Belief Revision
Christian Eichhorn, Gabriele Kern-Isberner and Katharina Diekmann

By using the concept of possible worlds as system states, it is possible to express a system's internal state with the configuration of the system's variables. In the same way, the (usually incomplete and not necessarily correct) belief of an intelligent agent about the system's state can be expressed by a set of possible worlds. If this belief is to be changed due to more accurate information about the system's true state, it is reasonable to incorporate the new information while at the same time abandon as few information as possible, that is, to minimally change the belief of the agent.

In this paper we define semantical distances between possible worlds based on the background beliefs of an agent which are represented as a conditional knowledge base, by defining distances on the syntax of the (semantical) conditional structure. With these distances, we instantiate AGM belief change operators that incorporate new information into the belief state and implement the principle of minimal change by selecting a set of worlds that are closest to the actual beliefs. We demonstrate that using the background knowledge to calculate distances allows us to change the belief state of the agent in a way that is semantically more correct than using, e.g., Dalal's distance. We finally discuss that defining the distances on a syntactical distance on conditional structures allows us to implement the resulting belief change operators more efficiently.

Some Entertainments Dealing with Three Valued Logic
Itziar García-Honrado

This chapter is devoted to do some calculations from the tables of the operations $(+, \cdot, ')$ in the three valued logic of Łukasiewicz, Gödel, Kleene, Bochvar and Post. The chapter is divided in two parts: the first one dealing with the Sheffer stroke, and the second one deals with models of Conjectures from a Basic Flexible Algebra.

On the Ability of Automatic Generation Control to Manage Critical Situations in Power Systems with Participation of Wind Power Plants Parks
Suad S. Halilčević and Claudio Moraga

The purpose of this work is design an automatic starter for the synchronizing equipment (*SE*) in power systems. Such a starter leads to a faster and secure decision concerning the introduction (i.e., a parallel switch) of a ready reserve generator to the power system as a type of ancillary service. The applied method is based on a hybrid neural model (*HNM*). The *HNM* consists of a feedforward, three-layer neural network using neurons with a sigmoid activation, and a perceptron with a biased hard limiter. The adopted *HNM* is excited by signals of the generator's operating status, current load regime, and an available power of the wind power plant park (*WPPP*). The logical decision-making is used to find out the actual load regime and the available power from *WPPP* relevant to building of *HNM*'s input. The automatic starter of the *SE* enables a reduction of the time spent in seeing whether or not the rescue action will imply resorting to the ready reserve power. Such a reduction is certainly a contribution to the efforts of preserving a power system's integrity during the critical situations (e.g. generating unit/area outages). *HNM* has the ability to recognize the crisis symptoms immediately, and to consequently suggest an introduction of the ready-reserve (*RR*) generator (a supplemental reserve) through *SE*.

The way to the BliZ
Erdmuthe Meyer zu Bexten

In this essay I talk about my professional career in which Claudio Moraga had a huge part since we first met in 1986. Under his tutelage I found my way into the field of medicinal engineering and in the end led me to establish the BliZ at the THM in Gießen, a project I am very proud of. The BliZ (*Zentrum für blinde und sehbehinderte Studierende*) is a facility specially designed to aid blind and visual impaired students throughout their course of studies and beyond by developing new technologies in the medicinal field with various partners.

Some Reflections on the Use of Interval Fuzzy Sets for Dealing with Fuzzy Deformable Prototypes
José A. Olivas

In this homage to Prof. Moraga, firstly a short introduction and definition of Fuzzy Deformable Prototypes, introduced by the author in 2000, referring some of the most interesting applications of this concept, mainly concerning prediction systems is presented. Then, there is a short introduction to interval fuzzy sets with the aim of showing some reflections on why it could be interesting to use interval fuzzy sets instead of standard ones for dealing with Fuzzy Deformable Prototypes and some guidelines for the representation and inference mechanisms required for applications.

Milestones of Information Technology—A Survey
Franz Pichler

In the following we give a survey on important epochs in the development of the electrical means for transmission and processing of information between humans and machines. The goal is to contribute to a kind of holistic knowledge of the function of the different complex systems and devices of modern information technology. We are convinced that a historical projection on the different stages of development can be of great help. However, by the limited space of this paper we can do this only by a sketch. The list of relevant literature which we give at the end may help to get the desirable deeper knowledge.

Logics—Many Values, Probabilities, Quantum States, and Fuzzy Sets
Rudolf Seising

This contribution is just a brief historical survey on logical calculus which lay in Caludio Moraga's interest, classical logic, multi-valued logics, fuzzy logic and quantum logic. After the appearance of Russell and Whitehead's *Principia Mathematica* the papers on multi-valued logics by Łukasiewicz an Post were published in 1920 and 1921. Kleene was interested in the Polish logic systems and Zadeh learned logic from Kleene. Quantum logic arose in the mid-thirties with a paper of Birkhoff and von Neumann. Then, some researchers worked in the area to combine these logics with probability concepts and others preferred fuzzy sets.

The Reed-Muller-Fourier Transform—Computing Methods and Factorizations
Radomir S. Stanković

Reed-Muller (RM) expressions are an important class of functional expressions for binary-valued (Boolean) functions which have a doube interpretation, as analogues to both Taylor series or Fourier series in classical mathematical analysis. In matrix notation, the set of basic functions in terms of which they are defined can be represented by a binary triangular matrix. Fourier (RMF) expressions are a generalisation of RM expressions to multiple-valued functions preserving properties of RM expressions including the triangular structure of the transform matrix. In this paper, we discuss different methods for computing RMF coefficients over different data structure efficiently in terms of space and time. In particular, we consider algorithms. corresponding to Cooley-Tukey and constant geometry algorithms for Fast Fourier transform. We also consider algorithms based on various decompositions borrowed from the decomposition of the Pascal matrix and related computing algorithms.

From Boolean to Multi-valued Bent Functions
Bernd Steinbach

Bent functions are functions that have the largest distance to all linear functions. Due to this property bent functions hedge statistical attacks against cryptosystems. This contribution reflects some steps on the way of the specification of Boolean bent functions by O.S. Rothaus, over the description of such bent functions using Boolean differential equations by Bernd Steinbach, the enumeration of bent functions by Natalia Tokareva, the evaluation of classes of bent functions using the Special Normal Form (SNF) by Bernd Steinbach and Christian Posthoff, the embedding of bent functions into other properties needed in cryptosystems by Jon T. Buttler and Tsutomu Sasao, the extension of such properties by Chunhui Wu and Bernd Steinbach, to the generalization to multi-valued bent function by Claudio Moraga et al.

Fuzziness as an Experimental Science: An Homage to Claudio Moraga
Marco Elio Tabacchi and Settimo Termini

In this contribution we collect a few considerations and remarks on such apparently unrelated topics as: an early paper by Norbert Wiener on the Nature of Mathematics; mathematical logic's heritage on the formalization of reasoning; cognitive aspects on the modalities of drawing conclusions. We hope that reading the present paper will show that they are, nevertheless, related in some way at least for what regards the problem of reasoning in the presence of uncertainty, showing a network of concepts that can help considering again the innovating aspects of fuzziness—in our opinion a more than fit homage to Claudio Moraga's interdisciplinary approach to fuzziness.

Sequential Bayesian Estimation of Recurrent Neural Networks
Branimir Todorović, Claudio Moraga and Miomir Stanković

This is short overview of the authors' research in the area of the sequential or recursive Bayesian estimation of recurrent neural networks. Our approach is founded on the joint estimation of synaptic weights, neuron outputs and structure of the recurrent neural networks. Joint estimation enables generalization of the training heuristic known as teacher forcing, which improves the training speed, to the sequential training on noisy data. By applying Gaussian mixture approximation of relevant probability density functions, we have derived training algorithms capable to deal with non-Gaussian (multi modal or heavy tailed) noise on training samples. Finally, we have used statistics, recursively updated during sequential Bayesian estimation, to derive criteria for growing and pruning of synaptic connections and hidden neurons in recurrent neural networks.

A Dialogue Concerning Contradiction and Reasoning
Enric Trillas

What follows is a virtual discussion between two imaginary characters, Carla and Karl, concerning the widespread idea that, with fuzzy sets, the Aristotle's principle of contradiction fails. Most thinkers see this principle as a guarantee for reasoning on solid grounds, and look at its failure with a suspicion of heterodoxy.

Recent Advances in High-Dimensional Clustering for Text Data
Juan Zamora

Clustering has become an important tool for every data scientist as it allows to perform exploratory data analysis and summarize large amounts of data. Specifically for text data, clustering faces other challenges derived from the high-dimensional space into which the data is represented. Furthermore and in spite of the fact that important contributions have already been made, scalability presents an important challenge when the whole-data-in-memory approach is no longer valid for real scenarios where data is collected in massive volumes. This chapter reviews the recent contributions on high-dimensional text data clustering with particular emphasis on scalability issues and also on the impact of the curse of dimensionality over the distance-based clustering methods.

A Hierarchical Distributed Linear Evolutionary System for the Synthesis of 4-bit Reversible Circuits
Fatima Zohra Hadjam and Claudio Moraga

Even limited to 4-bits reversible functions, the synthesis of optimal reversible circuits becomes an arduous task owing to the extremely large problem space. The current paper tries to answer the following question: is it possible to implement optimal 4-bit reversible circuits without relying on existing partial solutions libraries? A distributed linear genetic programming based-approach (DRIMEP2) is presented. It consists of a hierarchical topology with a new communication policy to allow the evolutionary algorithm to explore and exploit the search space in an efficient way. To test the effectivity and the efficiency of the proposed system, the design of 69 benchmarks (4-bits reversible functions) was performed. With respect to good results available in the literature, a gate count reduction up to 60 % was achieved with an average of 16.82 % (for the two first benchmark groups where the gate count of the circuit was considered by the reference authors) and a quantum cost reduction up to 62.71 % was reached with an average of 10.79 % (for the two remaining benchmark groups where the quantum cost of the circuit was considered by the reference authors).

Authors

Igor Aizenberg is Professor and Chair of the Department of Computer Science at *Manhattan College* in New York City. Before joining this institution in 2016 he was Professor of *Texas A&M University-Texarkana* in 2006–2016, research fellow in TU Dortmund and *Tampere University of Technology* in 2002–2006, VP Research in NNT Ltd, Associate Professor in *Uzhhorod National University* (Ukraine) and researcher in KU Leuven (Belgium). His main research interests are complex-valued neural networks, their applications, multiple-valued logic and image processing.

Héctor Allende, received the Ph.D. degree in Statistics in 1988 from *Universidad de Dortmund*, Germany. He is currently a Professor at the Computer Science Department of the *Federico Santa María Technical University*, Chile and Professor at the Engineering and Science Faculty of the *Adolfo Ibáñez University*, Chile. He works primarily in the broad areas of Computational Statistics, Machine Learning, Image Processing, and Pattern Recognition with applications to Time series. His research interests include classification algorithms, ensemble neural networks, intelligent data analysis, and incremental learning. He has published more than 50 journal articles, 2 books, and other books chapters, conferences and research reports.

Héctor Allende-Cid, Ph.D. in Computer Science Engineering. He is currently Assistant Professor at the *Pontificia Universidad Católica de Valparaíso* in Chile. He graduated in 2009 as a Computer Science Engineer and Master in Computer Science at the *Universidad Técnica Federico Santa María* in Chile. In 2015, he received his Ph.D. Degree at the same university.

Dr. Allende-Cid has published 22 papers in international conferences and journals. His main research lines are Machine Learning, Pattern Recognition and Time Series Analysis.

Luis Argüelles holds an MSc degree in Astronomy (High Distinction) from the *Swinburne University of Technology*, Melbourne, Australia. In March, 2010, he received a Master of Business Administration degree (High Distinction) from the *ITEAP Institute*, Malaga, Spain. He has worked as the R+D department's director in a private coal

R. Seising and H. Allende-Cid (eds.), *Claudio Moraga: A Passion for Multi-Valued Logic and Soft Computing*, Studies in Fuzziness and Soft Computing 349, DOI 10.1007/978-3-319-48317-7

mining company in Spain along 20 years, directing and/or coordinating more than 30 research projects, covering regional, national and European frames of research. Inside these projects he has developed discrete event simulation based models, virtual reality systems and Soft Computing based models for geology and mining engineering. He has also contributed as Affiliated Researcher to the works developed at the *European Centre for Soft Computing* in Mieres, Spain. Luis is a pioneer in developing intelligent models for double stars in observational astronomy, having explored also the utility of fuzzy-sets theory in galactic morphology. His fields of interest cover the use and application of fuzzy-logic based techniques in industrial processes and organizational management, especially in the field of total quality management (TQM). Actually he is interested in spreading Fuzzy Logic theories after writing the book *A Practical Introduction to Fuzzy Logic using Lisp*.

Senén Barro is a Full Professor of Computer Science and Artificial Intelligence and affiliated researcher at the *Research Center in Information Technologies* (CiTIUS), both at the *University of Santiago de Compostela*. He heads the Intelligent Systems Group and is author of more than 300 scientific papers in this area.

Katharina Behring obtained her master's degree in computer science from the *TU Dortmund University* in 2014 where she worked as a student research assistant on rationality and commonsense reasoning with counterfactual conditionals as well as on the topic of confidentiality in database systems. She currently works as a consultant for the topic of Product Lifecycle Management.

Henrik Björklund got his Ph.D. in Computer Science from *Uppsala University*, Sweden, in 2005. After a short postdoctoral stay at the *RWTH Aachen* he then worked as a researcher at the *TU Dortmund*, in the group of Thomas Schwentick. Since 2009, he is employed at *Umeå University*, Sweden, where he holds a position as associate professor. His research interests include automata theory, complexity, logic, and the foundations of natural language processing.

Alberto Bugarín is a Full Professor of Computer Science and Artificial Intelligence at the *University of Santiago de Compostela* and affiliated researcher at its *Research Center in Information Technologies* (CiTIUS). He co-authored 200 papers on fuzzy knowledge representation and reasoning and Natural Language Generation and their applications, among other areas.

Tomasa Calvo obtained her Master degree in Sciences (Mathematics) from the *University of Valencia* (1977) and the Ph.D. in Computer Science from *Polytechnic University of Madrid* (1989). She started her professional life at the University of the Balearic Islands (1977–2000), and since 2000 she is a Full Professor at the *University of Alcalá*, Madrid. Her major research interests are in the area of uncertainty modeling, including mathematical fuzzy logic, several types of aggregation techniques, non-additive measures and integral theory, preference structures, multi-distances and functional equations. She is a faithful collaborator of the LOBFI research group.

Christian Eichhorn received his diploma degree in Computer Science from the *TU Dortmund University* in 2011. He is currently enrolled as a Ph.D. student at the

Technical University of Dortmund. His research work is focused on formal properties of nonmonotonic logics and qualitative approaches to nonmonotonic inference, in particular, on the application and generation of ranking functions for this behalf.

Itziar García Honrado obtained a M.Sc. degree in Mathematics with specialization in Statistic at the *University of Oviedo* in 2007. That year she joined the Unit of Fundamentals of Soft Computing at the *European Centre of Soft Computing* where she carried out the research for her Ph.D. Dissertation entitled "Contribution to the study of Ordinary Reasoning and Computing With Words" presented in 2011 at the *University of León*. Currently, she is a member of the department of Statistics, Operational Research and Mathematics Education at the *University of Oviedo*, and teaches in the section of Mathematical Education. She has published around 40 papers in international journals, international and national conferences, and several book chapters. These publications mainly deal with models of Conjectures, Hypotheses and Consequences (CHC models), mathematical representations of linguistic imprecise terms and with some educational aspects in the context of fuzzy logic.

Suad S. Halilčević received the B.Sc. and Ph.D. degrees from the *University of Tuzla*, Bosnia and Herzegovina, and M.Sc. degree from the *University of Zagreb*, Croatia. He spent his research time with the Universities of Ljubljana, Dortmund (DAAD) and Aachen (DAAD). He joined the Faculty of Electrical Engineering, *University of Tuzla* in 1994, where he is currently a Full Professor. In 2004/2005, he was awarded the Fulbright scholarship at the *Iowa State University*. He spent eleven years at the Electric Companies EP BiH, d.d. Sarajevo and EP HZHB, d.d. Mostar. His special fields of interest are power system operation and control and energy management.

Gabriele Kern-Isberner is professor of Information Engineering at the Faculty of Computer Science at the *TU Dortmund University*. Her research work is focused on qualitative and quantitative approaches to knowledge representation, such as default and non-monotonic logics, uncertain reasoning, probabilistic reasoning, belief revision, and argumentation, as well as multi-agent systems and knowledge discovery.

Gaspar Mayor (Mallorca, 1946) received the B.S. degree in Mathematics from the *Universitat de Barcelona* in 1971 and the Ph.D. degree in Mathematics from the *Universitat de les Illes Balears* (UIB) in 1985. He is currently a full Professor at the UIB where he is the Head of the LOBFI research group. His main fields of research include aggregation functions, functional equations, discrete t-norms and copulas, preference modeling and multi-argument distances.

Erdmuthe Meyer zu Bexten, Prof. Dr. rer. nat. She studied computer sciences with theoretical medical science as minor field of study. After that she was a scientific assistant at the *Frauenhofer Institut for Microelectronic Circuits and Systems*, Duisburg where she graduated as Dr. rer. nat. in 1992. In 1994 she worked at the TU Dortmund at the Department Informatik Lehrstuhl 1. In 1996 she was appointed professor in Gießen. Since 1998 she has the function of managing director of the BliZ (Zentrum für blinde und sehbehinderte Studierende). Her main fields of interests lie in the inclusion of handicapped students and the medical computer sciences.

José A. Olivas, Head of the SMILe research team (Soft Management of Internet and Learning), Assistant Director of the Department of Information Technologies and Systems and Director of the Computer Science Ph.D. and Master Programs at the *University of Castilla-La Mancha*. Visiting Scholar at Lotfi A. Zadeh's BISC (*Berkeley Initiative in Soft Computing*), his current main research interests are in the field of Soft Computing for Information Retrieval and Knowledge Engineering applications. He received the Environment Research Award 2002 from the Madrid Council.

Franz Pichler, born 1936 (Salzburg), technician with the Austrian Post-and Telegraph Company (1954–1967), studied Mathematics and Physics at the University Innsbruck (1962–1967), Ph.D. thesis on the Theory of Walsh functions, professor of Systems Theory at the *Johannes Kepler University Linz* (1973–2004).

Thomas Schwentick works in several areas where Mathematical Logic meets Computer Science, particularly in Database Theory. He often studies questions that concern the expressiveness of logical languages together with the computational complexity of related algorithmic problems.

He is a professor for Theoretical Computer Science at *TU Dortmund* University (Germany). He graduated in Mathematics at the *Johannes Gutenberg University* in Mainz (Germany), where he also finished his Ph.D. and his habilitation. Before he moved to his current position, he had professor positions in Jena and Marburg. He was PC chair of PODS 2011 and PC co-chair of ICDT 2007, STACS 2010 and STACS 2011. He is a member of the ICDT Council, the EATCS Council, and the Academia Europaeae.

Rudolf Seising, Ph.D. in Philosophy of Science and the habilitation degree in History of Science from the Ludwig-Maximilians University of Munich. He was acting as professor for history of science and assistant professor for computer sciences and medical computer sciences at various universities and held positions as Visiting Researcher (2009–2011) and as Adjoint Researcher (2011–2014) at the *European Centre for Soft Computing* (ECSC) in Mieres (Asturias). He is co-editor in chief of the *Archives for Philosophy and History of Soft Computing*.

Alejandro Sobrino is an Associated Professor at the Department of Logic and Moral Philosophy of the *University of Santiago de Compostela*. His research interests are approximate causality and neuro-philosophical foundations of vague language. He is the Secretary of the journal "Ágora" and Assistant editor of the *Archives for the Philosophy and History of Soft Computing*.

Miomir Stanković received the Ph.D. degree from the Faculty of Electronic Engineering, *University of Nis*, Serbia, 1979. He is Full Professor in the Department of Mathematics, Faculty of Occupational Safety, *University of Nis*, Serbia.

Radomir S. Stanković received the B.Sc. degree in electronic engineering from the faculty of Electronics, *University of Niš*, Serbia, in 1976, and M.Sc. and Ph.D. degrees in applied mathematics from the Faculty of Electrical Engineering, *University of Belgrade*, Serbia, in 1984 and 1986, respectively. Currently, he is a Professor at the department of Computer Science, Faculty of Electronics, *University of Niš*, Serbia. In 1997, he was awarded by the *Kyushu Institute of Technology Fellowship*, Japan, and in 2000 by the Nokia Professorship by Nokia, Finland. From 1999 to 2014, he worked in part at the *Tampere International Center for Signal Processing*, Department of Signal Processing, Faculty of Computing and Electrical Engineering, *Tampere University of Technology*, Tampere, Finland, where he is currently an adjunct professor.

Bernd Steinbach studied Information Technology at the *University of Technology in Chemnitz* (Germany) and graduated with an M.Sc. in 1973. He graduated with a Ph.D. and with a Dr. sc. techn. (Doctor scientiae technicarum) for his second doctoral thesis from the Faculty of Electrical Engineering of the *Chemnitz University of Technology* in 1981 and 1984, respectively. In 1991 he obtained the Habilitation (Dr.-Ing. habil.) from the same faculty.

He was working in industry as an Electrician, there he had tested professional controlling systems at the Niles Company. Since 1992 he is a Full Professor of Computer Science/Software Engineering and Programming at the *Freiberg University of Mining and Technology*, Department of Computer Science. He has served as Head of the Department of Computer Science and Vice-Dean of the Faculty of Mathematics and Computer Science. He published more than 250 chapters in books, complete issues of journals, and papers in journals and proceedings. He received the Barkhausen Award from the University of Technology Dresden in 1983.

Jaume Suñer received the B.S. degree in Mathematics from the *Universitat Auténoma de Barcelona* in 1980, the M.Sc. degree in Communication Engineering from the *Imperial College of Science, Technology and Medicine* of London in 1990, and the Ph.D. degree in Computer Science from the *Universitat de les Illes Balears* in 2002. He is currently a University Lecturer at the *Universitat de les Illes Balears*. He is a member of the LOBFI (Fuzzy Logic and Information Fusion) research group, and his current research interests include aggregation operators, fuzzy connectives and functional equations.

Marco Elio Tabacchi is a Computational Intelligence Cognitivist. Currently the Scientific Director at Istituto Nazionale di Ricerche Demopolis, and a Research Assistant with SCo2 Research Unit at DMI—*Universitá degli Studi di Palermo*. Has co-authored more than 100 papers on a wide range of subjects, is a member of the Steering Committee at AISC (Italian Association for Cognitive Science) and co-editor in chief of the *Archives for Philosophy and History of Soft Computing*.

Settimo Termini has been professor of Cybernetics and Theoretical Computer Science at the Universities of Perugia and Palermo and, previously, researcher at the *Istituto di Cibernetica "Eduardo Caianiell"* of the CNR (National Research Council), which he directed from 2002 to 2009. Fellow of IFSA and of the Accademia nazionale di Scienze, Lettere e Arti of Palermo. A physicist by training, interested on the problem of the presence of uncertainty in different kind of systems, has introduced the theory of measures of fuzziness and studied the epistemological implications of the presence of fuzziness and uncertainty in scientific theories as well as their use as a bridge between Hard sciences and Humanities. Recently, he has been interested in the connections between "curiosity driven" research and technological innovations. Among his contributions, recall the volumes *Aspects of vagueness*, edited with H. J. Skala and E. Trillas, Kluwer, (1984), *Imagination and rigor*, Springer (2006), *Contro il declino*, coauthored with Pietro Greco, Codice edizioni (2007).

Branimir Todorovic is associate professor at Computer Science Department, Faculty of Mathematics and Sciences, *University of Nis*. He received his Doctor of Science degree from Faculty of Electrical Engineering, *University of Belgrade*. His research interest include sequential Bayesian training of feed forward and recurrent neural networks, blind source separation and deconvolution, on line training of structural classifiers, active and semi-supervised learning algorithms.

Enric Trillas (Barcelona, 1940), Dr. Sc. by the *University of Barcelona*, currently is an Honorary Emeritus Professor at the *University of Oviedo*. Did teach at the universities of Barcelona, Technical of Catalonia, Technical of Madrid, and has been an Emeritus Researcher at the *European Center for Soft Computing*. Served in the Spanish Government in several positions, holds two "Honoris Causa" doctorates, as well as several awards, medals, and civil and military decorations.

Carlos Valle received the Ph.D. degree in computer science from *Federico Santa María Technical University*, Chile in 2014. Currently he is working as a researcher in the Computer Science Department of the *Federico Santa María Technical University*, Chile. His research interest include Machine Learning, Ensemble Learning, Pattern Recognition with applications to Time series and Intelligent Distributed Systems.

Juan Zamora received his Ph.D. in Informatics Engineering at the *Universidad Técnica Federico Santa María* in 2016. His research is focused on clustering large document collections. He worked with Professor Moraga at the *European Centre for Soft Computing* (ECSC) in Mieres (Asturias) in 2012. He is currently at the *Universidad de Valparaíso of Chile* developing computational methods with applications in precision medicine.

Fatima Zohra Hadjam was born in Sidi Bel Abbes, Algeria, in 1970. She received an Engineering degree in Informatics in 1993 and a Magister in Computer Science in 2001 from the *University Djillali Liabes of Sidi Bel Abbes*. She did specialization studies in France and in Germany. In 2008 she obtained her Doctorate in Artificial Intelligence at the same University, where she became Assistant Professor of Computer Science.

In December 2011 she obtained a Postdoctoral position at the *European Centre for Soft Computing*, where she was working on the evolutionary design of reversible computing circuits at the Unit Fundamentals of Soft Computing until 2015. Dr. Hadjam occupies now, a position of professor assistant at the *University of Djillali Liabes*, Faculty of Exact Science, Computer Science Department. Dr. Hadjam has published 15 papers and has made presentations at different international conferences and journals. Her main research lines are artificial intelligence, evolutionary computations and distributed processing applied to the design of digital and reversible circuits.

Printed in the United States
By Bookmasters